# SUSTAINABLE PRACTICES IN SURFACE AND SUBSURFACE MICRO IRRIGATION

*Research Advances in Sustainable Micro Irrigation*

## VOLUME 2

# SUSTAINABLE PRACTICES IN SURFACE AND SUBSURFACE MICRO IRRIGATION

*Edited by*
**Megh R. Goyal, PhD, PE**

## Apple Academic Press

TORONTO      NEW JERSEY

Apple Academic Press Inc. | Apple Academic Press Inc.
3333 Mistwell Crescent | 9 Spinnaker Way
Oakville, ON L6L 0A2 | Waretown, NJ 08758
Canada | USA

ISBN 13: 978-1-77463-338-0 (pbk)
ISBN 13: 978-1-77188-017-6 (hbk)

**Library of Congress Control Number: 2014941332**

---

**Library and Archives Canada Cataloguing in Publication**

---

Sustainable practices in surface and subsurface micro irrigation/edited by Megh R. Goyal, PhD, PE.

(Research advances in sustainable micro irrigation; volume 2)

Includes bibliographical references and index.
ISBN 978-1-77188-017-6 (bound)
1. Microirrigation. 2. Microirrigation--Management.
3. Sustainable agriculture. I. Goyal, Megh Raj, editor
II. Series: Research advances in sustainable micro irrigation ; v. 2

S619.T74S996 2014          631.5'87          C2014-903543-8

---

# CONTENTS

## PART I: SUBSURFACE MICRO IRRIGATION

## PART II: MICRO IRRIGATION RESEARCH ADVANCES AND APPLICATIONS

# LIST OF CONTRIBUTORS

**Abedin, El**
Agricultural and Biosystems Engineering Department, Faculty of Agriculture, Shatby, Alexandria, 21545 Egypt

**Almodovar, Carlos**
Research Associate, Fortuna Agricultural Experiment Station, University of Puerto Rico, Mayaguez Campus, Juana Diaz, PR

**Chavan, Vishal Keshavao**
Assistant Professor and Senior Research Fellow in SWE, Agriculture University, Akola, Maharashtra, Website: www.pdkv.ac.in; Email: vchavan2@gmail.com

**Coleman, Bruce**
BECA Irrigation Supplier, Inc., Coamo, Puerto Rico

**DeValerio, Jim**
University of Florida, Institute of Food and Agricultural Science, Bradford County Extension Service, 2266 N Temple Ave, Starke, FL 32091, USA

**Devasirvatham, Viola**
Lecturer, University of Sydney, School of Natural Sciences, Sydney, AU; and Postgraduate student, School of Natural Sciences, University of Western Sydney, NSW, Australia

**El-nesr, Mohammad A B**
Assistant Professor, Water and Irrigation Systems Engineering, Department of Agricultural Engineering, College of Food and Agricultural Sciences, King Saud University, P.O. Box 2460 Riyadh 11451, Email: drnesr@gmail.com

**Gazula, Aparna**
University of Florida, Institute of Food and Agricultural Science, Alachua County Extension Service, 2800 NE 39th Ave, Gainesville, FL 32609, USA

**Goenaga, Ricardo**
Research Plant Physiologist and Research Leader, USDA-ARS Tropical Agriculture Research Station, 2200 P.A. Campos Ave., Suite 201, Mayaguez, Puerto Rico 00680-5470, Tel.: 787-831-3435 ext. 226, Fax: 787-831-3386, E-mail: ricardo.goenaga@ars.usda.gov

**Gonzalez, Eladio**
Associate Professor, Mountain College of Agricultural Sciences, University of Puerto Rico, Utuado Campus, Utuado, PR. Email: eladio.gonzalez1@upr.edu

**Goyal, Megh R.**
Retired Professor in Agricultural and Biological Engineering, University of Puerto Rico – Mayaguez Campus, PO Box 86, Rincon, PR, 00677; and Senior Acquisitions Editor and Senior Editor-in-Chief, Apple Academic Press Inc., New Jersey. Email: goyalmegh@gmail.com

**Harmsen, Eric W.**
Professor, Department of Agricultural and Biosystems Engineering, University of Puerto Rico, Mayagüez, PR 00681, USA. Email: harmsen1000@gmail.com

**Hochmuth, Robert**
University of Florida, Institute of Food and Agricultural Science, Florida Cooperative Extension Service, Suwannee Valley Agricultural Extension Center, 7580 CR 136, Live Oak, FL 32060-7434, USA

**Irizarry, Heber**
Research Horticulturist, USDA-ARS Tropical Agriculture Research Station, 2200 P.A. Campos Ave., Suite 201, Mayaguez, Puerto Rico 00680-5470

**Ismail, Samir M.**
Professor of Irrigation Systems, Agricultural and Biosystems Engineering Dept., Faculty of Agriculture, Shatby, Alexandria, 21545 Egypt, Fax: (+203) 5922780, Mobile: (+203) 01001281641, Email: drsamir.ismail@gmail.com

**Jensen, Marvin E.**
Retired Research Leader USDA-ARS, 1207 Springwood Drive, Fort Collins, CO, 80525, Email: mjensen419@aol.com

**Kumar, Pradeep**
Scientist, National Institute of Hydrology, Roorkee, India

**Muñoz-Muñoz, Miguel A.**
Ex-President of University of Puerto Rico, University of Puerto Rico, Mayaguez Campus, College of Agriculture Sciences, Call Box 9000, Mayagüez, PR. 00681-9000. Tel. 787-265-3871, Email: miguel.munoz3@upr.edu

**Ortiz, Eulalio**
Certified Public Accountant and Professor, College of Business Administration, University of Puerto Rico, Mayaguez, Puerto Rico

**Patel, Neelam**
Senior Scientist, Water Technology Centre (WTC), Indian Agricultural Research Institute (IARI), New Delhi, Delhi 110012, India, Email: neelam@iari.res.in

**Porch, Timothy G.**
Genetics Researcher, USDA-ARS Tropical Agriculture Research Station, 2200 P.A. Campos Ave., Suite 201, Mayaguez, Puerto Rico 00680-5470. Email: Timothy.Porch@ars.usda.gov

**Rajput, T.B.S.**
Principal Scientist, Water Technology Centre (WTC), Indian Agricultural Research Institute (IARI), New Delhi, Delhi 110012, India, Email: tbsraj@iari.res.in

**Ramirez Builes, Victor H.**
Agroclimatology and Crop Science Researcher, National Coffee Research Center (Cenicafe), Chinchina (Caldas, Colombia). Email: victor.ramirez@cafedecolombia.com

**Rivera, Edmundo**
Agronomist, USDA-ARS Tropical Agriculture Research Station, 2200 P.A. Campos Ave., Suite 201, Mayaguez, Puerto Rico 00680-5470.

**Simonne, Eric**
University of Florida, Institute of Food and Agricultural Science, Horticultural Sciences Department, 1052 McCarty Hall, Gainesville, FL 32611-0220, USA

**Singh, Gajendra**
Former Vice President, Asian Institute of Technology, Thailand. C-86, Millennium Apartments, Plot E-10A, Sector 61, NOIDA U.P. – 201301, INDIA, Mobile: (011)-(91)-997-108-7591, Email: prof.gsingh@gmail.com

**Sivanappan, R. K.**
Former Dean and Professor, Tamil Nadu Agricultural University at Coimbatore, Tamil Nadu, India; and Consultant, 14, Bharathi Park, 4th Cross Road, Coimbatore – 641 043, India. Email: sivanappan@hotmail.com

**Tripathi, Vinod Kumar**
Assistant Professor, Centre for Water Engineering and Management, Central University of Jharkhand, Ranchi, India, E mail: tripathiwtcer@gmail.com

**Wassif, M. A.**
Agricultural and Biosystems Engineering Department, Faculty of Agriculture, Shatby, Alexandria, 21545 Egypt

**Zien, T. K.**
Agricultural and Biosystems Engineering Department, Faculty of Agriculture, Shatby, Alexandria, 21545 Egypt

# LIST OF ABBREVIATIONS

| | |
|---|---|
| AALCS | Australian Agricultural Land Classification System |
| AAP | Apple Academic Press Inc. |
| AC-FT | acre foot |
| ANOVA | analysis of variance |
| ASABE | American Society of Agricultural and Biological Engineers |
| ASW | available soil water |
| AWC | available water content |
| B-C | Blaney-Criddle |
| CDGG | cumulative degree days |
| cfs | cubic feet per second |
| cfsm | water depth, cubic feet per second per square mile |
| CGDD | cumulative growing degree days |
| CPFDR | capacitance probe/frequency domain reflectometer |
| CRCIF | Cooperative Research Center for Irrigation Futures – Australia |
| CRZI | capillary root zone irrigation system |
| CU | coefficient of uniformity |
| CWSI | crop water stress index |
| DAP | days after planting |
| DAT | days after transplanting |
| DOY | day of the year |
| Dr | root zone depletion |
| DTPA | diethylenetriamine pentaacetic acid |
| DU | distribution uniformity |
| EDTA | ethylenediaminetetraacetic acid |
| EPAN | pan evaporation |
| ET | evapotranspiration |
| ETc | crop evapotranspiration |
| ETr | reference evapotranspiration |
| FAO | Food and Agricultural Organization, Rome |
| FC | field capacity |
| fc | fraction of the soil covered by vegetation |
| FDR | frequency domain reflectometer |
| gph | gallons per hour |
| gpm | gallons per minute |
| HRG | Hargreaves |
| ICAR | Indian Council of Agriculture Research |
| ISAE | Indian Society of Agricultural Engineers |
| Kc | crop coefficient |
| Kcb | transpiration coefficient |

| | |
|---|---|
| Ke | evaporation coefficient |
| KISSS™ | Kapillary Irrigation Subsurface System |
| Kp | pan coefficient |
| Ks | drought stress coefficient |
| KSA | Kingdom of Saudi Arabia |
| LAI | leaf area index |
| LAR | leaf appearance rate |
| LEPA | low energy pressure irrigation system |
| lps | liters per second |
| MAD | maximum allowable depletion |
| MSL | mean sea level |
| NSW | New South Wales |
| PC | plant crop |
| PE | potential evapotranspiration |
| PET | potential evapotranspiration |
| pH | acidity/alkalinity measurement scale |
| PM | Penman-Monteith |
| ppb | one part per billion |
| ppm | one part per million |
| psi | pounds per square inch |
| PVC | poly vinyl chloride |
| PWP | permanent wilting point |
| R1 | first ratoon (crop) |
| R2 | second ratoon (crop) |
| R3 | third ratoon crop |
| RA | extraterrestrial radiation |
| RH | relative humidity |
| RMAX | maximum relative humidity |
| RMIN | minimum relative humidity |
| RMSE | root mean squared error |
| RS | solar radiation |
| SAR | sodium absorption rate |
| SCS-BC | SCS Blaney-Criddle |
| SDI | subsurface drip irrigation |
| SPAC | soil-plant-atmosphere continuum |
| SWB | soil water balance |
| SWP | soil water potential |
| TDR | time domain reflectometer |
| TE | transpiration efficiency |
| TEW | total evaporable water |
| TMAX | maximum temperature |
| TMIN | minimum temperature |
| TR | temperature range |
| TUE | transpiration use efficiency |
| USDA | US Department of Agriculture |

| USDA-SCS | US Department of Agriculture-Soil Conservation Service |
| UWS | University of Western Sydney |
| VPD | vapor pressure deficit |
| VSWC | volumetric soil water content |
| VWC | volumetric water content |
| WATBAL | water balance |
| WISP | wind speed |
| WS | weather Stations |
| WSEE | weighed standard error of estimate |
| WUE | water use efficiency |
| YTD | yield threshold depletion |
| µg/g | micrograms per gram |
| µg/L | micrograms per liter |

# LIST OF SYMBOLS

| | |
|---|---|
| $\mathfrak{R}_i^{(v)}$ | error matrix |
| $\mathfrak{R}_i^{(v)}$ | Jacobian matrix |
| $\theta$ | moisture content |
| $\Upsilon$ | pore-connectivity |
| $\theta_d$ | desired soil moisture content |
| $\tau_j$ | time-step |
| $\theta_r$ | residual water content |
| $\Delta$ | slope of saturation vapor pressure curve |
| a | emitter spacing |
| b | maximum monolayer adsorption |
| c | residual solution metal concentration |
| $C_d$ | denominator constant |
| $C_n$ | numerator constant |
| Cp | specific heat of air |
| D | diffusivity |
| d | zero displacement height |
| $D_r$ | root zone depletion |
| E | evaporation |
| $e_a$ | actual vapor pressure |
| $e_s$ | saturation vapor pressure |
| $ET_c$ | crop evapotranspiration |
| $ET_o$ | standardized reference crop |
| $ET_{os}$ | evapotranspiration for short |
| $f_c$ | function of vegetation |
| $f_{ew}$ | irrigation |
| G | soil heat flux density |
| $\lambda$ | grid spacing |
| $g_s$ | saturated grid length |
| h | canopy height |
| H | heat flux |
| h | pressure head |
| $I_i$ | irrigation depth |
| j | time counter |
| k | affinity |
| K | soil hydraulic conductivity |
| $K_{bc}$ | basal crop coefficient |
| $K_{c\ dev}$ | development period |
| $K_c$ | crop coefficient |

| | |
|---|---|
| $K_{c,end}$ | late season period |
| $K_{c,mid}$ | the midseason period |
| $K_{cb}$ | basal crop coefficients |
| $K_e$ | soil evaporation coefficient |
| $K_s$ | stress coefficient |
| $l$ | current grid location |
| $q$ | emitter discharge |
| $q_{var}$ | emitter flow variation |
| $R$ | radius |
| $r_a$ | aerodynamic resistance |
| $r_L$ | stomatal resistance |
| $R_n$ | net radiation |
| $R_o$ | surface runoff |
| $r_s$ | high surface resistances |
| $S$ | matric flux potential |
| $T_{max}$ | maximum daily temperature |
| $T_{min}$ | minimum daily temperature |
| $t$ | simulation time |
| $T_b$ | base temperature |
| $T_{dew}$ | average dew point temperature |
| $T_{min}$ | minimum air temperatures |
| $u$ | integration temporary variable |
| $u_z$ | horizontal wind speed |
| $V$ | calculated volume |
| $V_a$ | actual volume |
| $v_z$ | water vertical velocity |
| $x$ | amount of adsorbed cations |
| $z$ | direction |
| $Z$ | thickness of soil layer |
| $z_h$ | height of the humidity measurements |
| $Z_m$ | height of wind measurements |
| $Z_t$ | rooting depth |
| $\gamma$ | psychometric constant |
| $\theta_{FC}$ | water content at field capacity |
| $\theta_g$ | gravimetric water content |
| $\theta_{v,I}$ | volumetric soil moisture |
| $\theta_{WP}$ | water content at wilting point |
| $\lambda$ | length of the grid |
| $\lambda E$ | latent heat flux |
| $\pi$ | osmotic potential |
| $\rho$ | bulk density |
| $P$ | pressure or turgor |
| $\tau$ | soil matric potential |
| $\psi$ | water potential in the plant |
| $\psi_g$ | gravitational potential energy |

$\psi_m$        matric potential due to capillary pressure
$\psi_o$        osmotic potential due to salts
$\psi_p$        pressure potential
$\psi_t$        total soil water potential energy

# PREFACE

Due to increased agricultural production, irrigated land has increased in the arid and subhumid zones around the world. Agriculture has started to compete for the water use with industries, municipalities and other sectors. This increasing demand along with increments in water and energy costs have made it necessary to develop new technologies for the adequate management of water. The intelligent use of water for crops requires understanding of evapotranspiration processes and use of efficient irrigation methods.

The "http://newindianexpress.com/cities/bangalore/Micro-irrigation-to-be-promoted/2013/08/17/" weblink published an article on the importance of micro irrigation in India. Every day, similar news appear all-round the world indicating that Government agencies at central/state/local level, research and educational institutions, industry, sellers and others are aware of the urgent need to adopt micro irrigation technology that can have an irrigation efficiency up to 90% compared to 30–40% for the conventional irrigation systems. I here share with the readers the news on 17 August of 2013 by Indian Express Newspaper, "*In its efforts to increase the irrigated area by efficiently distributing the available water in the Cauvery basin, The Cauvery Neeravari Nigama Limited (CNNL) is planning to undertake pilot projects on micro irrigation at four places. The CNNL Managing Director Kapil Mohan said, 'the Cauvery water disputes tribunal has permitted the state to irrigate up to 18.85 lakh acres of land in the Cauvery basin. Therefore, we have to judiciously use the available water to increase the irrigated area. In the conventional irrigation method, a lot of water is required to irrigate even a small piece of land. Therefore, we are planning to undertake pilot projects to introduce micro irrigation in four or five places in the Cauvery basin.' Kapil further said that unless the farmers are willing to embrace micro irrigation, it would be difficult for the project to succeed. Therefore, the CNNL is holding discussions with the farmers in different villages of the basin to select the villages in which the project would be undertaken. The CNNL is also in the process of finalizing the technology that should be adopted while undertaking the pilot project. 'If everything goes as planned we should implement the pilot project within this financial year. If the project yields the desired result, we will think of extending it to the other areas in the basin,' Kapil added. According to the official sources, water would be supplied through micro sprinklers instead of canals in the micro irrigation system. Therefore, one can irrigate more than two acres of land through the system with the water that is used to irrigate one acre of land in the conventional canal irrigation system.*"

Evapotranspiration (ET) is a combination of two processes: evaporation and transpiration. Evaporation is a physical process that involves conversion of liquid water into water vapor and then into the atmosphere. Evaporation of water into the atmosphere occurs on the surface of rivers, lakes, soils and vegetation. Transpiration is a physical process that involves flow of liquid water from the soil (root zone) through the trunk, branches and surface of leaves through the stomates. An energy gradient is created during the evaporation of water, which causes the water movement into and out of the plant stomates. In the majority of green plants, stomates remain open during the day and stay closed during the night. If the soil is too dry, the stomates will remain closed during the day in order to slow down the transpiration.

Evaporation, transpiration and ET processes are important for estimating crop water requirements and for irrigation scheduling. To determine crop water requirements, it is necessary to estimate ET by on site measurements or by using meteorological data. On site measurements are very costly and are mostly employed to calibrate ET methods using climatological data. There are a number of proposed mathematical equations that require meteorological data and are used to estimate the ET for periods of one day or more. Potential ET is the ET from a well-watered crop, which completely covers the surface. Meteorological processes determine the ET of a crop. Closing of stomates and reduction in transpiration are usually important only under drought or under stress conditions of a plant. The ET depends on four factors: (1) climate, (2) vegetation, (3) water availability in the soil and (4) behavior of stomates. Vegetation affects the ET in various ways. It affects the ability of the soil surface to reflect light. The vegetation affects the amount of energy absorbed by the soil surface. Soil properties, including soil moisture, also affect the amount of energy that flows through the soil. The height and density of vegetation influence efficiency of the turbulent heat interchange and the water vapor of the foliage.

Micro irrigation, also known as trickle irrigation or drip irrigation or localized irrigation or high frequency or pressurized irrigation, is an irrigation method that saves water and fertilizer by allowing water to drip slowly to the roots of plants, either onto the soil surface or directly onto the root zone, through a network of valves, pipes, tubing, and emitters. It is done through narrow tubes that deliver water directly to the base of the plant. It is a system of crop irrigation involving the controlled delivery of water directly to individual plants and can be installed on the soil surface or subsurface. Micro irrigation systems are often used in farms and large gardens, but are equally effective in the home garden or even for houseplants or lawns. They are easily customizable and can be set up even by inexperienced gardeners. Putting a drip system into the garden is a great do-it-yourself project that will ultimately save the time and help the plants grow. It is equally used in landscaping and in green cities.

The mission of this compendium is to serve as a text book or a reference manual for graduate and under graduate students of agricultural, biological and civil engineering; horticulture, soil science, crop science and agronomy. I hope that it will be a valuable reference for professionals that work with micro irrigation and water management; for professional training institutes, technical agricultural centers, irrigation centers, Agricultural Extension Service, and other agencies that work with micro irrigation programs.

After my first textbook on *Drip/Trickle or Micro Irrigation Management* by Apple Academic Press Inc. and response from international readers, I was motivated to bring out for the world community this series on *Research Advances in Sustainable Micro Irrigation*. This book series will complement other books on micro irrigation that are currently available on the market, and my intention is not to replace anyone of these.

This book series is unique because it is complete and simple, a one stop manual, with worldwide applicability to irrigation management in agriculture. Its coverage of the field of micro irrigation includes, historical review; current status and potential; basic principles and applications; research results for vegetable/row/tree crops; research studies from Chile, Colombia, Egypt, India, Mexico, Puerto Rico, Saudi Arabia, Spain, and U.S.A.; research results on simulation of micro irrigation and wetting patterns; development of software for micro irrigation design; micro irrigation for small farms and marginal farmers; studies related to agronomical crops in arid, humid, semiarid, and tropical climates; and methods and techniques that can be easily applied to other locations (not included in this book). This book offers basic principles, knowledge and techniques of micro irrigation management, that are necessary to understand before designing/developing and evaluating an agricultural irrigation management system. This book is a must for those interested in irrigation planning and management, namely, researchers, scientists, educators and students.

Volume 1 in this book series is titled *Sustainable Micro Irrigation: Principles and Practices*, and includes 16 chapters.

Volume 2 in this book series is titled *Sustainable Practies in Surface and Subsurface Micro Irrigation*, and includes 16 chapters consisting of: Wetting Pattern Simulation of Subsurface Micro Irrigation: Part I, Model Development by *M.N. El-nesr, S.M. Ismail, T.K. Zien El-Abedin, and M.A. Wassif*; Wetting Pattern Simulation of Subsurface Micro Irrigation: Part II, Model Validation by *M.N. El-nesr, S.M. Ismail, T.K. Zien El-Abedin, and M.A. Wassif*; Micro Irrigation in Egyptian Sandy Soil: Hydraulic Barrier Technique by *M.N. El-nesr, S.M. Ismail, T.K. Zien El-Abedin, and M.A. Wassif*; Micro irrigation Design using MicroCAD by *M.N. El-nesr, S.M. Ismail, T.K. Zien El-Abedin, and M.A. Wassif*; Subsurface Drip Irrigation in Australia: Vegetables by *Viola Devasirvatham;* Mechanics of Clogging in Microirrigation System by *Vishal keshavrao Chavan, P. Balakrishnan, Santosh Deshmukh, and M.B. Nagdeve*; Water Movement in Drip Irrigated Sandy Soils by *Eric Simonne, Aparna Gazula, Robert Hochmuth, and Jim DeValerio*; Crop Coefficients: Trickle Irrigated Common Beans by *Victor H. Ramirez Builes, Eric W. Harmsen and Timothy G. Porch*; Water Requirements for Papaya on a Mollisol Soil by *Ricardo Goenaga, Edmundo Rivera, and Carlos* Almodovar; Water Requirements for Tanier (*Xanthosoma* spp.) by *Ricardo Goenaga*; Water Requirements for Tanier (*Xanthosoma* spp.) on a Mollisol Soil by *Ricardo Goenaga*; Water Requirements for Banana on a Mollisol Soil by *Ricardo Goenaga, and Heber Irizarry*; Water Requirements for Banana on an Oxisol Soil by *Ricardo Goenaga, and Heber Irizarry*; Water Requirements for Plantains on a Mollisol Soil by *Ricardo Goenaga, Heber Irizarry, and Eladio Gonzalez*; Drip Irrigation Management: Plantain and Banana by *Ricardo Goenaga, Heber Irizarry, Bruce Coleman and Eulalio Ortiz*; Biometric Response of Eggplant under Sustainable Micro Irrigation with Municipal Wastewater *Vinod Kumar Tripathi, T.B.S. Rajput, Neelam Patel, and Pradeep Kumar;* and Appendices.

Volume 3 in this book series is titled *Sustainable Micro Irrigation Management for Trees and Vines*.

Volume 4 in this book series is titled *Management, Performance, and Applications of Micro Irrigation*.

The contribution by all cooperating authors to this book series has been most valuable in the compilation of this three-volume compendium. Their names are mentioned in each chapter. This book would not have been written without the valuable cooperation of these investigators, many of them are renowned scientists who have worked in the field of evapotranspiration throughout their professional careers.

I would like to thank the editorial staff, Sandy Jones Sickels, Vice President, and Ashish Kumar, Publisher and President at Apple Academic Press, Inc., (http://appleacademicpress.com) for making every effort to publish the book when the diminishing water resources is a major issue worldwide. Special thanks are due to the AAP Production Staff for typesetting the entire manuscript and for the quality production of this book. We request the reader to offer us your constructive suggestions that may help to improve the next edition. The reader can order a copy of this book for the library, the institute or for a gift from CRC Press (Taylor & Francis Group), 6000 Broken Sound Parkway, NW Suite 300, Boca Raton, FL, 33487, USA; Tel.: 800-272-7737 or search at Weblink: http://www.crcpress.com.

I express my deep admiration to my family for understanding and collaboration during the preparation of this three volume book series. With our whole heart and best affection, I dedicate this book series to my wife, Subhadra Devi Goyal, who has supported me during the last 44 years. We both have been trickling on to add our drop to the ocean of service to the world of humanity. Without her patience and dedication, I would not have been a teacher with vocation and zeal for service to others. I present here the Hymn on Micro Irrigation by my students. As an educator, there is a piece of advice to one and all in the world: *"Permit that our almighty God, our Creator and excellent Teacher, irrigate the life with His Grace of rain trickle by trickle, because our life must continue trickling on..."*

**—Megh R. Goyal, PhD, PE, Senior Editor-in-Chief**
July 14, 2014

# FOREWORD

Since 1978, I have been a research assistant at Agricultural Experiment Substation – Juana Diaz, soil scientist, Chairman of Department of Agronomy and Soils in the College of Agricultural Sciences at the University of Puerto Rico – Mayaguez Campus; and President of University of Puerto Rico (February 2011 to June 2013). I was also an Under-Secretary (1993–1997) and Secretary of the Puerto Rico Agriculture Department (1997–2000). I am privileged to write a foreword for Goyal's three volume book series that is titled *Micro Irrigation Research Advances and Applications*.

I have known Dr. Megh R. Goyal since 1st of October of 1979 when he came from Columbus – Ohio (later I went to study at the OSU to complete my MSc and PhD in Soil Fertility during 1981–1988) to Puerto Rico with his wife and three children. According to his oral story, he had job offers from Texas A&M, Kenya – Nigeria, University of Guelph and my university. He accepted the lowest paid job in Puerto Rico. When I asked why he did so. His straight-forward reply was challenges in drip irrigation offered by this job. With no knowledge of Spanish, Megh survived. He also started learning Spanish language and tasting Puerto Rican food (of course no meat, as he with his family is vegetarian till today).

Within four months of his arrival in Puerto Rico, first drip irrigation system in our university for research on water requirements of vegetable crops was in action. Soon, he formed State Drip Irrigation Committee consisting of experts from university, suppliers and farmers. He published his first 22-page Spanish publication titled, "Tensiometers: Use, service and maintenance for drip irrigation." Soon, he would have graduate students for their MSc research from our College of Agricultural Sciences. I saw him working in the field and laboratory hand in hand with his students. These students would later collaborate with Megh to produce a Spanish book on *Drip Irrigation Management* in 1990. I have personally read this book and have found that it can be easily adopted by different groups of readers with a high school diploma or a PhD degree: farmers, technicians, agronomists, drip irrigation suppliers and designers, extension workers, scientists. Great contribution for Spanish speaking users!

Megh is a fluent writer. His research studies and results started giving fruits with at least one peer-reviewed publication on drip irrigation per month. Soon, our university researchers will have available basic information on drip irrigation in vegetable and tree crops so that they could design their field experiments. Megh produced research publications not only on different aspects of drip irrigation, but also on crop evapotranspiration estimations, crop coefficients, agroclimatic data, crop water requirements, etc. I had a chance to review his two latest books by Apple Academic Press Inc.: *Management of Drip/Trickle or Micro Irrigation*, published (2013) and *Evapotranspiration: Research Advances and Applications for Water Management*, (2014) and wrote a foreword for both books. I am impressed with professional organization of the contents in each book that indicates his relationship with the world educational community. Now he has written three - volume series on "Research Advances in Sustainable Micro Irrigation," that is due for out of press in the middle of next year. My appreciation to Megh for his good work and contribution on micro irrigation; and for this he will always be remembered among the educational fraternity today, tomorrow and forever.

Conventional irrigation systems, such as gravity irrigation and flooding tend to waste water as large quantities are supplied to the field in one go, most of which just flows over the crop and runs away without being taken up by the plants. Micro irrigation keeps the water demand to a minimum. It has been driven by commercial farmers in arid regions of the United States of America and Israel in farming areas where water is scarce. Micro irrigation, also known as trickle irrigation or drip irrigation or pressurized irrigation, saves water and fertilizer by allowing water to drip slowly to the roots of plants, either onto the soil surface or directly near the root zone, through a network of valves, pipes, tubing, and emitters. It is done through narrow tubes that deliver water directly to the base of the plant. Primitive drip irrigation has been used since ancient times.

Ancient Methods: Centuries ago in the Middle East, farmers developed an efficient way of irrigating trees in desert soil with a minimum of water. If poured directly on the ground, much water flows away from the plant and seeps beyond the reach of the roots. To control the flow of water, farmers buried special unglazed pots near the trees and periodically filled them with water. The water seeped through the clay walls slowly, creating a pocket of wet soil around the tree. Trees grew as well with the pot method as in orchards watered by trench irrigation.

Clay Pipes: Predecessors of today's drip irrigation systems included experiments with unglazed clay pipe systems in Afghanistan in the late 1800s. Research conducted by E. B. House at Colorado State University in 1913 showed that slow irrigation could target the root zone of plants. In Germany during the 1920s, researchers devised controlled irrigation systems based on perforated pipe. None of these systems proved as efficient as modern drip irrigation technology.

Plastics: In the 1950s, plastics molding techniques and cheap polyethylene tubing made micro irrigation systems possible for the first time. Though researchers in both England and France experimented with controlled irrigation, the greatest advancements came from the work of a retired British Water Agency employee—Symcha Blass. In Israel, Blass found inspiration in a dripping faucet near a thriving tree and applied his knowledge of microtubing to an improved drip method. The Blass system overcame clogging of low volume water emitters by adding wider and longer passageways or labyrinths to the tubing. Patented in 1959 in partnership with Kibbutz Harzerim in Israel, the Blass emitter became the first efficient drip irrigation method.

Expansion: By the late 1960s, many farmers in Americas and Australia shifted to this new drip irrigation technology. Typical water consumption decreased from 30 percent to 50 percent. In the 1980s, drip irrigation saw use in commercial landscaping applications. Because the drip irrigation emitter technology focused water below ground in the root zone, these systems saved labor costs by reducing weed growth. With drip systems yards and gardens flourished without sprinklers or manual watering.

Plantations: One of drip irrigation's bigger success stories involves the sugar plantations of Hawaii. Sugar cane fields require irrigation for two years before harvesting, and in Hawaii, the hillside fields make ditch irrigation impractical. Producers abandoned sprinkler systems in favor of the low-flow drip irrigation methods—a conversion which took 16 years. In 1986, 11 sugar plantations in Hawaii had completely shifted to drip irrigation. One plantation spanned 37,000 acres of drip-irrigated sugar cane fields. The total cost of the conversion reached $30 million. Typically, these commercial irrigation systems consist of a surface or buried pipe distribution network using emitters supplying water directly to the soil at regular intervals along the pipework. They can be permanent or portable.

Many parts of the world are now using micro irrigation technology. The systems used by large commercial companies are generally quite complex with an emphasis on reducing the amount of labor involved. Small-scale farmers in developing countries have been reluctant to take up micro irrigation methods due to the initial high investment required for the equipment.

Today, drip irrigation is used by farms, commercial greenhouses, and residential gardeners. Drip irrigation is adopted extensively in areas of acute water scarcity and especially for crops such as coconuts, containerized landscape trees, grapes, bananas, berries, eggplant, citrus, strawberries, sugarcane, cotton, maize, and tomatoes.

Properly designed, installed, and managed, drip irrigation may help achieve water conservation by reducing evaporation and deep drainage when compared to other types of irrigation such as flood or overhead sprinklers since water can be more precisely applied to the plant roots. Some advantages of drip irrigation system are: Fertilizer and nutrient loss is minimized due to localized application and reduced leaching; water application efficiency is high; field leveling is not necessary; fields with irregular shapes are easily accommodated; recycled nonpotable water can be safely used; moisture within the root zone can be maintained at field capacity; soil type plays less important role in frequency of irrigation; soil erosion is minimized; weed growth is minimized; water distribution is highly uniform, controlled by output of each nozzle; labor cost is less than other irrigation methods; variation in supply can be regulated by regulating the valves and drippers; fertigation can easily be included with minimal waste of fertilizers; foliage remains dry, reducing the risk of disease; usually operated at lower pressure than other types of pressurized irrigation, reducing energy costs.

The disadvantages of drip irrigation are: Initial cost can be more than overhead systems; the sun can affect the tubes used for drip irrigation, shortening their usable life; if the water is not properly filtered and the equipment not properly maintained, it can result in clogging; drip irrigation might be unsatisfactory if herbicides or top dressed fertilizers need sprinkler irrigation for activation; drip tape causes extra cleanup costs after harvest; and users need to plan for drip tape winding, disposal, recycling or reuse; waste of water, time and harvest, if not installed properly; highly technical; in lighter soils subsurface drip may be unable to wet the soil surface for germination; requires careful consideration of the installation depth; and the PVC pipes often suffer from rodent damage, requiring replacement of the entire tube and increasing expenses.

Modern drip irrigation has arguably become the world's most valued innovation in agriculture. Drip irrigation may also use devices called microspray heads, which spray water in a small area, instead of emitters. These are generally used on tree and vine crops with wider root zones. Subsurface drip irrigation (SDI) uses permanently or temporarily buried dripper-line or drip tape located at or below the plant roots. It is becoming popular for row crop irrigation, especially in areas where water supplies are limited or recycled water is used for irrigation. Careful study of all the relevant factors like land topography, soil, water, crop and agro-climatic conditions are needed to determine the most suitable drip irrigation system and components to be used in a specific installation.

The main purpose of drip irrigation is to reduce the water consumption by reducing the leaching factor. However, when the available water is of high salinity or alkalinity, the field soil becomes gradually unsuitable for cultivation due to high salinity or poor infiltration of the soil. Thus drip irrigation converts fields in to fallow lands when natural leaching by rain water is not adequate in semiarid and arid regions. Most drip systems are designed for high efficiency, meaning little or no leaching fraction. Without sufficient leaching, salts applied with the irrigation water may build up in the root zone, usually at the edge of the wetting pattern. On the other hand, drip irrigation avoids the high capillary potential of traditional surface-applied irrigation, which can draw salt deposits up from deposits below.

This three-volume compendium brings academia, researchers, suppliers and industry partners together to present micro irrigation technology to partially solve water scarcity problems in agriculture sector. The compendium includes key aspects of micro irrigation principles and applications. I find it user-friendly and easy-to-read and recommend being to be on shelf of each library. My hats high to Apple Academic Press Inc. and Dr. Megh R. Goyal, my longtime colleague.

**Miguel A Muñoz-Muñoz, PhD**
Ex-President of University of Puerto Rico, USA
Professor and Soil Scientist
University of Puerto Rico – Mayaguez Campus
Call Box 9000
Mayaguez, P.R., 00681-9000, USA
Email: miguel.munoz3@upr.edu

June 14, 2014

# FOREWORD

With only a small portion of cultivated area under irrigation and with the scope of the additional area that can be brought under irrigation, it is clear that the most critical input for agriculture today is water. It is important that all available supplies of water should be used intelligently to the best possible advantage. Recent research around the world has shown that the yields per unit quantity of water can be increased if the fields are properly leveled, the water requirements of the crops as well as the characteristics of the soil are known, and the correct methods of irrigation are followed. Significant gains can also be made if the cropping patterns are changed so as to minimize storage during the hot summer months when evaporation losses are high, if seepage losses during conveyance are reduced, and if water is applied at critical times when it is most useful for plant growth.

Irrigation is mentioned in the Holy Bible and in the old documents of Syria, Persia, India, China, Java, and Italy. The importance of irrigation in our times has been defined appropriately by N.D Gulati: "In many countries irrigation is an old art, as much as the civilization, but for humanity it is a science, the one to survive." The need for additional food for the world's population has spurred rapid development of irrigated land throughout the world. Vitally important in arid regions, irrigation is also an important improvement in many circumstances in humid regions. Unfortunately, often less than half the water applied is used by the crop – irrigation water may be lost through runoff, which may also cause damaging soil erosion, deep percolation beyond that required for leaching to maintain a favorable salt balance. New irrigation systems, design and selection techniques are continually being developed and examined in an effort to obtain high practically attainable efficiency of water application.

The main objective of irrigation is to provide plants with sufficient water to prevent stress that may reduce the yield. The frequency and quantity of water depends upon local climatic conditions, crop and stage of growth, and soil-moisture-plant characteristics. Need for irrigation can be determined in several ways that do not require knowledge of evapotranspiration (ET) rates. One way is to observe crop indicators such as change of color or leaf angle, but this information may appear too late to avoid reduction in the crop yield or quality. Other similar methods of scheduling include determination of the plant water stress, soil moisture status, or soil water potential. Methods of estimating crop water requirements using ET and combined with soil characteristics have the advantage of not only being useful in determining when to irrigate, but also enables us to know the quantity of water needed. ET estimates have not been made for the developing countries though basic information on weather data is available. This has contributed to one of the existing problems that the vegetable crops are over irrigated and tree crops are under irrigated.

Water supply in the world is dwindling because of luxury use of sources; competition for domestic, municipal, and industrial demands; declining water quality; and losses through seepage, runoff, and evaporation. Water rather than land is one of the limiting factors in our goal for self-sufficiency in agriculture. Intelligent use of water will avoid problem of sea water seeping into aquifers. Introduction of new irrigation methods has encouraged marginal farmers to adopt these methods without taking into consideration economic benefits of conventional, overhead, and drip irrigation systems. What is important is "net in the pocket" under limited available resources. Irrigation of crops in tropics requires appropriately tailored working principles for the effective use of all resources peculiar to the local conditions. Irrigation methods include border-, furrow-, subsurface-, sprinkler-, sprinkler, micro, and drip/trickle, and xylem irrigation.

Drip irrigation is an application of water in combination with fertilizers within the vicinity of plant root in predetermined quantities at a specified time interval. The application of water is by means of drippers, which are located at desired spacing on a lateral line. The emitted water moves due to an unsaturated soil. Thus, favorable conditions of soil moisture in the root zone are maintained. This causes an optimum development of the crop. Drip/micro or trickle irrigation is convenient for vineyards, tree orchards, and row crops. The principal limitation is the high initial cost of the system that can be very high for crops with very narrow planting distances. Forage crops may not be irrigated economically with drip irrigation. Drip irrigation is adaptable for almost all soils. In very fine textured soils, the intensity of water application can cause problems of aeration. In heavy soils, the lateral movement of the water is limited, thus more emitters per plant are needed to wet the desired area. With adequate design, use of pressure compensating drippers and pressure regulating valves, drip irrigation can be adapted to almost any topography. In some areas, drip irrigation is used successfully on steep slopes. In subsurface drip irrigation, laterals with drippers are buried at about 45 cm depth, with an objective to avoid the costs of transportation, installation, and dismantling of the system at the end of a crop. When it is located permanently, it does not harm the crop and solve the problem of installation and annual or periodic movement of the laterals. A carefully installed system can last for about 10 years.

The publication of this book series and this volume is an indication that things are beginning to change, that we are beginning to realize the importance of water conservation to minimize the hunger. It is hoped that the publisher will produce similar materials in other languages.

In providing this resource in micro irrigation, Megh Raj Goyal, as well as the Apple Academic Press, are rendering an important service to the farmers, and above all to the poor marginal farmers. Dr. Goyal, Father of Irrigation Engineering in Puerto Rico, has done an unselfish job in the presentation of this compendium that is simple and thorough. I have known Megh Raj since 1973 when we were working together at Haryana Agricultural University on an ICAR research project in "Cotton Mechanization in India."

**Gajendra Singh, PhD,**
prof.gsingh@gmail.com, Tel. +91 99 7108 7591
Adjunct Professor, Indian Agricultural Research Institute, New Delhi
Ex-President (2010-2012), Indian Society of Agricultural Engineers
Former Vice Chancellor, Doon University, Dehradun, India.
Former Deputy Director General (Engineering), Indian Council of Agricultural Research (ICAR), New Delhi.
Former Vice-President/Dean/Professor and Chairman, Asian Institute of Technology, Thailand

New Delhi
June 14, 2014

# FOREWORD

In the world, water resources are abundant. The available fresh water is sufficient even if the world population is increased by four times the present population, that is, about 25 billion. The total water present in the earth is about 1.41 billion $Km^3$ of which 97.5% is brackish and only about 2.5% is fresh water. Out of 2.5% of fresh water, 87% is in ice caps or glaciers, in the ground or deep inside the earth. According to Dr. Serageldin, 22 of the world's countries have renewable water supply of less than 1000 cubic meter per person per year. The World Bank estimates that by the year 2025, one person in three in other words 3.25 billion people in 52 countries will live in conditions of water shortage. In the last two centuries (1800–2000) the irrigated area in the world has increased from 8 million-ha to 260 million-ha for producing the required food for the growing population. At the same time, the demand of water for drinking and industries has increased tremendously. The amount of water used for agriculture, drinking, and industries in developed countries are 50% in each and in developing countries it is 90% and 10%, respectively. The average quantity of water is about 69% for agriculture and 31% for other purposes. Water scarcity is now the single threat to global food production. To overcome the problem, there is a compulsion to use the water efficiently and at the same time increase the productivity from unit area. It will involve spreading the whole spectrum of water thrifty technologies that enable farmers to get more crops per drop of water. This can be achieved only by introducing drip/trickle/micro irrigation in large scale throughout the world.

Micro irrigation is a method of irrigation with high frequency application of water in and around the root zone of plant (crop) and consists of a network of pipes with suitable emitting devices. It is suitable for all crops except rice especially for widely spaced horticultural crops. It can be extended to wastelands, hilly areas, coastal sandy belts, water scarcity areas, semi arid zones, and well-irrigated lands. By using micro irrigation, the water saving compared to conventional surface irrigation is about 40–60% and the yield can be increased up to 100%. The overall irrigation efficiency is 30–40% for surface irrigation, 60-70% for sprinkler irrigation, and 85–95% for micro irrigation. Apart from this, one has the advantage of saving of costs related to labor and fertilizer, and weed control. The studies conducted and information gathered from various farmers in India has revealed that micro irrigation is technically feasible, economically viable, and socially acceptable. Since the allotment of water is going to be reduced for agriculture, there is a compulsion to change the irrigation method to provide more area under irrigation and to increase the required food for the growing population.

The farmers in the developing countries are poor and hence it is not possible for them to adopt/install the micro irrigation with fertigation though it is economically viable and profitable. In Tamil Nadu – India, the number of marginal farmers (holding less than 1.0 hectare) and small farmers (holding 1 to 2 ha) has increased from 50,76,915 in 1967–1968 to 71,84,940 in 1995–1996 and area owned by them has also decreased in the same period from 0.63 ha to 0.55 ha. In addition, the small farmers category is about 89.68% in 1995–1996 of the total farmers in the state. At the same time if micro irrigation is used in all crops, yield can be increased and water saving will be 50%. In the case of sugarcane crop, the yield can be increased to 250 tons/ha from the present average yield of 100 tons/ha, which is highest at present in India. Therefore, to popularize the micro irrigation system among this group of farmers, more books like this, not only in English but also in the respective national languages, should be published.

Volumes 1 and 2 in this book series cover micro irrigation status and potentials, reviews of the system, principles of micro irrigation, the experience of micro irrigation in desert region—mainly in the Middle East, and application in the field for various crops, especially in water requirements, like banana, papaya, plantations, tanier, etc. The chapters are written by experienced scientists from various parts of the world bringing their findings, which will be useful for all the micro irrigation farmers in the world in the coming years. I must congratulate Dr. Goyal for taking trouble in contacting and collecting papers from experts on their subjects and publishing nicely in a short time.

Professor Megh R. Goyal is a reputed agricultural engineer in the world and has wide knowledge and experience in soil and water conservation engineering, particularly micro irrigation. After a big success for his first book titled, *Management of Drip/Trickle or Micro Irrigation* by Apple Academic Press Inc., this compendium is unique. Dr Goyal, Senior Editor-in-Chief of this book series, has taken into account the fate of marginal farmers and is thus serving the poor. He has contacted/consulted many experts who are involved in the subject matter to bring the experience and knowledge about micro irrigation to this book. He has also given many figures, illustrations and tables to understand the subject. I congratulate the author for writing this valuable book series. The information provided in this book series will go a long way in bringing micro irrigation the world especially in water scarcity countries. On behalf of Indian scientists and agricultural engineers on micro irrigation, I am indebted to Dr. Megh R. Goyal and Apple Academic Press for undertaking this project.

**Professor (Dr.) R. K. Sivanappan,**
Email: sivanappan@hotmail.com
Former Dean-cum-Professor of College of Agricultural
Engineering and Founding Director of Water Technology
Centre at Tamil Nadu Agricultural University [TAMU],
Coimbatore – India. Ex-member of Tamil Nadu State
Planning Commission (2005–2006).
Father of Micro irrigation in India as mentioned by Mrs.
Sandra Postel in her book *Pillar of Sand — Can the
Irrigation Miracle Last* by W. W. Norton and Company –
New York.
Recipient of Honorary PhD. degree by Linkoping University
– Sweden; and conferment of the honorary DSc degree by the
TAMU-India.

June 14, 2014
Coimbatore—
India

# FOREWORD

The micro irrigation system, more commonly known as the drip irrigation system, was one of the greatest advancements in irrigation system technology developed over the past half century. The system delivers water directly to individual vines or to plant rows as needed for transpiration. The system tubing may be attached to vines, placed on or buried below the soil surface.

This book, written by experienced system designers/scientists, describes various systems that are being used around the world, the principles of micro irrigation, chemigation, filtration systems, water movement in soils, soil-wetting patterns, crop water requirements and crop coefficients for a number of crops. It also includes chapters on hydraulic design, emitter discharge and variability, and pumping station. Irrigation engineers will find this book to be a valuable reference.

**Marvin E. Jensen, PhD, PE**
Retired Research Program Leader at USDA-ARS; and
Irrigation Consultant
1207 Spring Wood Drive, Fort Collins, Colorado 80525, USA.
Email: mjensen419@aol.com

June 14, 2014

# BOOK SERIES: RESEARCH ADVANCES IN SUSTAINABLE MICRO IRRIGATION

**Volume 1: Sustainable Micro Irrigation: Principles and Practices**
Senior Editor-in-Chief: Megh R. Goyal, PhD, PE

**Volume 2: Sustainable Practices in Surface and Subsurface Micro Irrigation**
Senior Editor-in-Chief: Megh R. Goyal, PhD, PE

**Volume 3: Sustainable Micro Irrigation Management for Trees and Vines**
Senior Editor-in-Chief: Megh R. Goyal, PhD, PE

**Volume 4: Management, Performance, and Applications of Micro Irrigation**
Senior Editor-in-Chief: Megh R. Goyal, PhD, PE

# ABOUT THE SENIOR EDITOR-IN-CHIEF

Megh R. Goyal received his BSc degree in Engineering in 1971 from Punjab Agricultural University, Ludhiana, India; his MSc degree in 1977 and PhD degree in 1979 from the Ohio State University, Columbus; his Master of Divinity degree in 2001 from Puerto Rico Evangelical Seminary, Hato Rey, Puerto Rico, USA. He spent a one-year sabbatical leave in 2002–2003 at Biomedical Engineering Department, Florida International University, Miami, USA.

Since 1971, he has worked as Soil Conservation Inspector; Research Assistant at Haryana Agricultural University and the Ohio State University; and Research Agricultural Engineer at Agricultural Experiment Station of UPRM. At present, he is a Retired Professor in Agricultural and Biomedical Engineering in the College of Engineering at University of Puerto Rico – Mayaguez Campus; and Senior Acquisitions Editor and Senior Technical Editor-in-Chief in Agriculture and Biomedical Engineering for Apple Academic Press, Inc.

He was the first agricultural engineer to receive the professional license in Agricultural Engineering in 1986 from the College of Engineers and Surveyors of Puerto Rico. On September 16, 2005, he was proclaimed as "Father of Irrigation Engineering in Puerto Rico for the twentieth century" by the ASABE, Puerto Rico Section, for his pioneer work on micro irrigation, evapotranspiration, agroclimatology, and soil and water engineering. During his professional career of 45 years, he has received awards such as: Scientist of the Year, Blue Ribbon Extension Award, Research Paper Award, Nolan Mitchell Young Extension Worker Award, Agricultural Engineer of the Year, Citations by Mayors of Juana Diaz and Ponce, Membership Grand Prize for ASAE Campaign, Felix Castro Rodriguez Academic Excellence, Rashtrya Ratan Award and Bharat Excellence Award and Gold Medal, Domingo Marrero Navarro Prize, Adopted son of Moca, Irrigation Protagonist of UPRM, Man of Drip Irrigation by Mayor of Municipalities of Mayaguez/Caguas/Ponce and Senate/Secretary of Agriculture of ELA, Puerto Rico.

He has authored more than 200 journal articles and textbooks including: *Elements of Agroclimatology* (Spanish) by UNISARC, Colombia; two *Bibliographies on Drip Irrigation*. Apple Academic Press Inc. (AAP) has published his books, namely, *Biofluid Dynamics of Human Body*, *Management of Drip/Trickle or Micro Irrigation*, *Evapotranspiration: Principles and Applications for Water Management*, and *Biomechanics of Artificial Organs and Prostheses*. With this volume, AAP will publish 10-volume set on *Research Advances in Sustainable Micro Irrigation* by Readers may contact him at: goyalmegh@gmail.com.

# WARNING/DISCLAIMER

The goal of this compendium is to guide the world community on how to manage the sustainable surface and subsurface drip/trickle or micro irrigation system efficiently for economical crop production. The reader must be aware that the dedication, commitment, honesty, and sincerity are the most important factors in a dynamic manner for a complete success. It is not a one-time reading of this compendium. Read and follow every time, that it is needed. To err is human. However, we must do our best. Always, there is a space for learning new experiences.

The editor, the contributing authors, the publisher and the printer have made every effort to make this book as complete and as accurate as possible. However, there still may be grammatical errors or mistakes in the content or typography. Therefore, the contents in this book should be considered as a general guide and not a complete solution to address any specific situation in irrigation. For example, one size of irrigation pump does not fit all sizes of agricultural land and to all crops.

The editor, the contributing authors, the publisher and the printer shall have neither liability nor responsibility to any person, any organization or entity with respect to any loss or damage caused, or alleged to have caused, directly or indirectly, by information or advice contained in this book. Therefore, the purchaser/reader must assume full responsibility for the use of the book or the information therein.

The mention of commercial brands and trade names are only for technical purposes. It does not mean that a particular product is endorsed over another product or equipment not mentioned.

All weblinks that are mentioned in this book were active on October 31, 2013. The editors, the contributing authors, the publisher and the printing company shall have neither liability nor responsibility if any of the weblinks is inactive at the time of reading of this book.

**PART I**
# SUBSURFACE MICRO IRRIGATION

# CHAPTER 1

# SUBSURFACE DRIP IRRIGATION: WETTING PATTERN SIMULATION PART I: MODEL DEVELOPMENT

M. N. EL-NESR, S. M. ISMAIL, T. K. ZIEN EL-ABEDIN, and M. A. WASSIF

## CONTENTS

*Partially printed from: "*El-Nesr, M. N., 2006. Subsurface Drip Irrigation System Development and Modeling of Wetting Pattern Distribution. Unpublished PhD Thesis, Faculty of Agriculture, Alexandria University, Egypt* <www.alexu.edu.eg/index.php/en>." Academic and research guidance/support by S.M. Ismail, T.K. Zien El-Abedin, and M.A. Wassif; and by Faculty of Agriculture at Alexandria University is fully acknowledged.

## 1.1   INTRODUCTION

Campbell [5] stated that unlike flood and sprinkler irrigation systems modeling, drip source infiltration modeling cannot be accepted in any way to be modeled in less than two dimensions (2D). However, the drip source infiltration can be represented in a very satisfactory way in 2D cylindrical coordinates for point source and 2D Cartesian coordinates for line source.

Several investigators had developed mathematical models to predict infiltration characteristics from water source. These models are either numerical or analytical, for solving steady or time dependent flow, of surface or subsurface point or line sources. Most of these models dealt only with the wetting pattern shape in bare soil, but some of them dealt also with solute transport or root uptake. Some of these models and their properties are listed in Table 1.

In this chapter, authors discuss the research results on development of a simulation model for both the surface and subsurface drip irrigation systems. The model includes effects of soil type, soil-water characteristics, lateral burying depth, lateral spacing, physical barrier, and multiple tubing.

## 1.2   MODEL DEVELOPMENT

### 1.2.1   GOVERNING EQUATIONS

Vadose zone of the soil is the shallow, unsaturated zone above water table. In the vadose zone of a soil, the soil-moisture is governed by the Darcy's law. The Darcy's law can be applied in the vertical direction ($z$) as follows:

**TABLE 1**   Selected two-dimensional models for water infiltration in soil.

| Model Name | nD | Loc. | Shape | Time | Method | Wf | Tc |
|---|---|---|---|---|---|---|---|
| Brandt et al. [3] | 2 | Sr | Point | Td | Nm | Wf | ADI |
| Bresler [4] | 2 | Sr | Point | Td | Nm | Wf+St | ADI |
| Raats [17] | 2 | Br | Point | Td | Nm | | |
| Raats [18] | 2 | Sr | Point | Td | Nm | | |
| Selim, and Kirkham [23] | 2 | Sr | Trench | Td | Nm | | ADI |
| Zachmann and Thomas [32] | 2 | Sr | Line | Ss | An | | D.A. |
| Warrick [30] | 2 | Sr | Point | Td | An | Wf | |
| Lomen, and Warrick [12] | 2 | Sr | Line | Td | An | | |
| Warrick and Lomen [31] | 2 | Sr | Disk | Td | An | | |
| Gilley, and Allred [8] | 2 | Br | Line | Ss | An | | |
| van Der Ploeg et al. [27] | n | Sr | Point | Td | Nm | | |
| Thomas et al. [24] | 2 | Br | Line | Ss | An | | |
| Raats [19] | 2 | Br | Both | Ss | An | | D.A. |

**TABLE 1** *(Continued)*

| Model Name | nD | Loc. | Shape | Time | Method | Wf | Tc |
|---|---|---|---|---|---|---|---|
| Thomas et al. [25] | 2 | Br | Point | Ss | An | | chart |
| Caussade et al. [6] | 2 | Sr | Point | Td | Nm | | ALN |
| Ahmad [1] | 2 | Br | Line | Td | Nm | | |
| Oron [16] | 2 | Br | Point | Ss | Nm | Ru | |
| Ragab et al. [20] | 2 | Sr | Line | Td | Nm | | |
| Ben-Asher et al. [2] | 2 | Sr | Point | Td | An | | |
| Zazueta et al. [33] | 2 | Sr | Point | Td | An | Wf | D.A. |
| Lafolie et al. [11] | 2 | Sr | Point | Td | Nm | Wf | |
| Morcos et al. [15] | 2 | Sr | Point | Td | Nm | | |
| Zin El-Abedin et al. [34] | 2 | Br | Point | Ss | Nm | Wf | |
| Revol et al. [21] | 2 | Sr | Point | Td | An | Wf | |
| Russo et al. [22] | 2 | Sr | Point | Td | Nm | St+Ru | |
| Šimůnek et al. [11] | 2 | SrBr | Point | Td | Nm | St+Ru | |
| Mmolawa, and Or [16] | 2 | SrBr | Point | Td | An | St | |
| Russo et al. [22] | 3 | Br | Point | Td | Nm | St | |
| Moncef et al. [14] | 2 | Sr | Point | Td | Nm | Wf | ADI |
| Khalifa et al. [9] | 2 | Sr | Point | Td | Nm | | ADI |

**An**: Analytical solution; **Br**: Buried source; **D.A.**: dimensional analysis; **nD**: number of dimensions; **Nm**: Numerical solution; **Ru**: Root uptake study; **Sr**: Surface source; **Ss**: Steady state solution; **St**: solute transport study; **Tc**: Solution technique; **Td**: Time dependent solution; **Wf**: Wetting front determine.

$$v_z = -K(\theta)\left(\frac{\partial h}{\partial \theta} \cdot \frac{\partial \theta}{\partial z} - 1\right) \tag{1}$$

where, $v_z$ is a water vertical velocity (LT$^{-1}$) in $z$ direction (L); K is a soil hydraulic conductivity (LT$^{-1}$) at soil moisture content $\theta$. Darcy's law can be represented in the diffusivity form as follows:

$$v_z = -D(\theta)\frac{\partial \theta}{\partial z} + K(\theta) \tag{2}$$

where, $D(\theta)$ is the water diffusivity of soil (L$^2$T$^{-1}$), defined by Eq. (3):

$$D(\theta) = K(\theta)\frac{\partial h}{\partial \theta} \tag{3}$$

Darcy's law is modified for the unsaturated flow in three dimensions to obtain a nonlinear partial differential equation (known as Richard's equation) as follows.

$$\therefore \frac{\partial \theta}{\partial t} = \frac{\partial}{\partial x}\left[D(\theta)\frac{\partial \theta}{\partial x}\right] + \frac{\partial}{\partial y}\left[D(\theta)\frac{\partial \theta}{\partial y}\right] + \frac{\partial}{\partial z}\left[D(\theta)\frac{\partial \theta}{\partial z}\right] - \frac{\partial}{\partial z}\left[K(\theta)\right] \tag{4}$$

Richard's equation can also been expressed in terms of the Matric Flux Potential (MFP or $S(\theta)$) term in order to make the equation linear. So the diffusivity form of MFP is:

$$S(\theta_d) = \int_{\theta_r}^{\theta_d} D(\theta) \cdot d\theta \tag{5}$$

where, $\theta_d$ is the desired soil moisture content which corresponds to the matric flux potential; $\theta_r$ is the residual water content of the soil.

$$\frac{\partial \theta}{\partial t} = \frac{\partial^2 S(\theta)}{\partial x^2} + \frac{\partial^2 S(\theta)}{\partial y^2} + \frac{\partial^2 S(\theta)}{\partial z^2} - \frac{\partial K(\theta)}{\partial z} \tag{6}$$

However, for a two-dimensional flow in cylindrical ($r$-$z$) coordinates, the Eq. (6) for a cylindrical radius, $r$, is expressed as follows:

$$\frac{\partial \theta}{\partial t} = \frac{\partial^2 S(\theta)}{\partial r^2} + \frac{1}{r}\frac{\partial S(\theta)}{\partial r} + \frac{\partial^2 S(\theta)}{\partial z^2} - \frac{\partial K(\theta)}{\partial z} \tag{7}$$

### 1.2.1.1. SOIL WATER CONDUCTIVITY

Several empirical functions have been proposed to describe the soil water retention curve. One of the most popular functions is a function defined by *Brook and Corey* (BC-equation), given below:

$$\Theta = \frac{\theta - \theta_r}{\theta_s - \theta_r} = \begin{cases} (\alpha h)^{-\lambda_{ps}} & \alpha h > 1 \\ 1 & \alpha h \leq 1 \end{cases} \tag{8}$$

where, $\Theta$ is the effective degree of saturation or the reduced water content; $\theta_r$ and $\theta_s$ are the residual and saturated water contents, respectively; $\alpha$ is an empirical parameter ($L^{-1}$) whose inverse is often referred to as the air entry value or bubbling pressure; and $\lambda_{ps}$ is a pore-size distribution parameter affecting the slope of the retention function; **h** is the soil water pressure head taken positive for unsaturated soils (denotes suction).

The residual water content, $\theta_r$ in Eq. (8) specifies the maximum amount of water in a soil that will not contribute to liquid flow because of blockage from the flow paths or strong adsorption onto the solid phase [13]. It is defined as the water content at which both $d\theta/dh$ and $K$ go to zero when **h** becomes large negative value.

The saturated water content, $\theta_s$ denotes the maximum volumetric water content of a soil. $\theta_s$ is generally about 5 to 10% smaller than the porosity because of entrapped or dissolved air.

The Eq. (9) by *van Genuchten* [VG-equation, 28] is the alternate smooth function for soil water retention curve:

$$\Theta = \frac{\theta - \theta_r}{\theta_s - \theta_r} = \frac{1}{\left(1 + (\alpha h)^n\right)^m} \tag{9}$$

where, $\alpha$, $n$ and $m$ are empirical constants affecting the shape of the retention curve; $m$ can be calculated as a function of $n$ according to $m=1-1/n$ for the Mualem conductivity model and $m=1-2/n$ for Burdine conductivity model. However, most of the soils have $1<n<2$, and $m$ must be positive. Unlike Mualems' model, Burdine's model cannot be used for most of the soil types.

van Genuchten, et al. [28] combined Mualems' conductivity model (MCM) and Burdine's model (BCM) with the VG and BC retention models. Therefore, they derived some smooth, simple, and, durable functions, which are defined below:

1. Hydraulic conductivity model combining MCM and VG models:

$$K\left(\Theta\right) = K_S \Theta^\Upsilon \left(1-\left(1-\Theta^{\frac{1}{m}}\right)^m\right)^2 \tag{10}$$

where, $\Upsilon$ is the pore-connectivity parameter $= 0.5$ as an average for many soils; $K_s$ is the saturated hydraulic conductivity.

2. Soil-water diffusivity $D(\Theta)$ by combining MCM and VG models:

$$D\left(\Theta\right) = \frac{\left(1-m\right)K_S \Theta^{\Upsilon-\frac{1}{m}}}{\alpha m \left(\theta_s - \theta_r\right)} \left(\left(1-\Theta^{\frac{1}{m}}\right)^{-m} + \left(1-\Theta^{\frac{1}{m}}\right)^m - 2\right) \tag{11}$$

Other combinations of the retention-conductivity models are defined by van Genuchten et al. [28].

## 1.2.2 MODELING PROCEDURE AND ASSUMPTIONS

### 1.2.2.1 MODELING TARGETS

The current model consists of several models of soil water relationships. Although, these models agree with the main equation, yet these are for different initial and boundary conditions. The models for the wetting pattern in this chapter were developed to simulate following conditions:

- Single surface dripper with/without a physical barrier. For simplification, these models are called GDF and GDN respectively; where, GD stands for ground dripper, F: with physical isolation, N: with no physical isolation.
- Single subsurface dripper with/without physical barrier, models BDF, and BDN respectively; where, BD stands for buried dripper.
- Double dripper lines, with/without physical barrier, models DDF and DDN respectively; where, DD stands for double dripper-lines. These dripper lines must have at least one of the laterals buried, i.e., either both lines are subsurface, or one surface and one subsurface.

### 1.2.2.2 THEORETICAL ASSUMPTIONS

- The soil is assumed uniform, homogeneous, and isotropic.
- The initial water content $\theta_{ini}$ should be uniform, and no sensible water movement initially in the soil $\theta = \theta_{ini}$.
- Darcy's law applies in both saturated and unsaturated zones.

- During the infiltration process: Soil water at any point in the system can either increase due to infiltration, or remain unchanged, i.e., it cannot decrease at any time.
- The hydraulic conductivity of the soil and all its derived functions: Are differentiable, continuous and single valued functions of the moisture content.

## 1.3   DEVELOPMENT OF THE BASIC WETTING PATTERN GDN MODEL

**For the simplest model "GDN":** Consider a field that is being irrigated with a set of emitters, spaced at $2x$, and $2y$ as shown in Fig. 1: where, $z$ direction is considered positive downward. Therefore, the element under consideration is bounded by the plans X=0, X=x, Y=0, Y=y, Z=0, and Z=z: where, $z$ is not a fixed boundary, but it should be taken far enough of the expected wetting front during the modeled experiment time.

To ensure non interference between emitters or any other wetting sources, the flow must vanish through the plans X=0, X=x, Y=0, Y=y, and Z=z (not Z=0). Therefore, at time $t{>}0$, we have: $\partial\theta/\partial x = 0$ at X plans, $\partial\theta/\partial y = 0$ at Y plans, and $\partial\theta/\partial z = 0$ at Z=z plan.

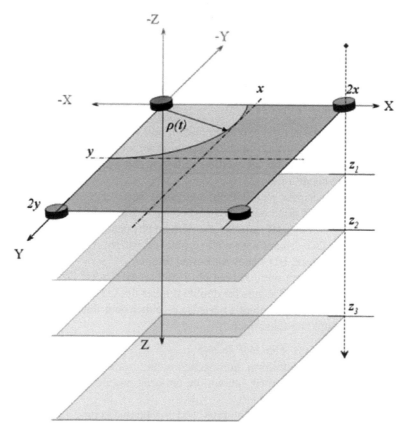

**FIGURE 1**    The elemental scheme under study for ground drip model.

To define the boundary conditions for plan Z=0, several considerations should be taken into account, while developing the model:

- This plan is divided into two zones, the first is the wetted zone, from the emitter (X=0, Y=0) to a moving boundary, which increases with time $\rho(t)$, has a wetting radius $\rho(t) = \sqrt{x^2 + y^2}$. The second zone is the dry zone from $\rho(t)$ to the end of the grid X=x, and Y=y.
- Water flows through the emitter discharge $q$ should instantly infiltrate into the soil, or evaporates to air, hence a film of water pond occur with no runoff, so, the water content within the wetted zone is equal to the water content at saturation $\theta = \theta_s$.
- Outside the wetted zone, no water occurs at surface, so either no water reaches this zone, or the infiltrated water is exactly equals the evaporated water. From Darcy's Eq. (2) for downward water movement, it can be said that: $E = -D(\theta) \cdot \partial \theta / \partial z + K(\theta)$ , where $E$ is the evaporation (LT$^{-1}$), from Eq. (5) $\therefore E = -\partial S / \partial z + K(\theta)$.
- Inside the wetting zone, the integration over horizontal plan of the net water, which evaporates and infiltrates into the soil is equals to the corresponding emitter flow.

$$2\pi \int_0^{\rho(t)} \left[ K + E - \frac{\partial S}{\partial z} \right] r \cdot dr = q \tag{12}$$

### 1.3.1  MODEL SOLUTION METHOD

The model equations should be solved for all the wetting pattern (θ) values throughout the soil matrix. However, Richard equation is an implicit equation of the second order, and it can only be solved by using the numerical methods. Many investigators have found that the best numerical-solution model of the diffusivity equation is the alternative direction implicit difference method (ADI). The ADI method solves the two-dimensional equation in two stages:

In the first stage: Advancing from time $t$ (j) to time $t$ (j+½), instead of $t$ (j+1), the $r$ direction variables are considered explicit, and the other $z$ direction variables are considered implicit, thus the equation can simply be solved.

The resulted values of the first stage are entered to the **second stage** but by inverting the implicit-explicit order, and advancing from time $t$ (j+½), to time $t$ (j+1), i.e., $r$ considered implicit, and $z$ considered explicit.

This technique is very durable, accurate, and efficient In order to solve by ADI, several auxiliary techniques are used. **The first method** involves the simplification of each stage-equation by analytical methods, which can be done by approximating some variables, or variable groups into simpler variables, or constants, and then the equation is solved. These approximations reduce the accuracy of the method but increase the ease and the speed of the solution.

Another method is to combine the iterative Newton-Raphson method with the noniterative ADI method to find the solution of each stage of the latter. This combination has been used by Brandt et al. [3], and will be used to develop the model for the wetting pattern in this chapter.

## 1.3.2 THE NUMERICAL GRID

The square grid is used to solve the finite difference model. The dynamic/static nature of the grid, and its dimensions specifies the accuracy of the solution. Figure 2 shows the grid of the model, given in this chapter. The grid is divided into squares in $r$ and $z$ directions:

$r_i = (i - 0.5)\lambda$   $i$=0, 1, 2, ..., N+1; (integer)

$z_l = (l)\lambda$   $l$=0, 1, 2, ..., M+1; (integer)

$t^j = \sum_{\delta=1}^{j} \tau_\delta$   $j$>=1, $t^0$=0; (real)

where, $\lambda$ is the length of the grid unit so that: $R_{max} \approx N\lambda$; $Z_{max} = \lambda M$; M, and N are integers representing number of grid units in $Z$ and $R$ directions, respectively; and $\tau_\delta$= variable time step.

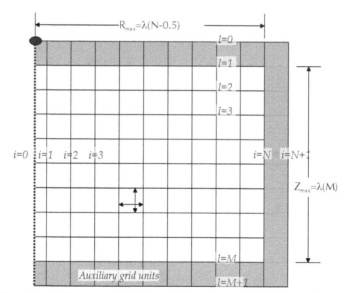

**FIGURE 2**   Finite difference grid for the wetting pattern model in this study.

Moisture content during the simulation process is denoted by location and time. Location coordinates are put in the subscript position, while time is put in the superscript position. For example: $\theta_{i,l}^{j}$, denotes the water content in the horizontal position $i$, and vertical position $l$, on time-step $j$. Sometimes, for brief, the sub and super scripts may be omitted.

Normally, the time step is considered fixed, the smaller the time step is, the more accurate is the results, and the less the rapidity of the solution. Campbell [5] reported that the selection of the time step is a balance between accuracy and speed. In the iterative Newton-Raphson technique (named after Isaac Newton and Joseph Raphson): There is no reason to worry about accuracy, the convergence of the answer will not be achieved unless it is within an acceptable range of error. If the answer converges at a large time step, the final answer will be achieved faster with a satisfactory accuracy. However, since Newton-Raphson method is used in this chapter, a variable time step was used. A medium time-step was assumed first, if convergence occurred, and it is accepted. Otherwise, a smaller time-step will be assumed and so on.

### 1.3.3 THE FIRST STAGE OF ADI METHOD

The main equation of the model in Eq. (7) for **the first stage**, can be transformed to finite difference format as shown in Eq. (13). Note that $r$ groups are explicit while the $z$ groups are implicit.

The corresponding boundary conditions to be satisfied in the same time step are the conditions in which the $\partial/\partial z$ terms appear or in which $z$ is given a certain value not a range, These conditions, as well as the second stage BCs are converted to Finite Difference Model (FDM) Eqs. (14)–(20), as shown below:

$$\frac{\theta_{i,l}^{j+\frac{1}{2}} - \theta_{i,l}^{j}}{\tau_j/2} = \underbrace{\frac{S_{i+1,l}^{j} - 2S_{i,l}^{j} + S_{i-1,l}^{j}}{\lambda^2} + \frac{S_{i+1,l}^{j} - S_{i-1,l}^{j}}{(2i-1)\lambda^2}}_{r\,\text{groups(explicit)}}$$

$$+ \underbrace{\frac{S_{i,l+1}^{j+\frac{1}{2}} - 2S_{i,l}^{j+\frac{1}{2}} + S_{i,l-1}^{j+\frac{1}{2}}}{\lambda^2} - \frac{K_{i,l+1}^{j+\frac{1}{2}} - K_{i,l-1}^{j+\frac{1}{2}}}{2\lambda}}_{z\,\text{groups(implicit)}}$$

(13)

| Format of the Equation | $i$ | $l$ | $j$ | Eq.# |
|---|---|---|---|---|
| $\theta_{i,j}^{0} = \theta_{ini}$ | 0 to N+1 | 0 to M+1 | 0 | (14) |
| $\theta_{0,l}^{j+1} = \theta_{1,l}^{j+1}$ or $S_{0,l}^{j+1} = S_{1,l}^{j+1}$ | 1 | 1 to M | 0 to T | (15) |
| $\theta_{N,l}^{j+1} = \theta_{N+1,l}^{j+1}$ or $S_{N,l}^{j+1} = S_{N+1,l}^{j+1}$ | N | 1 to M | 0 to T | (16) |
| $\theta_{i,M+1}^{j+\frac{1}{2}} = \theta_{i,M}^{j+\frac{1}{2}}$ or $S_{i,M+1}^{j+\frac{1}{2}} = S_{i,M}^{j+\frac{1}{2}}$ | 1 to N | M | 0 to T | (17) |
| $\theta_{i,0}^{j+\frac{1}{2}} = \theta_s$ | 1 to $i_S - 1$ | 0 | 0 to T | (18) |
| $F_i = 2\pi \cdot r\lambda \left( K_{i,0}^{j+\frac{1}{2}} + E_{i,0}^{j+\frac{1}{2}} - \Delta_z S_{i,0}^{j+\frac{1}{2}} / 2\lambda \right) = 0$ | $i_S + 1$ to N | 0 | >0 to T | (19) |
| where, at soil surface, $\Delta_z S_{i,0}^{j+\frac{1}{2}} = 4S_{i,1}^{j+\frac{1}{2}} - 3S_{i,0}^{j+\frac{1}{2}} - S_{i,2}^{j+\frac{1}{2}}$ | | | | |

$$F_{i_s} = q\left(t^{j+\frac{1}{2}}\right) -$$

$$\sum_{i=1}^{i_s-1} 2\pi \cdot r \lambda \left( K_{i,0}^{j+\frac{1}{2}} + E_{i,0}^{j+\frac{1}{2}} - \Delta_z S_{i,0}^{j+\frac{1}{2}} / 2\lambda \right)$$

| | $i_s$ | 0 | >0 to T | (20) |

where: $i_s = \rho(t)/\lambda$;

The system of equations for the first stage is a closed system for each $i$. The number of equations to be solved is $M+2$ equations in $M+2$ unknowns. The unknowns here are $\theta_{i,l}^{j+\frac{1}{2}}$ for fixed $i$ and $j+\frac{1}{2}$, and for $l=0, 1, 2, 3, \ldots M, M+1$). Therefore, for each time-step, there are N systems of the M+2 equations. Brandt, et al. [3], suggested an approximation of the solution by means of vectors, i.e., for each time step. The complete system of equations can be written in a reduced format as follows:

$$\Phi_i\left(\Omega_i\right) = 0 \tag{21}$$

where, $\Omega_i = \left(\theta_{i,0}^{j+\frac{1}{2}}, \theta_{i,1}^{j+\frac{1}{2}}, \theta_{i,2}^{j+\frac{1}{2}}, \theta_{i,3}^{j+\frac{1}{2}}, \cdots \theta_{i,M}^{j+\frac{1}{2}}, \theta_{i,M+1}^{j+\frac{1}{2}}\right)$ = Matrix for unknowns; and,

$\Phi_i = \left(\varsigma_{i,0}, \varsigma_{i,1}, \varsigma_{i,2}, \varsigma_{i,3}, \cdots \varsigma_{i,M}, \varsigma_{i,M+1}\right)$ = Matrix for equations.

The subscripts in the unknowns' matrix denote the location of the variable $\theta$ in the grid, while the subscripts in the equations matrix denote an equation of the system corresponding to such location. For example, $\varsigma_{i,M+1}$ denotes the boundary condition defined in Eq. (17).

Initially all values of $\theta$ should be set to the given initial soil wetness $\theta_{ini}$, therefore, this will be the first guess of $\Omega_i^{(v)}$ for the iteration number, $v$. This means that the initial iteration is denoted by $\Omega_i^{(1)}$. The above system (Matrix system) can be solved using matrices algebra by the solving the following matrix system:

$$\Re_i^{(v)} \bullet \omega_i^{(v)} = -\Phi_i^{(v)} \tag{22}$$

where, $\omega_i^{(v)}$ is the error matrix, $\Re_i^{(v)}$ is the Jacobian matrix. $\Re$ is the $(M+2) \times (M+2)$. Jacobian matrix denotes the partial derivative of each element with respect to its position, that is, $\Re(p, g) = \dfrac{\partial \varsigma_{i,p}}{\partial \theta_{i,g}}$, where: p, g denotes the horizontal and vertical position in the Jacobian matrix, respectively, each equals 0, 1, 2, ... M+1. The details of the Jacobian matrix will be discussed in details in this chapter for each model.

After each time-step, the value of $\Phi_i^{(v)}$ of the next time step equals:

$$\Phi_i^{(v+1)} = \Phi_i^{(v)} + \omega_i^{(v)} \tag{23}$$

where, $\omega_i^{(v)}$ is the error matrix of (M+2) unknowns, however, this is the target of each time step iteration. The matrix system in Eq. (22) needs to be solved to find the error matrix $\omega_i^{(v)}$. If the absolute value of all elements of it are below the allowable error ($\varepsilon$), then the time-step has ended, and the values of the $\Phi_i^{(v+1)}$ matrix contains the proper values of $\theta_{i,l}^{j+\frac{1}{2}}$. Otherwise, if the absolute value of any element of the $\omega_i^{(v)}$ matrix

has exceeded ε, the full time-step should be repeated with the new value of the $\Phi_i^{(v+1)}$.
Assuming convergence of the diffusivity function, the operation will complete within
about seven iteration steps.

### 1.3.3.1 THE JACOBIAN MATRIX $\Re$

The Jacobian matrix in this chapter is a backbone of the model for the wetting pattern
in this chapter. Nonetheless, any differences in the boundary conditions from model
to model will be reflected mainly in this matrix. However, as mentioned before, it is
the local partial differentiation of the zeta function $\zeta_{i,l}$ with respect to the moisture
content variable. For more clarification, Zeta function of the GDN model will be discussed in details here:

For any point in the grid, with the coordinates (i, l), Eq. (13) can be applied except
for the points whose coordinates belong to the range of any of the boundary conditions
(14) to (20). The derivation of the zeta function of the point that only belongs to Eq.
(13) is defined as: zeta equals the difference between the right and the left hand side of
the equation, as shown below:

$$\zeta_{i,l} = \frac{\theta_{i,l}^{j+\frac{1}{2}} - \theta_{i,l}^j}{\tau_j/2} - \left( \frac{S_{i+1,l}^j - 2S_{i,l}^j + S_{i-1,l}^j}{\lambda^2} + \frac{S_{i+1,l}^j - S_{i-1,l}^j}{(2i-1)\lambda^2} \right) K$$
$$- \left( \frac{S_{i,l+1}^{j+\frac{1}{2}} - 2S_{i,l}^{j+\frac{1}{2}} + S_{i,l-1}^{j+\frac{1}{2}}}{\lambda^2} - \frac{K_{i,l+1}^{j+\frac{1}{2}} - K_{i,l-1}^{j+\frac{1}{2}}}{2\lambda} \right)$$

(24)

Multiplying Eq. (24) by $\tau.\lambda^2/2$, we get:

$$\zeta_{i,l} = \lambda^2 \left( \theta_{i,l}^{j+\frac{1}{2}} - \theta_{i,l}^j \right) - \frac{\tau_j}{2} \left[ \left( S_{i+1,l}^j - 2S_{i,l}^j + S_{i-1,l}^j + \frac{S_{i+1,l}^j - S_{i-1,l}^j}{(2i-1)} \right) + \left( S_{i,l+1}^{j+\frac{1}{2}} - 2S_{i,l}^{j+\frac{1}{2}} + S_{i,l-1}^{j+\frac{1}{2}} - \frac{\lambda}{2} \left( K_{i,l+1}^{j+\frac{1}{2}} - K_{i,l-1}^{j+\frac{1}{2}} \right) \right) \right]$$

(25)

For $1 < i < i_s - 1$, and $j = 0$ (at soil surface): Eq. (13) combined with the boundary condition
BC (18) is used. This combination can be transformed into zeta format as shown in
Eq. (26). Zeta function for the rest of the boundary conditions combined with the main
equation, as well as the stated equations, can be summarized in the case of $\zeta_{i,l}$ where
$l = 0, 1, 2, ..., M+1$:

If ($l = 0$, $1 < i < i_s - 1$):  $\zeta_{i,l} = \lambda^2 \left( \theta_{i,0}^{j+\frac{1}{2}} - \theta_s \right)$

(26)

If ($l = 0$, $i = i_s$):

$$\zeta_{i,l} = -\frac{\tau_j}{2} \left[ \left( 2\lambda K_{i_s,0}^{j+\frac{1}{2}} + 2\lambda E_{i_s,0}^{j+\frac{1}{2}} + 3S_{i_s,0}^{j+\frac{1}{2}} - 4S_{i_s,1}^{j+\frac{1}{2}} + S_{i_s,2}^{j+\frac{1}{2}} \right) - \frac{q\left(t^{j+\frac{1}{2}}\right) - \sigma_{i-1}}{\pi r_{i_s}} \right]$$

(27)

where, $\sigma_{i-1} = \sum_{i=1}^{i_s-1} 2\pi \cdot r\lambda \left( K_{i,0}^{j+\frac{1}{2}} + E_{i,0}^{j+\frac{1}{2}} - \frac{\Delta_z S_{i,0}^{j+\frac{1}{2}}}{2\lambda} \right)$, see Eq. (19).

If ($l$=0, N>$i$>$i_s$):

$$\zeta_{i,l} = -\frac{\tau_j}{2}\left(2\lambda K_{i,0}^{j+\frac{1}{2}} + 2\lambda E_{i,0}^{j+\frac{1}{2}} + 3S_{i,0}^{j+\frac{1}{2}} - 4S_{i,1}^{j+\frac{1}{2}} + S_{i,2}^{j+\frac{1}{2}}\right) \tag{28}$$

If (M+1>$l$>0, i≠0):

$$\zeta_{i,l} = \lambda^2\left(\theta_{i,l}^{j+\frac{1}{2}} - \theta_{i,l}^{j}\right) - \frac{\tau_j}{2}\left[\begin{array}{l} S_{i,l+1}^{j+\frac{1}{2}} - 2S_{i,l}^{j+\frac{1}{2}} + S_{i,l-1}^{j+\frac{1}{2}} - \frac{\lambda}{2}\left(K_{i,l+1}^{j+\frac{1}{2}} - K_{i,l-1}^{j+\frac{1}{2}}\right) + \\ S_{i+1,l}^{j} - 2S_{i,l}^{j} + S_{i-1,l}^{j} + \frac{1}{2i-1}\left(S_{i+1,l}^{j} - S_{i-1,l}^{j}\right) \end{array}\right] \tag{29}$$

If ($l$=M+1, i≠0): $\quad \zeta_{i,l} = -\frac{\tau_j}{2}\left(S_{i,M}^{j+\frac{1}{2}} - S_{i,M+1}^{j+\frac{1}{2}}\right)$ (30)

It should be mentioned that these equations are only applied to the GDN model, and may vary totally or partially if the model changed.

### 1.3.3.1.1 THE COMPONENTS OF THE JACOBIAN MATRIX

The Jacobian matrix components are the partial derivative of the zeta function with respect to the moisture content, where:

$$\left[\Re_i^{(v)}\right]_{p,g}^{j+\frac{1}{2}} = \frac{\partial\zeta_{i,p}}{\partial\theta_{i,g}^{j+\frac{1}{2}}} \tag{31}$$

where, $v$ is the number of iteration trial, $p$ and $g$ are the position of the matrix, each having the values of $l$ = 0, 1, 2, ..., M, M+1. To understand how to evaluate the Jacobian matrix, consider an example and let the Jacobian for the first stage is 9×9 matrix (for a 7×7 grid), starting from row 0 to row 8, and from column 0 to column 8:

$$\left[\Re_i^{(v)}\right]_{0,0}^{j+\frac{1}{2}} = \frac{\partial\zeta_{i,0}}{\partial\theta_{i,0}^{j+\frac{1}{2}}}$$

Notice that mismatching means that the implicit suffix, so that the function is differentiated to, does not equal the corresponding suffix in the equation, i.e.: in the previous equation, differentiation is performed with respect to $\theta_{i,l=0}$ so that $l$ value here is fixed to zero. Therefore, any variable containing $\theta$ with any other value of $l$ other than the zero is considered a fixed value (not variable) so that its derivative is zero. Other examples of the Jacobean can be found in El-Nesr [7].

$$\therefore \left[ \mathfrak{R}_i^{(v)} \right]_{0,0}^{j+\frac{1}{2}} = \frac{\partial \zeta_{i,0}}{\partial \theta_{i,0}^{j+\frac{1}{2}}} = \begin{cases} \lambda^2 & i < i_S \\ \dfrac{-\tau}{2}\left(2\lambda\left[\dfrac{\partial K}{\partial \theta} + \dfrac{\partial E}{\partial \theta}\right] + 3\dfrac{\partial S}{\partial \theta}\right) & i \geq i_S \end{cases}$$

It can be observed that the major components of the Jacobian matrix are zero values. However, the first stage Jacobian matrix is semi-tri-diagonal matrix (banded matrix), which can be solved by some methods other than normal gauss-related methods. The full Jacobian of the first stage is shown in the matrix in Fig. 3, where,

$$D = \frac{\partial S}{\partial \theta}, \hat{K} = \frac{\partial K}{\partial \theta}, \hat{E} = \frac{\partial E}{\partial \theta}.$$

### 1.3.4 THE SECOND STAGE OF ADI METHOD

The finite difference transformation of the equation for the second stage is shown below:

$$\underbrace{\frac{\theta_{i,l}^{j+1} - \theta_{i,l}^{j+\frac{1}{2}}}{\tau_j/2} = \frac{S_{i+1,l}^{j+1} - 2S_{i,l}^{j+1} + S_{i-1,l}^{j+1}}{\lambda^2} + \frac{S_{i+1,l}^{j+1} - S_{i-1,l}^{j+1}}{(2i-1)\lambda^2}}_{r \text{ groups (implicit)}} + \underbrace{\frac{S_{i,l+1}^{j+\frac{1}{2}} - 2S_{i,l}^{j+\frac{1}{2}} + S_{i,l-1}^{j+\frac{1}{2}}}{\lambda^2} - \frac{K_{i,l+1}^{j+\frac{1}{2}} - K_{i,l-1}^{j+\frac{1}{2}}}{2\lambda}}_{z \text{ groups (explicit)}} \qquad (32)$$

In Eq. (32), it can be noticed that $r$ groups are implicit while the $z$ groups are explicit. In addition, the current known time-step is $j+\frac{1}{2}$, while the unknown time step is $j+1$.

The corresponding boundary conditions to be satisfied in the same time step are the conditions in which the $\partial/\partial r$ terms appear or in which $r$ is given a certain value not a range. These conditions can be converted to Finite Difference Model FDM as shown in Eqs. (15), and (16).

The system of equations for the second stage is a closed system for each $l$. The number of equations to be solved is $N+2$ equations for $N+2$ unknowns. The unknowns here are $\theta_{i,l}^{j+1}$ for fixed $l$ and $j+1$, and $i=0, 1, 2, 3, \ldots N, N+1$). Thus for each time-step, there are M systems of the $N+2$ equations. Same reduced format of the first stage in Eq. (21)) can be applied.

$$\left[ \Phi_l^{(v)} \right]_p = \zeta_{p,l}\left( \Omega_l^{(v)} \right) \qquad (33)$$

where, $\Omega_l = \left( \theta_{0,l}^{j+1}, \theta_{1,l}^{j+1}, \theta_{2,l}^{j+1}, \theta_{3,l}^{j+1}, \cdots \theta_{N,l}^{j+1}, \theta_{N+1,l}^{j+1} \right)$ = Matrix for unknowns;

and: $\Phi_l = \left( \zeta_{0,l}, \zeta_{1,l}, \zeta_{2,l}, \zeta_{3,l}, \cdots \zeta_{N,l}, \zeta_{N+1,l} \right)$ = Matrix for equations.

As discussed in the first stage, zeta function can be expressed as follows:

• if $(p=0, l=1 \rightarrow M)$:

$$\zeta_{p,l} = -\tau_j \left( S_{1,l}^{j+1} - S_{0,l}^{j+1} \right)/2$$

- if $(p \neq 0, l \neq 0)$:

$$\zeta_{p,l} = \lambda^2 \left( \theta_{i,l}^{j+1} - \theta_{i,l}^{j+\frac{1}{2}} \right) - \frac{\tau_j}{2} \left[ \begin{array}{l} \left(1 + \frac{1}{2i-1}\right) S_{i+1,l}^{j+1} - 2S_{i,l}^{j+1} + \left(1 - \frac{1}{2i-1}\right) S_{i-1,l}^{j} \\ + S_{i,l+1}^{j+\frac{1}{2}} - 2S_{i,l}^{j+\frac{1}{2}} + S_{i,l-1}^{j+\frac{1}{2}} - \frac{\lambda}{2} \left( K_{i,l+1}^{j+\frac{1}{2}} - K_{i,l-1}^{j+\frac{1}{2}} \right) \end{array} \right]$$

- if $(p=N+1, l \neq 0)$:

$$\zeta_{p,l} = -\tau_j \left( S_{N,l}^{j+1} - S_{N+1,l}^{j+1} \right) / 2$$

## 1.4 DEVELOPMENT OF THE SUBSURFACE MODEL WITH NO PHYSICAL BARRIER (BDN)

The difference between the GDN and BDN models are:

- At the soil surface as $l=0$, no in-flow occurs. Thus only evaporation and upward water movement may occur.
- At the location of the emitter $l=z_e$, the flow occurs in all directions (up, down, left, right, and in-between). However, the flow from the source is theoretically equal in all directions assuming the soil is uniform and no gravity effect, actually, the gravity effect plays a big role of the deformation of the wetting bulb.
- A subsurface special condition must be defined instead of the expanding circle (Onion shape pattern) at the soil-surface of the ground drip modeling. This implies that to simulate subsurface water source in a grid, sum of the grid elements around source must be assumed "always-saturated" elements and these elements must not be less than the diameter of the dripper line (i.e., about 2 cm or one grid unit at least). During the development of the model in this chapter, this method succeeds, only when some of the conditions of the method are modified and developed. This significantly improved the accuracy of the simulation.
- Zeta functions of the first stage of the GDN model differs from that of the BDN model. However, the zeta function of the second stage is identical for all models, as the variation between model occurs in the implicit $z$ direction, hence occurs in the stage in which $z$-direction is implicit.

Similar to the GDN model, the zeta function of the first stage and the corresponding Jacobian matrix of the BDN are shown below, for: $g_s$ is the saturated grid length, and $p=0, 1, 2, ..., M+1$.

- if $(p=M+1, i \neq 0)$:

$$\zeta_{i,p} = -\tau_j \left( S_{i,M}^{j+\frac{1}{2}} - S_{i,M+1}^{j+\frac{1}{2}} \right) / 2$$

- if $(p=z_e$ or $p=z_e+1)$, and $(i \leq g_s)$:

$$\zeta_{i,p} = \tau_j \left( \theta_{i,z_e}^{j+\frac{1}{2}} - \theta_S \right) / 2$$

- if $(p=0$, any $i)$:

$$\zeta_{i,p} = -0.5 \pi \tau_j r_i \left( 2\lambda K_{i,0}^{j+\frac{1}{2}} + 2\lambda E_{i,0}^{j+\frac{1}{2}} + 3S_{i,0}^{j+\frac{1}{2}} - 4S_{i,1}^{j+\frac{1}{2}} + S_{i,2}^{j+\frac{1}{2}} \right)$$

- Another values of $p$ and $i$ other than above

$$\zeta_{i,p} = \lambda^2 \left( \theta_{i,l}^{j+\frac{1}{2}} - \theta_{i,l}^{j} \right) - \frac{\tau_j}{2} \left[ \begin{array}{l} S_{i,l+1}^{j+\frac{1}{2}} - 2S_{i,l}^{j+\frac{1}{2}} + S_{i,l-1}^{j+\frac{1}{2}} - 0.5\lambda \Big/ \left( K_{i,l+1}^{j+\frac{1}{2}} - K_{i,l-1}^{j+\frac{1}{2}} \right) + \\ S_{i+1,l}^{j} - 2S_{i,l}^{j} + S_{i-1,l}^{j} + \left( S_{i+1,l}^{j} - S_{i-1,l}^{j} \right) \Big/ (2i-1) \end{array} \right]$$

## 1.5   DEVELOPMENT OF THE BILATERAL MODEL WITH NO PHYSICAL BARRIER, DDN

The DDN model refers to two dripper lines, with similar discharge and at least one is buried. In case that one emitter is on ground and the other is buried, a mixed model of GDN and BDN is obtained.

- While dealing with the surface grid layer ($l$=0), equations of the GDN model are considered.
- For the grid layer ($l$=$z_e$), equations of the BDN model are considered.
- For other layers, both models are identical.
- In case that both drippers are buried, the BDN model only is considered. However, the ($l$=$z_e$) conditions are expanded applicable on the two layers in which any of the emitters is located.

## 1.6   SIMULATION FOR PRESENCE OF PHYSICAL BARRIER

Physical barrier was simulated in in this chapter by forcing all the grid points in the barrier's zone to be at $\theta_{begin}$ at the end of every time step. Therefore, no moisture could pass through and water accumulation occur.

### 1.6.1   THE PROCEDURE BACKBONE

The main procedure in any of the mentioned modules calculates the values of the matrix of $\theta_{i,l}^{j+1}$, when the values of the $\theta_{i,l}^{j}$ matrix are known. As described before in this chapter, this can be solved by the ADI method combined with Newton-Raphson iteration, in two stages: Each stage deals with only one variable as implicit, and the other is explicit. The main steps of the GDN model is given by El-Nesr [7].

## 1.7   MODELING DIFFICULTIES AND SPECIAL CONSIDERATIONS

In the previous sections of this chapter, the theoretical steps of model development are planned and shown. However, while translating these steps to numerical form or to computer language, some difficulties are encountered. These difficulties with solutions and special considerations are discussed in this section.

### 1.7.1   MATRICES SOLUTION

Most of the matrices used in this chapter are tridiagonal or semitridiagonal (banded). Both types can be converted to a compact form and solved in a BW M matrix instead of N M matrix, where BW is the band width of the matrix; M and N are number of rows and columns in the matrix. In the current model and any complicated finite difference models, many near-zero values occur in the calculations especially when using Newton iteration method combined with FDM. These near-zero values occurs mostly in the initial calculation steps, and can cause a "division-by-zero error" in the computer program even when using double precession variables, as the near-zero values in most are very close to zero i.e., about $10^{-30}$ or less. Under these conditions: Pivoting mostly causes the program to crash, even if not, it will lead to inaccurate results, due to division approximation. Unlike pivoting method of solving normal matrices, Walker et al. [29] showed a partial-pivoting (PP) solution of the banded systems derived from the LU decomposition method of matrices solution.

Thorson [26] showed two algorithms for solving banded systems, one of the algorithms is PP and the other is no-pivoting (NP) solution. The partial pivoting algorithms are almost the same, and lead to same results. The comparison was between NP and PP; and both methods were tested with the current model. The NP solution was better as it is more stable (cause of no divisions), more accurate (no truncation errors), less memory-consuming (PP requires 150% more memory), and faster (PP requires 225% more time) than the partial-pivoting algorithm. PP is more suitable for the solution of wave equations as suggested by Thorson [26].

### 1.7.2   MOISTURE FIELD OVER RELAXATION

In solving any system by Newton-Raphson method, the function must be continuous, smooth, and converging, as the method finds the desired root by convergence from guess to guess. Each guess is the root of the tangent of the previous guess. In some cases no solution can be obtained or the computer stops the execution (halts) process. This can happen because the formation of the function; or because the first-guess-point given leads to a second point, but the second point leads to the first point again and so on. In this situation, the function is called over-relaxed function at point $x$. The solution is obtained in this case by changing the first guess even slightly to escape from the over relaxation point.

In our model, to ensure no over relaxation occur in the function, every guess of the Newton Jacobian field is stored and compared to the next guesses in the same time-step. If no-change or circulatory-change occurs, the initial field is incremented uniformly by half percent of the previous succeeding time-step. Sometimes, the first-guess field is just **near** the over relaxation point, in this way the convergence will take much time, and may lead to unexpected results. This situation is considered in the volume balance check.

### 1.7.3   VOLUME BALANCE CHECK

After each time-step, the corresponding moisture pattern is identified and is rounded up to the difference in data type, as the computation is executed in double precession mode, while storage in single precession, sometimes errors occur due to matrices inversion and solution rounding too. These rounding errors may accumulate and hence generate an uncontrollable error. Thereby, before moving to next step, a **volume balance check** should be made, if the calculated volume-increase was not equal (with some tolerance) to the expected or actual volume, which equals emitter discharge multiplied by elapsed time, then the calculated wetting pattern should be adjusted to meet the actual volume. However, actual wetted volume is known in hole not in detailed pattern. Volume balance computation, comparison and adjustment are shown as follows:

$$V_a = q \cdot \sum_{j=0}^{t} \tau_j \tag{34}$$

$$V_c = \sum_{l=1}^{M-1} \sum_{i=1}^{N-1} \left( \theta_{i,l}^j + \theta_{i+1,l}^j + \theta_{i,l+1}^j + \theta_{i+1,l+1}^j - 4\theta_b \right) \cdot \frac{\lambda\pi}{4} \left( R_c^2(i) - R_c^2(i-1) \right) \tag{35}$$

where, $V_a$: actual volume, $q$: emitter discharge, $t$: simulation time, $j$: time counter, $\tau_j$: time-step, $V_c$: calculated volume, $i$, $l$: current grid location in horizontal and vertical directions respectively, $N$, $M$: end grid location in horizontal and vertical directions respectively, $\theta$: moisture content, $\theta_b$: beginning moisture content, $\lambda$: grid spacing, $R_c$: radius far from emitter.

*BALANCING PROCEDURE:*

$$VBR = V_c / V_a \tag{36}$$

$$UBR = \left( V_a - V_c \right) / V_a \tag{37}$$

$$MUR = \min\left( \frac{UBR}{2}, \varepsilon_b \right) \tag{38}$$

Do a loop from $l = 1$ to $l = M - 1$
    Do a loop from $i = N$ to $i = 1$
      $RR = 1 - \left( R_c(i) / R_c(N) \right)$
      $\theta_{i,l}^j = \min\left( \theta_{sat}, \ \theta_{i,l}^j + \left( \theta_{i-1,l}^j - \theta_b \right) \times MUR \times RR \right)$       (39)
    Continue loop $i$
Continue loop $l$

Do a loop from $i = 1$ to $i = N$
    $RR = 1 - \left( R_c(i) / R_c(N) \right)$
    Do a loop from $l = M$ to $l = 2$
      $\theta_{i,l}^j = \min\left( \theta_{sat}, \ \theta_{i,l}^j + \left( \theta_{i,l-1}^j - \theta_b \right) \times MUR \times RR \right)$       (40)
    Continue loop $l$
Continue loop $i$

where, $VBR$ = volume balance ratio; $UBR$ = unbalance ratio; $MUR$ = modified unbalance ratio; $\varepsilon_b$ = small decimal less than 0.1 to ensure gradualism of volume balance even if severe unbalance occur; $l$ = vertical direction grid numbering; $M$ = maximum grid number of depth; $i$ = horizontal grid numbering; $N$ = maximum grid spacing from emitter position; $RR$ = element radius ratio; $R_c(i)$ and $R_c(N)$ = radius from emitter of element at grid position $i$, $N$ respectively; and $\theta_{sat}$ = soil moisture content at saturation.

In the procedure of volume balance described here: the actual and calculated volumes are compared to both volume balance ratio, and unbalance ratio as presented in steps (36) and (37). Sometimes unbalance ratio reaches big value (up to ±20%), and therefore, correcting these values using any algorithm may lead to lack of confidence in results of the model. However, the ratio of unbalance was restricted to a maximum value of $\varepsilon_b$, or half of *UBR* (which is less), as shown in Eq. (38). Normally $\varepsilon_b$ should not exceed 0.05 because the volume balance is made mainly to escape from the over-relaxation field, after applying such small volume-correction. The whole time-step is repeated to ensure that the output pattern is actually well-adjusted by ADI and the Newton-Raphson method.

As in the ADI method, volume balance modification occurs in two stages one on $Z$ direction, and the other on $R$ direction. The double loop in Eq. (39) applies corrections in $Z$ direction, while the double loop in Eq. (40) applies corrections in $R$ direction.

In each correction stage, the moisture content of the current element is raised or lowered by a small fraction equal to the moisture unbalance ratio *MUR* multiplied by the amount of moisture increase in the previous cell. Hence the errors occur mostly beside saturation, then the correction fraction is multiplied by radius ratio *RR*, which gives bigger corrections to element with less diameter (more close to the emission source). After applying the correction, the resultant moisture content must not exceed the saturation water content $\theta_{sat}$.

## 1.8  EFFECTS OF SOIL-WATER RELATIONSHIPS

Our model depends on three soil water characteristics: hydraulic conductivity $K$, diffusivity $D$, and matric flux potential $S$. For the first stage of the model construction, the equations by van Genuchten et al. [28] were used.

### 1.8.1  DIFFUSIVITY

The main problem of diffusivity is the extreme value: the diffusivity equations of VG retention model with either Mualem's or Burdines' conductivity models. In both models, the diffusivity tends to infinity or zero when the water content function tends to 0 or 1, respectively:

$$\lim_{\Theta \to 1} D = \infty, \qquad \lim_{\Theta \to 0} D = 0 \qquad (41)$$

The diffusivity is defined below:

$$D(\Theta) = \int_0^\Theta \max\left\{\varepsilon, \min\left\{\frac{1}{\varepsilon}, \frac{D(u)}{K_s}\right\}\right\} du. \qquad (42)$$

where. $\varepsilon$, is a small positive value, $u$ is an integration temporary variable represents the water content variable.

### 1.8.2  MATRIC FLUX POTENTIAL (MFP)

The matric flux potential is the limited integration of the diffusivity with respect to water content. The only way to determine MFP is a numerical integration. However, doing so in a model that uses thousands of iterations through Newton's iterative method, leads to billions of operations to reach only one step of the mentioned procedure. Therefore, thousands of numerical integration steps must be reduced through the use of curve fitting.

| | $g=0$ | 1 | 2 | 3 | 4 | □ □ □ | $M-2$ | $M-1$ | $M$ |
|---|---|---|---|---|---|---|---|---|---|
| $p=0$ | $\begin{bmatrix}\lambda^2 & i<i_s \\ -\tau\big(\lambda[\bar{K}+\bar{E}]+3D/2\big) & i\geq i_s\end{bmatrix}$ | $\begin{bmatrix}0 & i\leq i_s \\ 2\tau D & i>i_s\end{bmatrix}$ | $\begin{bmatrix}0 & i\leq i_s \\ -\tau D/2 & i>i_s\end{bmatrix}$ | 0 | 0 □ □ □ | | 0 | 0 | 0 |
| 1 | $\frac{-\tau}{4}(\lambda\bar{K}+2D)$ | $\lambda^2+\tau D$ | $\frac{\tau}{4}(\lambda\bar{K}-2D)$ | 0 | □ □ □ □ | | 0 | 0 | 0 |
| 2 | 0 | $\frac{-\tau}{4}(\lambda\bar{K}+2D)$ | $\lambda^2+\tau D$ | $\frac{\tau}{4}(\lambda\bar{K}-2D)$ | □ □ □ □ | | 0 | 0 | 0 |
| 3 | 0 | 0 | $\frac{-\tau}{4}(\lambda\bar{K}+2D)$ | $\lambda^2+\tau D$ | • □ □ □ | | □ | □ | □ |
| 4 | □ | □ | □ | $\frac{-\tau}{4}(\lambda\bar{K}+2D)$ | • • □ □ | | □ | □ | □ |
| □ | □ | □ | □ | □ | □ • • • □ | | □ | □ | □ |
| □ | □ | □ | □ | □ | □ • • • | | □ | □ | □ |
| □ | □ | □ | □ | □ | □ □ • • | $\frac{\tau}{4}(\lambda\bar{K}-2D)$ | □ | □ | |
| □ | □ | □ | □ | □ | □ □ □ • | $\lambda^2+\tau D$ | $\frac{\tau}{4}(\lambda\bar{K}-2D)$ | □ | |
| $M-1$ | 0 | 0 | 0 | 0 | □ □ □ □ | $\frac{-\tau}{4}(\lambda\bar{K}+2D)$ | $\lambda^2+\tau D$ | $\frac{\tau}{4}(\lambda\bar{K}-2D)$ | |
| $M$ | 0 | 0 | 0 | 0 | □ □ □ □ | 0 | $\frac{-\tau}{4}(\lambda\bar{K}+2D)$ | $\lambda^2+\tau D$ | $\frac{\tau}{4}$ |
| $M+1$ | 0 | 0 | 0 | 0 | □ □ □ □ | 0 | 0 | $\frac{-\tau}{2}D$ | |

**FIGURE 3**  The Jacobean of the first stage to the GDN model.

By inspection, the shape of the MFP function seems to be power shape, but fitting MFP to power curve fails because of the zero values at the beginning of the integration (at $\theta=\theta_r$). Removing the zero pair leads to high correlation coefficient and very low standard error value. The polynomial fit is an alternate acceptable fitting method that only occurs in the 19th degree polynomial. However, after fitting MFP to any of the mentioned models, unexpected results occur and some indefinite loops let the model to freeze. Moreover, some negative values of the fitted function appear besides zero due to the coarseness of the initial function. This leads to ignoring functional-fitting of the MFP, and considering the Bézier curve (splines) interpolation.

Spline is a polynomial function that can have a locally very simple form, globally flexible and smooth. Splines are very useful for modeling arbitrary and coarse functions. One of the most widely used type of splines is cubic splines. A cubic spline is a spline constructed of piecewise third-order polynomials which pass through a set of $m$ control points. The second derivative of each polynomial is commonly set to zero at the endpoints, since this provides a boundary condition that completes the system of $m-2$ equations. This produces a so-called "natural" cubic spline and leads to a simple tridiagonal system, which can be solved easily to give the coefficients of the polynomials.

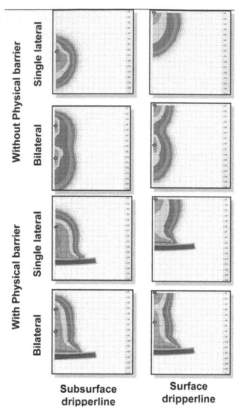

**FIGURE 4**    The interface of "Drip Chartist" model simulating surface, subsurface, and bilateral drip irrigation with and without physical barrier.

Cubic spline interpolation is a useful technique to interpolate between known data points due to its stable and smooth characteristics. Unfortunately, it does not prevent overshoot at intermediate points, which is essential for many engineering applications [10].

Kruger [10] suggested a solution of the intermediate points of the cubic splines. This technique is called constrained cubic splines. The fitting procedure starts after defining all soil data, then 20 points of moisture ratio ($\Theta$) were selected at equal intervals from 0 to 1, the corresponding MFP function was normally calculated, then 19 polynomial were fitted to the points. After fitting the MFP function using constrained cubic splines, no extreme values detected, no program hangings occurred, and the modeling time reduces to less than 10% of what it was before applying spline fittings.

Sample screen photos of the model are shown in Fig. 3. Drip Chartist succeeded to simulate surface and subsurface drip irrigation, with single and bilateral tubing, and with/without physical barrier as shown in Fig. 4.

## 1.9   CONCLUSIONS

The current model (called "Drip Chartist") was developed using the "Microsoft Visual Basic™ for Applications" language (VBA). However, this language is a special version of the original visual basic language, which is designed to work directly on any system containing Microsoft Office™ applications with no need of special dynamic links or registry entries for the program to work.

Drip Chartist succeeded to model surface and subsurface drip irrigation, with single and bilateral tubing, and with/without physical barrier. The model is fast and stable in all types of soil textures. It is suitable to monitor the effect of several design parameters, soil properties, and solution techniques on the wetting pattern shape. It was validated by field experiments and by comparing its results to other verified model. Model validation results are published in next chapter of this book.

## 1.10 SUMMARY

A computer model was developed to simulate surface and subsurface drip irrigation systems. The model combines the alternate direction implicit method (ADI) of solving two-dimensional linear partial differential equations with the iterative Newton-Raphson method to advance through variable time steps. After each time step, a volume balance check is applied to avoid error accumulation. Several techniques were selected to harmonize the auxiliary equations in the model. Cubic splines were used to smooth the diffusivity. A nonpivoting technique was used to solve banded and tridiagonal matrices. The model was developed using visual basic for applications (VBA) computer language, which runs under office environment. The model was named "Drip Chartist," and it was verified to run efficiently under any construction alternative of the drip irrigation, and under any soil texture.

## KEYWORDS

- **Alexandria University**
- **alternate direction implicit method (adi)**
- **barrier**
- **cubic spline**
- **diffusivity**
- **drip chart list**
- **dripper**
- **dripper spacing**
- **Egypt**
- **emitter**
- **emitter spacing**
- **finite difference method, FDM**
- **hydraulic conductivity**
- **Jacobian matrix**

- matrix flux potential
- micro irrigation
- Newton – Raphson method
- nonlinear differential equation
- onion shape
- polynomial fit
- Richards' equation
- soil moisture
- soil texture
- subsurface drip irrigation, SDI
- subsurface micro irrigation
- surface irrigation
- visual basic for applications, VBA
- wetting front
- wetting pattern

## REFERENCES

1. Ahmad, A. 1980. *Simulation of simultaneous heat and moisture transfer in soils heated by buried pipes.* Ph.D. thesis, The Ohio state university. 120 pp.
2. Ben-Asher J., Ch. Charach, and A. Zemel, 1986. *Infiltration and water extraction from trickle source: the effective hemisphere model.* Soil Sci. Soc. Am. J. Proc. 50: 882–887
3. Brandt A., E. Bresler, N. Dinar, I. Ben-Asher, J. Heller, and D. Goldberg, 1971. *Infiltration from trickle source: I. mathematical model.* Soil Sci. Soc. Am. J. Proc. 35: 675–682.
4. Bresler E, 1975. Two-dimensional transport of solutes during nonsteady infiltration from a trickle source, Soil Sci. Soc. Am. J. Proc. 39: 604–613.
5. Campbell, G. S., 1985, *Soil physics with BASIC, transport models for soil-plant systems*, Elsevier Pub., 150pp.
6. Caussade, B.H., G. Dourness, and G. Renard., 1979. *A new numerical solution of unsteady two-dimensional flow in unsaturated porous media*, Soil Science vol. 127(4): 193–201.
7. El-Nesr, M.N., 2006, *Subsurface drip irrigation system development and modeling of wetting pattern distribution.* Unpublished Ph.D. thesis, Fac. of Agric., Alex. Univ.
8. Gilley, J. R., and E. R. Allred 1974a. *Infiltration and root extraction from subsurface irrigation laterals.* Trans. ASAE 17(5): 927–933.
9. Khalifa, H.E., A.M. El-Gindy, G.A. Sharaf, and Kh. A. Allam, 2004. *Simulating water movement in sandy soil under surface point source emitter, i: model development*, Misr J. Ag. Eng., 21(2): 341–361
10. Kruger, C.J., 2001. *Constrained cubic spline interpolation for chemical engineering applications*, URL: http://www.korf.co.uk/spline.pdf
11. Lafolie, F., R. Geunnelon, and M. Th. van Genuchten, 1989a. *Analysis of water flow under trickle irrigation: i: theory and numerical solution.* Soil Sci. Soc. Am. J. Proc. 53: 1310 1318.
12. Lomen, D. O., and A. W. Warrick, 1974. *Time dependent linearized infiltration: ii. Line sources.* Soil Sci. Soc. Am. J. Proc. 38: 568–572.

13. Luckner, L., M. Th. van Genuchten, and D. R. Nielsen. 1989. *A consistent set of parametric models for the two-phase flow of immiscible fluids in the subsurface.* Water Resour. Res. 25: 2187–2193.
14. Moncef, H., D. Hedi, B. Jelloul, and M. Mohamed, 2002. *Approach for predicting the wetting front depth beneath a surface point source: theory and numerical aspect.* Irrig. And Drain. vol. 51: 347–360.
15. Morcos, M.A., M. Hanafy, M.F.H. Aly, 1994. *A mathematical model for predicting moisture distribution from trickle source under drip irrigation,* Misr. J. Ag. Eng. 11(4): 1151–1182.
16. Oron, G., 1981. *Simulation of water flow in the soil under subsurface trickle irrigation with water uptake by roots.* Agric. Wat. Manag. 3: 179–193.
17. Raats, P. A. C., 1971. *Steady infiltration from sources at arbitrary depth,* Soil Sci. Soc. Am. J. Proc. 36: 399–401.
18. Raats, P. A. C., 1972. *Steady infiltration from point sources, cavities and basins,* Soil Sci. Soc. Am. J. Proc. 36: 689–694.
19. Raats, P. A. C., 1977. *Laterally confined, steady flows of water from sources and to sinks in unsaturated soils,* Soil Sci. Soc. Am. J. Proc. 41: 294–304.
20. Ragab, R., J. Feyen, and D. Hellel, 1984. *Simulating infiltration into sand from a trickle line source using the matric flux potential concept.* Soil Science 137: 120–127.
21. Revol, P., B.E. Clothier, J.C. Mailhol, G. Vachaud, and M. Vaculin., 1997. *Infiltration from a surface point source and drip irrigation: 2. An approximate time-dependent solution for wet front position,* Wat. Resour. Res., 33(8): 1869–1874.
22. Russo, D., J. Zaideland, and A. Laufer, 1998. *Numerical analysis of flow and transport in a three dimensional partially saturated heterogeneous soil.* Wat. Resour. Res. vol. 34(6): 1451–1468.
23. Selim, H.M., and D. Kirkham, 1973. *Unsteady two-dimensional flow in unsaturated soils above an impervious barrier,* Soil Sci. Soc. Am. J. Proc. 37: 489–495.
24. Thomas, A.W., E.G. Kruse, and H.R. Duke, 1974. *Steady infiltration from line sources buried in soil.* Trans. ASAE 17: 125–133.
25. Thomas, A.W., H.R. Duke, and E.G. Kruse, 1977. *Capillary potential distribution in root zones using subsurface irrigation.* Trans. ASAE 20: 62–67.
26. Thorson, J., 2000. *Gaussian elimination on a banded matrix,* Stanford Exploration Project, Stanford University, 12 pp, Web Service, URL: http://sepwww.stanford.edu/oldreports/sep20/20_11.pdf.
27. van Der Ploeg, R., and P. Benecke, 1974. *Unsteady unsaturated, n-dimensional moisture flow in soil: a computer simulation program.* Soil Sci. Soc. Am. J. Proc. 38: 881–885.
28. van Genuchten, M. Th, F. J. Leij, and S. R. Yates, 1991. *The RETC code for quantifying the hydraulic functions of unsaturated soils,* U.S. Salinity Lab., U.S. Dept. of Agric., Agric. Res. Service, Riverside, California. 93pp
29. Walker, D.W., T. Aldcroft, A. Cisneros, G. C. Fox, and W. Furmanski, 1988, *LU decomposition of banded matrices and the solution of linear systems on hyper cubes,* report of the California Institute of Technology, Pasadena, CA, USA, 19pp
30. Warrick, A.W., 1974. *Time dependent linearized infiltration: i. Point sources.* Soil Sci. Soc. Am. J. Proc. 38: 383–386.
31. Warrick, A.W., and D.O. Lomen, 1976. *Time dependent linearized infiltration: iii. Strip and disk sources,* Soil Sci. Soc. Am. J. Proc. 40: 639–643.
32. Zachmann, D.W., and A.W. Thomas, 1973. *A mathematical investigation of steady infiltration from line sources.* Soil Sci. Soc. Am. J. Proc. 37(4): 495–500.
33. Zazueta, F. S., G.A. Clark, A.G. Smajstrla, and M. Carillo, 1995. *A simple equation to estimate soil-water movement from a drip irrigation source.* 5th International micro irrigation congress, Florida, ASAE, Proc.: 851–856.
34. Zin, El-Abedin T.K., G.A. Sharaf, and S.M. Ismail 1996. *Subsurface dripper line irrigation system, ii- modeling the soil moisture distribution.* Misr J. Agr. Eng, 13(3): 589–604.

# CHAPTER 2

# SUBSURFACE DRIP IRRIGATION: WETTING PATTERN SIMULATION PART II: MODEL VALIDATION

M. N. EL-NESR, S. M. ISMAIL, T. K. ZIEN EL-ABEDIN, and M. A. WASSIF

## CONTENTS

*Partially printed from: "*El-Nesr, M. N., 2006. Subsurface Drip Irrigation System Development and Modeling of Wetting Pattern Distribution. Unpublished PhD Thesis, Faculty of Agriculture, Alexandria University, Egypt* <www.alexu.edu.eg/index.php/en>." Academic and research guidance/support by S.M. Ismail, T.K. Zien El-Abedin, and M.A. Wassif; and by Faculty of Agriculture at Alexandria University is fully acknowledged.

## 2.1  INTRODUCTION

In Chapter 1 in this volume, a simulation model (Drip "Chartist") was developed for the wetting pattern under subsurface drip irrigation. In this chapter, we will discuss our results to validate our model to ensure its harmony for the predicted behavior based on three consequent operations: *verification, validation,* and *output analysis* [10]. Fishman and Kiviat [6] defined these operations as follows:

*"Verification"* operation determines whether a simulation model performs as intended, that is, debugging the computer program to compare step by step results to manual calculations for several program runs. The second operation *"Validation"* determines whether a simulation model, as opposed to the computer program, is an accurate representation of the real-world system under study. This can be evaluated by field and laboratory experiments, and by comparing these results with other trusted models, or by comparing these results with published cases. Finally, *"Output analysis"* is the operation in which the output of the model been revised for logic, harmony, and real situations. This analysis can be done by means of statistical and mathematical methods. After performing this step, the model is declared as reliable and ready to use.

The verification our "Drip Chartist" model was carried out while and after programming stage by debugging the program line by line to ensure it to be error-free: no errors like overflow (division by zero), undeclared variables, mistyped variable names, or mistyped equations. Several runs were performed and compared to manually solved calculations. Testing of extremes was done as well. The model was validated after it was found to be free from all programmatic errors and typos after thorough tests.

Model validation was done by two methods. The first method was comparing the results of our model with field-measured data, in a manner similar as was done by Levin et al. [11], who verified the model of Bresler [4] by using field experiments. The second validation method was to compare the current results of our model with the results of an alternate trustworthy model in a manner similar as was done by Ragab et al. [13].

## 2.2  MODEL VALIDATION

### 2.2.1  FIELD-MEASURED DATA VALIDATION

Field validation of the model was performed in the North Sinai research station of the Desert Research Center in "El-Shaikh Zowayed" city, 30 km from "El-Arish," and 12 km from "Rafah" on the Egyptian-Palestinians' borders. The soil texture was medium to fine sand as shown in Table 1. Soil moisture characteristic curve was determined according to FAO [5] and is listed in Table 2. Soil hydraulic conductivity was 24.6 m/d and was determined using a method by van Beers [15] method.

**TABLE 1**  Soil texture and particle size distribution percent at the experimental site.

| Particle size, mm | From To | >2.0 2.0 | 1.0 2.0 | 0.5 1.0 | 0.25 0.50 | 0.10 0.25 | 0.063 0.100 | <0.063 0.063 | Soil texture |
|---|---|---|---|---|---|---|---|---|---|
| Soil depth (cm) | 0 to 30 | 6.0 | 15.3 | 11.4 | 28.3 | 29.4 | 7.1 | 2.5 | medium to fine sand |
| | 30 to 60 | 6.3 | 12.1 | 10.4 | 32.7 | 28.6 | 6.9 | 3.0 | medium to fine sand |

**TABLE 2**  Values for soil moisture characteristic for the experimental site soil.

| Suction, bar | 0.10 | 0.50 | 1.00 | 5.00 | 10.00 | 15.00 |
|---|---|---|---|---|---|---|
| Water content, cm³/cm³ | 0.23 | 0.21 | 0.20 | 0.17 | 0.10 | 0.03 |

In "Drip Chartist" model, the soil was assumed to be physically uniform, and the initial water content was a constant value throughout the soil profile. This situation is theoretical and is hard to be established in the real situations. To overcome this problem, the comparison of wetting pattern was performed between the difference in soil moisture pattern after and before irrigation for model and for field.

"Drip Chartist" requires some soil parameters to define the simulated soil properly. In order to find these parameters, the laboratory-measured retention values (Table 2) were entered to the computer model RETC [16], which performs a neural-networks-based prediction of soil properties. The predicted properties are listed in Table 3.

**TABLE 3**  Soil properties at the experimental site.

| Water content | | van Genuchtin parameters | | Hydraulic conductivity |
|---|---|---|---|---|
| $q_r$ | $q_S$ | a | $n$ | $Ks$ |
| % Vol | %Vol | | | cm/min |
| 0.0507 | 0.3760 | 3.4400 | 4.4248 | 1.7083 |

The soil parameters (Table 3) were entered into our model (Drip Chartist) and a simulation was performed for two systems without physical barrier:

- System I: Single lateral line for surface dripper system; and
- System II: Bilateral system of 20, 40 cm buried lateral lines.

The model was allowed to simulate for an infiltration time of 45 and 20 min for the two systems, respectively. The output patterns of both systems were adjusted by subtracting the initial moisture content values from all the output to the values under field patterns.

In the field, and after harvesting all crops, the soil moisture was measured at soil depths of 10, 30, 50, and 70 cm using the neutron scattering probe (CPN HYDRO-PROBE, 503DR) to represent soil moisture at the depth range of 0–20, 20–40, 40–60, and 60–80 cm, respectively in the two systems under study. For the first and second systems, measurements were performed before irrigation and every 5 min after irrigation till the end time of each system. The measurements before irrigation were subtracted from the corresponding values of moisture after irrigation in all systems. Comparative profiles are presented in Fig. 1 for system I and Fig. 2 for system II, respectively.

Figures 1 and 2 reveal that our model appears to be very close to the measured values but with some under-estimations. The most under-estimated values in all charts were the top-layer points. This may be attributed to the inaccurate measurements of the neutron probe near soil surface due to the extent of roots tailings, herbs, and other organic substances, which confuse the hydro probe readings. To evaluate the overall amount of under-estimation, the 45° line was drawn to compare measured values with estimated (Fig. 3). In the surface system without the hydraulic barrier (Fig 3a), the model represents the real-situation by 94.99% (correlation coefficient of 0.9746) with under-estimation of 0.157. Whereas in the subsurface system with hydraulic barrier (Fig 3b), the model represents the real-situation by 81.96% (correlation coefficient of 0.9053) with under-estimation of 0.021. However, these values give high confidence in the results of our model in both surface and subsurface simulations.

## 2.2.2 COMPARATIVE VALIDATION WITH "HYDRUS 2D" MODEL

It is well known (in the field studies for soil-water) that the "Hydrus 2D" model [14] is the most trusted soil-water model in horizontal, and vertical plans (simulating infiltration and subsoil water movement as well as simulating rivers and surface runoff). However, many investigators tested this model and compared the results of the model with accurate laboratory measurements. Abbasi et al. [1], Assouline [2] and Li et al. [12] showed that Hydrus 2D simulations of water content and solute distributions are reasonably close to measured values. Therefore, we conducted a comparative case study to solve a standard problem that can be applicable for both models.

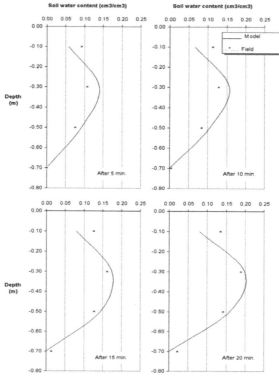

FIGURE 1    Soil moisture increment values under field conditions versus the predicted values with our model: For bilateral 20 and 40 cm treatments for an elapsed time of 5, 10, 15 and 20 min.

Fortunately, "Drip Chartist" and "Hydrus 2D" models use the same soil equations and input parameters. Thus, no modifications or conversions were performed. The main problem in comparing the results was that Hydrus2D deals only with flux from a point source, not with discharge rate like "Drip Chartist." However, the discharge can be converted to the flux by dividing the discharge by the infiltration area. However, in drip irrigation, the infiltration area is not fixed and it varies with time: The flux is inversely proportional to the time. To overcome this problem, several adjustments were performed such as:

- An adapted version of Hydrus2D was used similar to that used by Gärdenäs et al. [7].
- Variable boundary conditions were adjusted for decreasing flux with time.
- Volume balance (output from Hydrus2D) was checked to ensure that same volume of water was applied in both models.

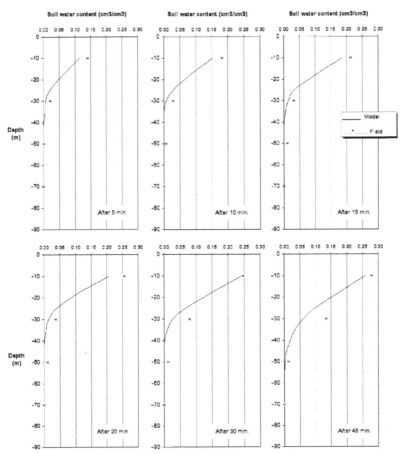

**FIGURE 2**  Soil moisture increment values under field conditions versus the predicted values with our model: For surface drip irrigation treatments for an elapsed time of 5, 10, 15, 20, 30 and 45 min.

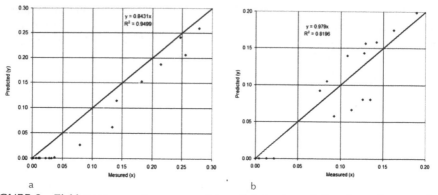

a                                                                b

**FIGURE 3**   Field measurements versus model-predicted values: Soil moisture increase after infiltration for: (Left, Fig. a) Single dripper line. (Right, Fig. b) Bilateral system.

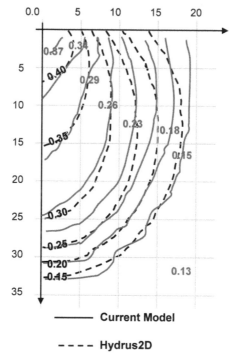

————— **Current Model**

- - - - **Hydrus2D**

**FIGURE 4**   A comparison of soil water content isolines between our model and Hydrus 2D: Output diagram. simulation water movement in a sandy soil using a 3 L/h emitter for 60 min (Cumulative volume of 3 L).

Figure 4 shows an isoline comparison between our model and Hydrus2D output diagram of simulating water movement through a 3L/h emitter for 60 min, in a sandy soil. The results from both models match very well, and almost coincide at boundaries. Although Hydrus2D is more accurate, and performs smoother isolines, but our model is quicker and simpler in interface.

## 2.3 ANALYSIS OF "DRIP CHATIST" MODEL

Several parameters of our model affect its output (the wetting pattern). Some of the parameters had more effect than the others. Several studies were performed to evaluate the effects of these parameters. Each study has special inputs and constrains but all of the studies have the same output-variables. The output-variables are the dimensions of the wetting pattern in some arbitrary isolines, and the simulation time consumed on the computer.

### 2.3.1 EFFECTS OF SOIL PHYSICAL PROPERTIES ON SOIL-WETTING PATTERNS

Within the same soil texture class, wetting pattern is affected by soil physical properties, the retention and conductivity parameters. However, we studied only the effects of saturated water content ($\theta_{sat}$), residual water content ($\theta_{res}$), initial water content ($\theta_b$), retention parameter (n), bubbling pressure inverse ($\alpha$, cm$^{-1}$), hydraulic conductivity ($K_s$, cm/min), and pore-connectivity parameter ($\Upsilon$). All studies were conducted for a sandy soil, with a profile of 50 cm depth, and 35 cm width (radius), with grid spacing of 2.5 cm, and cumulative water volume of 8 L.

Each variable under study was evaluated individually, while keeping the other variables as constant to the default values. Each variable has about ten levels, which were studied within the acceptable range of each parameter. The studied values of each variable are indicated in Table 4 (default values are in bold). When studying $\theta_{sat}$ for example, all other variables were fixed to the default values, and $\theta_{sat}$ value varies according to the listed values in the $\theta_{sat}$ column.

Figure 5 indicates effects of soil properties on the wetting pattern and simulation time for surface drip irrigation.

1. Saturated soil water content appears to affect the shape of wetting pattern only near the saturation zone (Fig. 5a). In the same way, the width of the wetting pattern is affected (Fig. 5b). In Fig. (5c), the simulation time and steps appear to increase in the extremes of the $\theta_{sat}$ values, while minimum time was achieved in the middle of the tested zone at $\theta_{sat}$=0.385.

**TABLE 4** The studied values for each soil parameter, where default values of each parameter are in bold/italics.

| Case study | $\theta_{sat}$ | $\theta_{res}$ | $\theta_{begin}$ | n | $\alpha$ (cm$^{-1}$) | $K_s$ (cm/min) | $\Upsilon$ |
|---|---|---|---|---|---|---|---|
| 1 | 0.3450 | 0.0526 | 0.065 | 2.70 | 0.03530 | 0.047 | 0.05 |
| 2 | 0.3550 | 0.0626 | 0.075 | 2.80 | 0.04500 | 0.095 | 0.20 |
| 3 | 0.3650 | *0.0726* | *0.085* | *2.93* | 0.06000 | 0.145 | 0.30 |
| 4 | *0.3759* | 0.0826 | 0.095 | 3.00 | 0.07500 | 0.215 | 0.40 |
| 5 | 0.3850 | 0.0926 | 0.105 | 3.10 | *0.09015* | 0.285 | *0.50* |
| 6 | 0.3950 | 0.1026 | 0.115 | 3.18 | 0.10500 | 0.295 | 0.60 |
| 7 | 0.4050 | 0.1126 | 0.125 | | 0.12000 | 0.345 | 0.70 |
| 8 | 0.4150 | 0.1226 | 0.135 | | 0.14500 | 0.395 | 0.95 |
| 9 | 0.4250 | 0.1326 | 0.145 | | | 0.445 | |
| 10 | | | 0.155 | | | *0.495* | |

2.  Residual soil water content affects wetting pattern shape mostly in the near
    saturation zone as shown in Figs. 5d, 5e, and 5f. Depth of the near saturation
    isoline increases with the increment of $\theta_{res}$ while other isolines are nearly not
    affected except the 0.11 isoline which had a jump after the value of $\theta_{res}=0.10$.
    The widths of the isolines show an increasing trend with a strange jump in the
    0.926 isoline. However, this jump may be due to some cumulative over shoot-
    ing in the van Genuchten model (16). Simulation time of these cases is directly
    proportional to the $\theta_{res}$ value.
3.  Soil initial wetness, before irrigation or the initial water content ($\theta_{begin}$), affects
    wetting pattern as shown in Figs. 5a, 5b, and 5c. Excluding the near satura-
    tion isoline, all the isolines locations move towards the increment direction of
    depth and width, i.e., the wetting pattern area increases with the increment of
    $\theta_{begin}$. The simulation time trend tends to increase with the increment of $\theta_{begin}$.
4.  In Figs. 5d, 3e, and 5f, soil hydraulic conductivity at saturation ($K_s$) affects
    the wetting area profile widely. However, the more the $K_s$ value, the more the
    depth of the specified isolines. In contrast, $K_s$ is inversely proportional to the
    width of isolines. Explicitly, the increment of the $K_s$ causes wetting area to be
    narrower in width and longer in depth. Therefore, for design of drip irrigation
    systems in case of soils with higher conductivity, the emitters must be more
    close to each other, and irrigation should be managed so as to give smaller
    amounts of water on shorter frequencies to avoid deep percolation due to elon-
    gation of the wetting pattern.

**TABLE 5**    Soil properties for selected soil texture classes for case studies in our model.

| Texture Class | Symbol | $\theta_{sat}$ | $\theta_{res}$ | $\theta_{begin}$ | n | $\alpha$ | $K_s$ | $\Upsilon$ |
|---|---|---|---|---|---|---|---|---|
| | | | | | | $cm^{-1}$ | $cm/min$ | |
| Sand | S | 0.376 | 0.073 | 0.085 | 2.930 | 0.090 | 0.495 | 0.5 |
| Loamy sand | L Sa | 0.387 | 0.081 | 0.109 | 2.013 | 0.079 | 0.243 | 0.5 |
| Sandy loam | Sa L | 0.413 | 0.090 | 0.132 | 1.669 | 0.051 | 0.074 | 0.5 |
| Loam | L | 0.443 | 0.122 | 0.156 | 1.517 | 0.024 | 0.047 | 0.5 |
| Silt | Si | 0.429 | 0.116 | 0.095 | 1.523 | 0.011 | 0.058 | 0.5 |
| Silty loam | Si L | 0.453 | 0.137 | 0.197 | 1.536 | 0.013 | 0.061 | 0.5 |
| Sandy clay loam | Sa C L | 0.450 | 0.125 | 0.175 | 1.405 | 0.040 | 0.009 | 0.5 |
| Clay loam | C L | 0.479 | 0.154 | 0.200 | 1.362 | 0.017 | 0.006 | 0.5 |
| Silty clay loam | Si C L | 0.503 | 0.178 | 0.218 | 1.375 | 0.009 | 0.008 | 0.5 |
| Sandy clay | Sa C | 0.465 | 0.169 | 0.294 | 1.218 | 0.030 | 0.0078819 | 0.5 |
| Silty clay | Si C | 0.500 | 0.176 | 0.326 | 1.205 | 0.011 | 0.0066736 | 0.5 |
| Clay | C | 0.503 | 0.181 | 0.359 | 1.171 | 0.012 | 0.0102431 | 0.5 |

Inversely to the effect of $\theta_{sat}$ on simulation time, $K_s$ extreme values of the tested range lead to the least simulation time, while the peak simulation time was achieved in the middle value of 0.295 cm/min which took twice the time of the 0.047 cm/min as they was simulated in 46 and 23 seconds respectively.

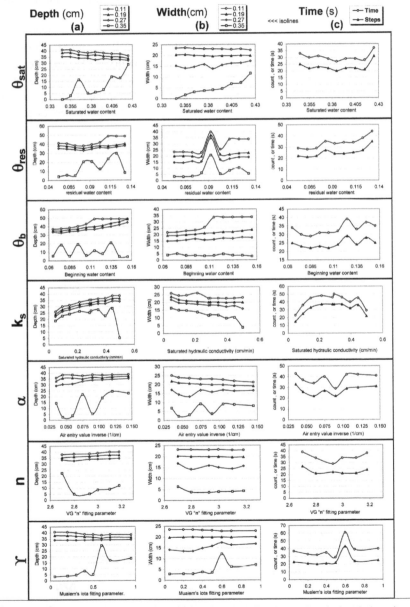

**FIGURE 5** Effects of seven soil properties on the wetting pattern and simulation time for surface drip irrigation.

5.  The results for conductivity and retention variables were plotted in Fig. 5. Effect of air entry inverse ($\alpha$) is nearly similar to the $\theta_{sat}$ effect on wetting pattern. It increases with the wetting depth and decreases with the wetting width. The least simulation time was obtained at the middle range values of $\alpha$ while peak time is obtained at the edges. The similarity of $\alpha$ plots to the $\theta_{sat}$ plots may be attributed to the direct physical relationship between these through the soil-water retention curve.

6.  The retention fitting parameter $n$ indicated no effect on the wetting pattern within the tested range values as shown in Figs. 5d, 5e, and 5f. However, these values were taken as the limits for the sand textured soil in the literature. The exception of the "n=2.68" can be due to the least value in range between sandy texture and the loamy sand texture. Therefore, the soil texture may be virtually changed to finer texture while it is fixed to "Sandy" in the rest of cases.

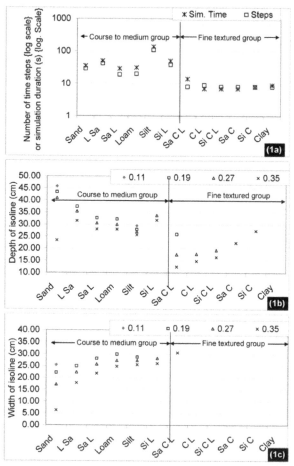

**FIGURE 6 (1a to 1c).** Effects of simulation time (A), isoline depth (B) and isoline width (C) for the "Drip Chartlist" model: Soil texture class studies.

7. The value of Mualem's fitting parameter (ϒ) ranged from 0 to 1 and about 0.5 for most of the soils. It was found to be ineffective to the wetting pattern and to the simulation time. However, a strange exception in the ϒ=0.6 value can be observed, as it spreads the saturation zone in depth and width, and doubles the simulation time. This can be due to instability of the equation at this value. The insignificant effect of on the wetting pattern supports the approximation of 0.5 in all soils reported by van Genuchten et al. [16].

The results for the seven properties under study are plotted in Figs. 6 (1a–3c) and 7 (1a–3a). Effect of each of these seven properties will be discussed in detail in this chapter.

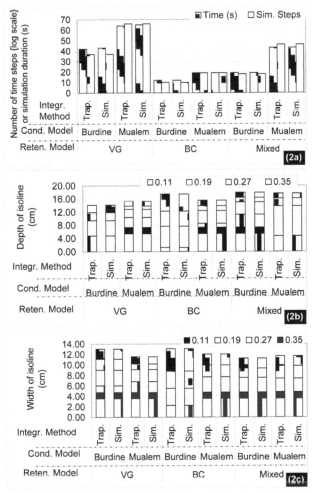

FIGURE 6 (2a to 2c)   Effects of simulation time, isoline depth and isoline width for the "Drip Chartlist" model: Soil moisture retention model, hydraulic conductivity model and numerical integration method.

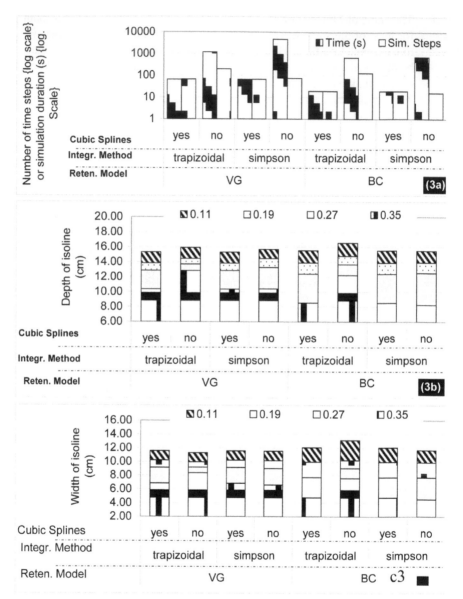

**FIGURE 6 (3a to 3c)**   Effects of simulation time, isoline depth and isoline width for the "Drip Chartlist" model: Cubic splines analysis.

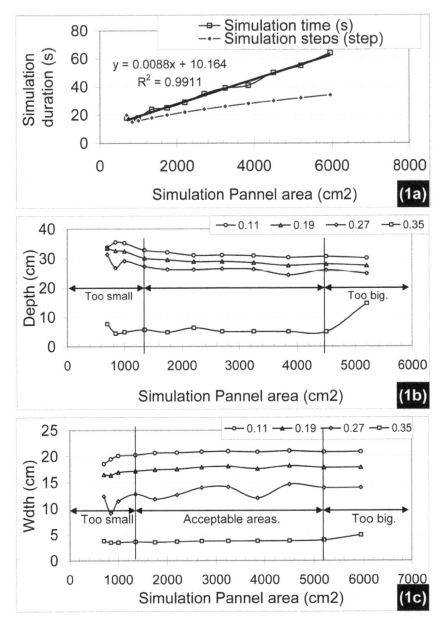

**FIGURE 7 (1a to 1c)**   Effects of simulation time (A), isoline depth (B) and isoline width (C) for the "Drip Chartlist" model: Simulation panel analysis.

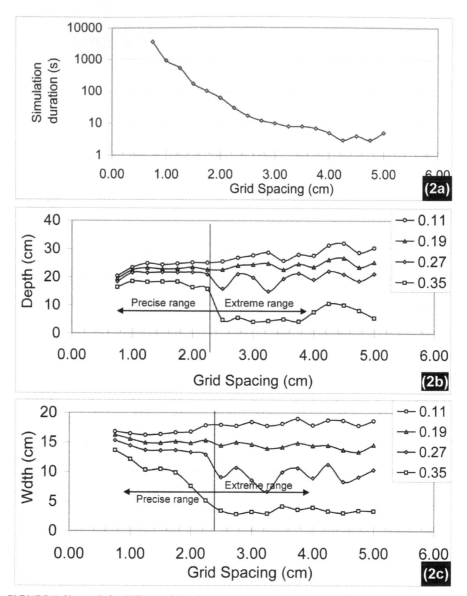

**FIGURE 7 (2a to 2c)**   Effects of simulation time (A), isoline depth (B) and isoline width (C) for the "Drip Chartlist" model: Grid spacing analysis.

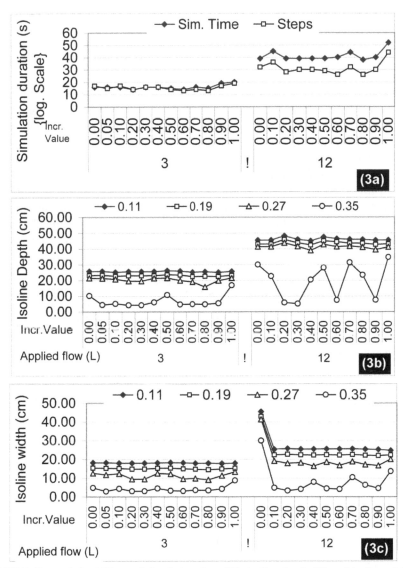

**FIGURE 7 (3a to 3c)**    Effects of simulation time (A), isoline depth (B) and isoline width (C) for the "Drip Chartlist" model: Time increment value analysis.

## 2.3.2   EFFECT OF SOIL TEXTURE ON SOIL WETTING PATTERN

Generally, the wetting pattern is deep and narrow in sandy soils; and shallow and wide in clayey soils. A study was done on the 12 main textures to find out the location of the previously considered values of isolines (0.11, 0.19, 0.27, and 0.35), though some of these values may not be applicable in some texture classes. However, these values were considered for harmony in all studies.

All case studies were applied to a soil profile of 50 cm depth, and 35 cm width (radius), with grid spacing of 2.5 cm, and cumulative water volume of 12 L. However, default values for soil physical properties varied according to each soil texture class. The 12 main texture classes were studied (Table 5). Each texture has its own soil properties.

The results of the case study were plotted in Fig. 6 (1a–1c). ; which concludes It can be concluded that heavy textured soils can be simulated faster than light texture soils. The silt texture soil was simulated in longer time compared to other texture classes. The reader will observe that the time axis scale is logarithmic.

In Fig. 6 (1b and 1c), the wetting pattern depth and width were plotted. In these figures, no trend was observed in all soil texture classes. However, some trends were noticed within the fine textured groups and within the coarse to medium group. In the coarse to medium texture group, the minimum depth of 0.35 isoline appears to be increasing in the direction from coarse to fine. One can observe that the 0.11 isoline disappears in most texture classes, because it is less than the $\theta_{begin}$ value of such a texture class. In addition, the 0.19 and 0.27 isolines decrease in the direction from sand to loam then increases again within the first group. In the heavy texture group, the 0.35 isoline moved deeper as the texture going finer. This may be due to that $\theta_{res}$ and $\theta_{begin}$ always increase when the texture goes finer. These isolines were logged in a far distance for finer textures till it disappears totally as in the texture class with $\theta_{begin}$=0.35.

However, the width of isolines increase in the coarse-to-fine direction as shown in Fig. 6 (1c), which shows the disappearance of the 0.11 isoline in almost all soil texture classes, and the vanishing of 0.19 and 0.27 isolines in most texture classes. The results in Fig. 6 (1b and 1c) may attribute the horizontal spread in the fine textured profiles, and the vertical spread in the coarse textured ones. However, the output charts of the "Drip Chartist" conclude that fine textured soils let the water spread horizontally than vertically, while the contrary is true for the coarse texture classes.

It can be concluded that soil texture class affects: Wetting pattern shape significantly in the coarse-to-fine direction; increase of width of an isoline; and decrease of depth of the isoline. However, more advanced study is suggested with the relative not absolute values of wetness isolines (isolines which are relative to $\theta_{begin}$ and $\theta_{sat}$).

### 2.3.3 EFFECTS OF METHODS TO ESTIMATE THE SOIL PROPERTIES ON WETTING PATTERN

The principal models for soil-water infiltration relationships are the retention models (by van Genuchten, VG, 16; and by Brookes and Corey's, BC), and the conductivity models by Mualem and Burdine. Wetting pattern was surely affected by either formula (each of these models), but to what degree? Here we will discuss this in detail. Another calculation method that is expected to affect wetting pattern is the matric flux potential (MFP is the integration of diffusivity with relative to moisture content) and integration method. Finally, the effect of cubic splines fitting on the variables under study is compared to direct integration methods.

All case studies were for a sandy soil, with a profile of 18.75 cm depth, and 13.75 cm width (radius), with grid spacing of 1.25 cm, and cumulative water volume of

0.75 L. However, these low volume and small grid area was selected because of the long time needed to finish some tests especially for the noncubic splines' fitted tests.

Three retention models were tested: VG, BC, and mixed model of VG/BC. Two conductivity models were tested: Mualem and Burdine. Also, two integration methods were examined: Trapezoidal and Simpson. Finally, cubic splines fitting usage were compared to direct integration methods.

The retention and conductivity models case study are plotted in Fig. 6 (2a–2c), where all plotted values represent cubic splines' fitted tests. The direct integration to cubic splines fitted comparisons are shown in Fig. 6 (3a–3c).

In Fig. 6 (2a), the fastest simulation was achieved through the combination BC retention model. With Burdine's conductivity model, the simulation time was 12 seconds. The slowest simulation was 64.5 seconds for VG-Mualems combination model. However, for the retention models, the sequence of speed of simulation was BC, Mixed, then VG models from faster to slower. For the conductivity models, Mualems' model appear to take more time than Burdines' in all benchmarks. However, no difference was for the simulation time between Simpson's and trapezoidal integration methods.

The effects of cubic splines effect are significant in the reduction of simulation time, as shown in Fig. 6 (3a–3c). However, because of the huge differences between simulation time values, the time axis is plotted in logarithmic scale. In the VG-Simpson's combination, the usage of direct integration requires 7153% more time than the usage of cubic splines fitting. These results display the importance of the usage of cubic splines fitting over direct integration.

The Fig. 6 (2b and 6.2c) show that changing the calculation method changes the shape of wetting pattern, especially in the depth of the saturated area (0.35 isoline), which fluctuates severely with the models. The outer boundary of the pattern did not change within the models (0.11 isoline) and it fluctuated four cm in depth and two cm in width. These results indicate that the models in this study affect the distribution of a wetting pattern within semifixed boundaries.

The corresponding charts in Fig. 6 (2a–2c) show that the usage of cubic splines does not affect, in most cases, the location of any of the isolines. It was observed that it didn't move any isoline>2 cm in width and>4 cm in depth. These results support the recommendation to use the cubic splines instead of direct integration.

From these results, it can be concluded that Mualem's conductivity model lead to wider and deeper profiles than the Burdine's model for both VG and BC retention relationships. This result is opposite for a mixed retention model. These findings are useful for validating the model under field and simulated conditions. However, if the results of the model underestimate or overestimate the field measurements, then the use of Cubic splines instead of direct integration is highly recommended due to significant reduction in simulation time, because this does not affect the output wetting pattern considerably.

### 2.3.4   EFFECTS OF FINITE DIFFERENCE GRID SETTINGS ON SOIL WETTING PATTERNS

Although simulation of soil infiltration under drip irrigation is performed using our "Drip Chartist" and other models to study the wetting patterns around a dripper. The boundaries for the spreading of wetting patterns must be expected and set before the simulation process is initiated. However, if these boundaries are less than the spread-flow in width or in depth, then the wetting pattern is distorted according to the boundary conditions with no flow outside boundaries. Therefore, the outer boundaries must be set as wide as possible before simulation initiation and this might need to increase simulation time. In this chapter, we will discuss results of a case study on the effects of the grid boundaries and grid spacing (unit length of the grid) on the simulation time and simulated pattern.

**TABLE 6**   Grid dimensions for each case study when the grid unit is fixed.

| Case | Grid dimensions | | | Grid Nodes | | |
|------|--------|--------|----------|----|----|-------|
|      | $R$ (cm) | $Z$ (cm) | $A$ (cm²) | $i$ | $l$ | *Total* |
| | 20.0 | 35.0 | 700 | 8 | 14 | 112 |
| | 22.5 | 37.5 | 844 | 9 | 15 | 135 |
| | 25.0 | 40.0 | 1000 | 10 | 16 | 160 |
| | 30.0 | 45.0 | 1350 | 12 | 18 | 216 |
| | 35.0 | 50.0 | 1750 | 14 | 20 | 280 |
| | 40.0 | 55.0 | 2200 | 16 | 22 | 352 |
| | 45.0 | 60.0 | 2700 | 18 | 24 | 432 |
| | 50.0 | 65.0 | 3250 | 20 | 26 | 520 |
| | 55.0 | 70.0 | 3850 | 22 | 28 | 616 |
| | 60.0 | 75.0 | 4500 | 24 | 30 | 720 |
| | 65.0 | 80.0 | 5200 | 26 | 32 | 832 |
| | 70.0 | 85.0 | 5950 | 28 | 34 | 952 |

Therefore, a starting grid of 20×35 (r×z), cm² was tested; the grid was increased by one or two unit grid (2.5 cm or 2.5×2 cm) in radius and in depth, till reaching a grid area of 70×85 cm². However, 12 grid areas were tested as shown in Table 6. To study the grid unit effect, a unified grid of 35×50 (r×z) cm² was used, whereas the grid unit started from 0.75 cm to 5 cm, as shown in Table 7.

The results are summarized in Fig. 7 (1a–3c). In these figures, the time was plotted in logarithmic scale to represent the larger and smaller values of simulation time. For the smaller values of the panel area, wetting pattern depth is larger than its value for bigger areas as shown in Fig. 7 (1a–1c). However, the changes in all isolines appears

**TABLE 7**  Grid dimensions for each case study when the grid area is fixed.

| Grid unit | 0.75 | 1.0 | 1.25 | 1.5 | 1.75 | 2.0 | 2.25 | 2.5 | 2.75 | 3.0 | 3.25 | 3.5 | 3.75 | 4.0 | 4.25 | 4.5 | 4.75 | 5.0 |
|---|---|---|---|---|---|---|---|---|---|---|---|---|---|---|---|---|---|---|
| $i$ | 46 | 35 | 28 | 23 | 20 | 17 | 15 | 14 | 12 | 11 | 10 | 10 | 9 | 8 | 8 | 7 | 7 | 7 |
| $l$ | 66 | 50 | 40 | 33 | 28 | 25 | 22 | 20 | 18 | 16 | 15 | 14 | 13 | 12 | 11 | 11 | 10 | 10 |
| *Total* | 3036 | 1750 | 1120 | 759 | 560 | 425 | 330 | 280 | 216 | 176 | 150 | 140 | 117 | 96 | 88 | 77 | 70 | 70 |

Nodes

to be almost negligible unless for very small and very big values of panel's area. In addition, the wetting pattern depth is more constant for the medium to large panel areas. The different results of the small panel areas are due to the fact that boundary conditions at borders (no side flow) distort the wetting pattern. The simulation time is directly proportional to grid area (with same grid spacing). However, the increment in simulation time is less than the increment in area as defined below (Fig. 7, 1a–1c):

$$\text{Time}_{(s)} = \{[0.0088 \times \text{Area}_{(cm^2)}] + 10.164\} \tag{1}$$

Grid spacing or the unit grid has more effect on both the wetting pattern and simulation time. Regarding the simulation time: Choosing smaller grid spacing increases the simulation time significantly. A nonlinear relationship was obtained between grid spacing and simulation time as given below:

$$Time_{(s)} = \min\left[2, 295.89 \times \left(GridSpacing_{(cm)} - 0.358\right)^{-2.665}\right]_{r^2=0.999, \ SE=34.0} \tag{2}$$

The Eq. (2) is valid only for the conditions of the current case study.

The location of isolines appears to be affected with grid spacing. The lower grid spacing increases the saturation area depth and width, as shown in Fig. 7 (2a–2c). This may be attributed to the larger matrices produced in the solution of the finite difference step, and to the shooting of accumulative error in the Newton-Raphson Jacobean solution. Although a volume balance check was performed, however, the wetting pattern distribution may vary within the same applied volume. In addition, the ADI technique for the model was developed and accepted to be unconditionally stable for all grid unit lengths. For this reason, it is advisable to use moderate grid spacing like 2.0 to 3.0 cm because most of the published soil-water simulation models do so. However, the verification of "Drip Chartist" as compared to "Hydrus 2D" shows almost coincidence in results when using the 2.5 cm grid spacing. This comparison is useful, because "Hydrus2D" with finite element method simulates the actual mesh of the real situations.

Based on these results, we recommend to use "moderate to large panel area initially," and the simulation should be repeated with larger area if the minimum isoline location is less than two grid units far from the grid boundaries. It is also advised to use a grid spacing of 2 to 3 cm to ensure perfect simulation and to reduce the simulation time.

### 2.3.5  EFFECTS OF ITERATIONS AND SHOOTING SETTINGS ON SOIL WETTING PATTERNS

Solving a complicated system of equations, to estimate of a set of values using Newton-Raphson technique includes: Estimation based on repeated calculations under certain conditions; modification of the basic variable slightly by the results of the Jacobean solution; the repetition of this iteration until the variable set is accepted according to an allowable error value, or till the maximum allowable number of iterations is reached. Using small allowable number of iterations (let it be called "v ") may lead to accepting incorrect values, but using big "v" may increase the simulation time.

To speed up the convergence of the solution, the new time step must be greater than the previous successful time step, using the following equation:

$$\tau_{j+1} = Max \left[ \tau_{MAX}, (1+\sigma)\tau_j \right]$$ (3)

where, $\tau_{MAX}$ is the maximum allowable time-step; $\tau_j$ is a decimal number from 0 to 1, mostly close to 0.75), however, the value of $\tau$ seems to affect simulation time.

Number of iterations used are $v$ values: 1–15, 20, and 25, with accumulative water volume of 5 L. Time increment factor values are: 0.00, 0.05, 0.1, 0.2, 0.3, 0.4, 0.5, 0.6, 0.7, 0.8, 0.9, and 1.0 for two accumulative volumes of 3 and 12 L. All case studies were for a sandy soil, with a profile area of 35×50 cm$^2$ with 2.5 cm grid spacing, and all "Drip chartist" default values were used.

Small allowable number of iterations noticeably increases the saturated isoline location, as shown in Fig. 7 (1a–3c). Approximate stability was established after $v$ =7, while absolute stability occurred after $v$ =11. However, this stability increases with $v$; which leads to a result that we have inaccurate simulation results for a value of $v$ <11. Some investigators including Brandt et al. [3] conclude that the maximum iterations used in their ADI scheme was 7 to 10 that matches with recommendation in our results. On the other hand, and unlike the expected effect on simulation time, using smaller $v$ values rises the simulation time till $v$=8; after which the simulation time go fixed.

The time increment value affects the near saturation isoline depth and width as shown in Fig. 7 (3a–3c), especially when large volume is being applied. However, the default value of the "Drip Chartist" model was 0.79, which was tested and accepted to give the most accurate results in all soil texture classes. Unlike expected, $\tau$ has no noticeable effect of the simulation time. It is advised to use the value of $v$ =11 at least to ensure accurate simulations.

### 2.3.6 EFFECTS OF BILATERAL GAP ON SOIL WETTING PATTERNS

Ismail et al. [8, 9] studied the hydraulic barricading of water through a secondary buried dripper line. This vertical space between the bilateral dripper lines is called "Bilateral Gap" that will be analyzed in this chapter.

Bilateral gap was varied from 4 cm to 32 cm with 4 cm increments. Each level was evaluated just after emission was stopped (before redistribution), and after 6 h of redistribution (the experiment time started from the initiation of infiltration process. The case studies were for a sandy soil, with a profile of 50 cm depth, and 35 cm width (radius), with grid spacing of 2.5 cm. Each case was repeated at two times: One after cumulative volume of 2 L, and the other after emission stopped and redistribution action of 6 h.

Redistribution was modeled the same way as infiltration; however, with the emission source discharge set to zero. The upper dripper line was laid on soil surface, while the secondary line was buried at different depths.

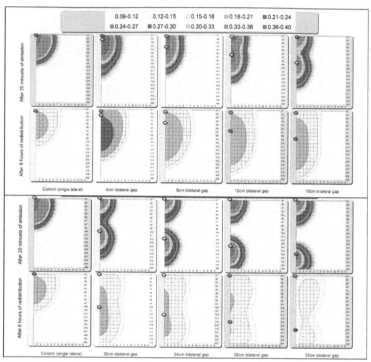

**FIGURE 8**   Drip Chartist output for bilateral gap case study.

Figure 8 shows sample output of "Drip Chartist" for eight bilateral-gap spaces, in addition to the control treatment of single lateral dripper line. All charts in Fig. 8 indicate that with increase in the bilateral gap, the wetting patterns of the "application" stage (the upper patterns) was spreaded more in the vertical direction with a throttle appearance as the gap exceeded from 12 cm till it reached a value of 24 cm (where the wetting pattern of either emitter has been totally separated with no overlapping). This shows that the barricading effect of the second dripper line vanishes after 20 cm gap space (Only for current case study constrains such as soil and emitter discharge, etc....).

**TABLE 8**   Simulation time, and simulation steps in application and redistribution for different gap spacings.

| Gap (cm) | Time, seconds | | | Num. of steps | | |
|---|---|---|---|---|---|---|
| | Application | Redistribution | | Application | Redistribution | |
| | | All | Difference | | All | Difference |
| control | 33 | 36 | 6 | 23 | 26 | 3 |
| 4 | 74 | 75 | 1 | 39 | 43 | 4 |
| 8 | 31 | 43 | 12 | 20 | 29 | 9 |
| 12 | 33 | 40 | 13 | 21 | 27 | 8 |
| 16 | 22 | 36 | 14 | 15 | 24 | 9 |

**TABLE 8**    *(Continued)*

| Gap (cm) | Time, seconds | | | Num. of steps | | |
|---|---|---|---|---|---|---|
| | Application | Redistribution | | Application | Redistribution | |
| | | All | Difference | | All | Difference |
| 20 | 23 | 33 | 10 | 15 | 22 | 7 |
| 24 | 22 | 33 | 11 | 15 | 22 | 7 |
| 28 | 24 | 29 | 5 | 16 | 24 | 8 |
| 32 | 22 | 36 | 14 | 14 | 27 | 13 |

"Redistribution" patterns show the state of water distribution of soil profile, six hours after opening the irrigation valve. The results of these wetting patterns due to bilateral gap are summarized as follows:

1. The effect of bilateral system is highly noticeable when comparing any of the redistribution patterns to a single emitter. This implies that although the same amount of water was applied in all patterns, yet the moisture content in the bilateral tubing patterns is more than its value when not using it.
2. With increase in the bilateral gap (especially after 20 cm gap space), the wetting pattern in redistribution spreaded more in horizontal direction but with low moisture content values (0.12 to 0.18).
3. Field capacity wetting range (0.18-0.21) appeared in the shallow root zone only below gap space of 20 cm.
4. Although the highest isoline occurred in the 4 cm gap-space (although it is not practical), yet the 8 to 16 cm gap-space seems to have very good wetting patterns in the root zone.

Simulation time of the case studies was only affected by the lower values of the gap space, because of more interference by dripping sources; causing partial instability of the model so that it requires more time to get the accurate solution for each time-step. Nonetheless, redistribution converges very rapidly than application. However, Table 8 shows that redistribution steps finished at very short time compared to application steps: In the 8 cm gap space, redistribution simulation took 9 more steps and finished in 4 more seconds, in addition these 4 seconds simulate 340 min of redistribution while the "application" in 31 seconds simulate only 20 min of application.

It is advised to use gap size of 8 cm to 16 cm in order to enhance wetting pattern distribution in root zone. Further studies are needed to set the optimum gap space with relation to soil type, emitter discharge, and upper dripper line's location.

## 2.3.6 EFFECTS OF UPPER LATERAL LOCATION ON SOIL WETTING PATTERNS

Bilateral method of subsurface drip irrigation has two main variables: The location of each dripper line or the location of the upper one; and the gap spacing between both. In Section 2.3.5, the gap spacing was studied. In this section, the upper dripper line location will be studied.

Upper emitter location varied as 0, 6, 10, 14, and 20 cm. Bilateral gap levels were 6, 10, and 14 cm. Each case was evaluated just after the emission stopped (before redistribution), and after 6 h of redistribution (the experiment time started from the infiltration initiation).

Figure 9 shows the sample output of "Drip Chartist" for this case study. In charts of Fig. 9, it can be observed that in the "application" stage, the wetting pattern moves downward with the increment of upper lateral depth. On the other hand, the redistribution patterns was affected widely except in the smaller gap spaces, possibly due to no-flow boundary conditions, which prohibit flow through lower boundary as well as side boundaries so that the same amount of water is forced to be redistributed in the soil profile.

Sufficient water in the root zone was established after 6 h of redistribution when the depth of upper dripper line was up to 10 cm, for any of the tested gap spaces, that is, up to 14 cm gap space. In the real situations, redistribution patterns can be different especially in the "20 cm upper lateral – 10 cm gap space," "14–14," and "20–14" cases, because the wetting front reaches the lower boundary, therefore, grid area must be expanded in these cases to ensure actual representation of the reality.

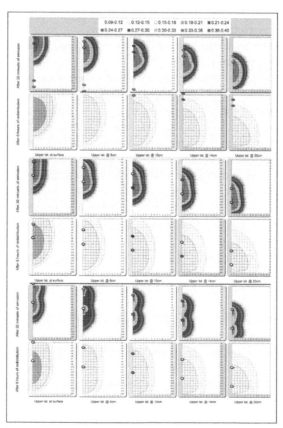

**FIGURE 9**    Drip Chartist output of upper lateral location case study.

It is advised to use shallower upper dripper line (from 0 to 10 cm) with gap size of about 10 cm in order to enhance wetting pattern distribution in root zone. Further studies are needed to set the optimum values with relation to soil type, and emitter discharge.

## 2.3.7 EFFECTS OF PHYSICAL BARRIER ON SOIL WETTING PATTERNS

There are four design parameters for the physical barrier, namely: Depth, width, thickness, and forming shape. However, in our model, only the depth and width variables were considered. Because the thickness of a barrier can be assumed as negligible; and the forming shape requires a flexible mesh system, which is out of scope for the research in this chapter. Moreover, depth and width parameters are very important considerations in the construction economy where the trench cost depends on the soil type, soil profile, mechanization availability, and trench dimensions. Determining the optimum parameters of the physical barrier depends on gain to losses balance. However, the "Drip Chartist" can help determining gains, while trenching costs should be calculated separately.

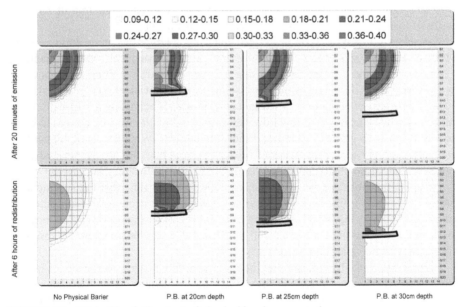

**FIGURE 10** "Drip Chartist" output of physical barrier case study.

Nine combinations of width and depth of the barrier were studied: Three widths (30, 40, and 50 cm), and three depths (20, 25, and 30 cm), in addition to a control case with no barrier.

Figure 10 shows the sample output for "Drip Chartist" for three of the nine studied cases, compared with the control case in each figure.

The effect of the physical barrier is significantly noticeable for both application and redistribution phases. However, the effect is predominant in the shallower depths of the barrier. In addition, no effect was noticed at the deepest barrier (30 cm depth) during application phase, while was noticeable during redistribution until the water reached the barrier. The 20 cm depth cases always resulted in wider wetting patterns than other cases. However, the 25 cm depth resulted in moisture distribution in the root zone.

Width of the barrier was not so effective, within the studied widths. However, the 40 cm width was slightly better that the 30 cm width, while the 50 cm width is not effective at all compared to the 40 cm. Therefore, we recommend using narrower barrier within the case studies in this research.

Based on the case studies, it is recommended to put a 30 cm width physical barrier at 25 cm depth; that is, smaller gutter should be trenched. Further studies are needed to identify the optimum dimensions and properties of the barriers as affected by soil, emitter discharge, hydraulic barrier, and application time.

## 2.3.8   EFFECTS OF EMITTER DISCHARGE ON SOIL WETTING PATTERNS

In this chapter, the emitter discharge rate varied from 0.25 $l$/h to 12 $l$/h. However, 12 application rates were studied for a sandy soil, with a profile of 50 cm depth, and 35 cm width (radius), with grid spacing of 2.5 cm, and cumulative volume of 5 L. The results of this case study are shown in four charts of Fig. 11.

Considering Fig. 11a, it can be noticed that using a larger discharge emitter lowers the location of the 0.11 isoline, while raises the near saturation front (the 0.35 isoline).

In Fig. 11b, one can notice that using a larger discharge emitter does not affect the 0.11 isoline while it spread the 0.35 isoline to a distant location. This means that the whole pattern is being condensed in a smaller area when using a larger discharge emitter. In other words, using low flow rate emitter allow the wetting pattern to cover more area but with gradual decrease of moisture content. Whereas, using higher flow rate emitter allow the wetting pattern to cover less area but with almost near saturated zone. This could be attributed to the limitation of soil infiltration rate, as the higher flow rate emitter pushes a large amount of water in small time that the soil moisture cannot redistribute to the surrounding areas. Hence, water accumulates and saturated condition occurs in this small spreading area. On the other hand, lower flow rates allow lateral distribution of water as well as vertical distribution and hence it results in more area with no saturation occurrence.

Figures 11c and 11d indicate that the faster simulated case was for the 3 l/h emitter discharge, however, simulation speed and number of steps were higher before and after this value as simulation time increases up to 400% of lowest value. This may be due to the saturation state occurring for the higher emitter discharge, which tends to the instability of the system of equations due to the usage of the van Genuchten equation to calculate water diffusivity and matric flux potential. Equation by van Genuchten [16] is undetermined at saturation; hence convergence cannot be achieved.

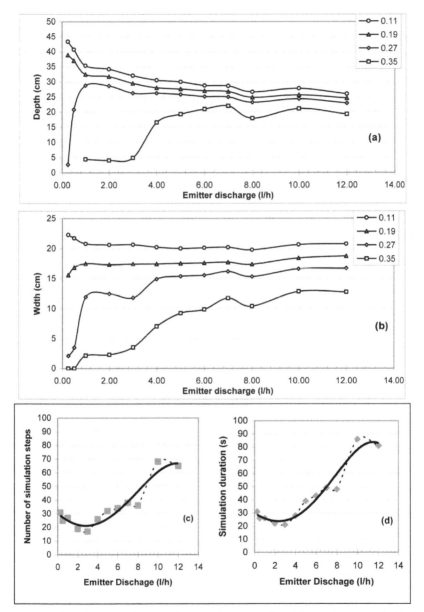

**FIGURE 11** Effects of emitter discharge rates on soil wetting patterns and simulation characteristics: (a) Depth of moisture content isoline; (b) Width of moisture content isoline; (c) Number of simulation steps; and (d) Simulation duration on the computer.

It is advised to use higher emitter discharge rates to achieve less simulation time, and thus less pumping costs. On the other hand, lower emitter discharge rates lead to gradual distribution of moisture and more wetting pattern area.

## 2.4   CONCLUSIONS

After field and model-based comparisons: the "Drip Chartist" model was validated to be a reliable simulation model for surface and subsurface drip irrigation especially in light textured soils. It is also proved to be able to simulate two sources of wetting patterns working simultaneously one above the other.

Several model-based studies were performed to benchmark our model. The studies resulted in several recommendations for the optimum values of the model parameters, formulas, and settings to ensure accurate, reliable, and fast modeling of surface and subsurface drip irrigation system.

## 2.5   SUMMARY

The developed model "Drip Chartist" was described in the previous chapter of this book. We verified and validated our model under field conditions; and compared it with other related models, such as: "Hydrus 2D."

## KEYWORDS

- **Alexandria University**
- **Bar**
- **Drip Chartist model**
- **Finite difference method**
- **Hydraulic conductivity**
- **Hydrus 2D**
- **Isoline**
- **Newton – Raphson method**
- **Physical barrier**
- **Richard's equation**
- **Saturated soil**
- **Simpson method**
- **Soil depth**
- **Soil moisture**
- **Soil profile**
- **Subsurface drip irrigation**
- **Suction**
- **Surface drip irrigation**
- **Wetting pattern**
- **Unsaturated soil**

## REFERENCES

1. Abbasi, F., Simunek, J., Feyn, J., van Genuchten, M. Th. Shouse, P.J., (2003b). Simultaneous inverse estimation of soil hydraulic and solute transport parameters from transient field experiments: homogeneous soil. Trans ASAE 46 (4): 1085–1095

2. Assouline, S., (2002). The effects of microdrip and conventional drip irrigation on water distribution and uptake. Soil Sci. Soc. Am. J. Proc. 66: 1630–1636.

3. Brandt, A., E. Bresler, N. Dinar, I. Ben-Asher, J. Heller, and D. Goldberg, (1971). Infiltration from trickle source: I. mathematical model. Soil Sci. Soc. Am. J. Proc. 35: 675–682.

4. Bresler, E., (1975). Two-dimensional transport of solutes during nonsteady infiltration from a trickle source. Soil Sci. Soc. Am. J. Proc. 39: 604–613.

5. FAO, (1970). *Methods of Soil and Water Analyzes*. Soil Bulletin # (10), FAO, Rome, Italy.

6. Fishman, G. S., and P. J. Kiviat, (1968). The statistics of discrete-event simulation. Simulation J. 10: 185–195. quoted from Law, and Kelton, 1982.

7. Gärdenäs, A.I., J. W. Hopmans, B. R. Hanson, J. Simunek, (2005). Two-dimensional modeling of nitrate leaching for various fertigation scenarios under micro irrigation. J. Agr. Wat. Manag. 74: 219–242.

8. Ismail, S.M., Zien El-Abedin, T.K., Wassif, M.A., El-Nesr, M.N., (2006a) *Wetting Pattern Simulation Of Surface And Subsurface Drip Irrigation Systems I- Model Development* (The previous chapter in this book).

9. Ismail, S.M., Zien El-Abedin, T.K., Wassif, M.A., El-Nesr, M.N., (2006b) *Drip Irrigation Systems In Sandy Soil Using Physical And Hydraulic Barriers.* (Chapter by authors in this book)

10. Law, A. M., and W. D. Kelton, (1982). *Simulation Modeling and Analysis*. McGraw-Hill Book Company, USA.

11. Levin, I., P.C. van Rooyen, and F.C. van Rooyen, (1979). The effect of discharge rate and intermittent water application by point-source irrigation on the soil moisture distribution pattern. Soil Sci. Soc. Am. J. Proc. 43: 8–16.

12. Li, J., Zhang, J., Rao, M., (2004). Wetting patterns and nitrogen distributions as affected by fertigation strategies from a surface point source. Agric. Wat. Manag. 67: 89–104.

13. Ragab, R., J. Feyen, and D. Hellel, (1984). Simulating infiltration into sand from a trickle line source using the matric flux potential concept. Soil Science 137: 120–127.

14. Šimůnek, J., K. Huang, and M.Th. van Genuchten, (1999). *The HYDRUS code for simulating the two-dimensional movement of water, heat, and multiple solutes in variably saturated soils v. 2.0*, US-Salinity Lab., Agric. Res. service, U.S. department of agriculture, Riverside, California. Research report #144, 253pp.

15. van Beers, W. F. J., (1976). *The auger hole method*. International Institute for Land Reclamation. ILRI, Netherlands.

16. van Genuchten, M. Th, F. J. Leij, and S. R. Yates, (1991). *The RETC code for quantifying the hydraulic functions of unsaturated soils*, U.S. Salinity Lab., U.S. Dept. of Agric., Agric. Res. Service, Riverside, California. 93pp.

# CHAPTER 3

# MICRO IRRIGATION AND HYDRAULIC BARRIER TECHNIQUE

M. N. EL-NESR, S. M. ISMAIL, T. K. ZIEN EL-ABEDIN, and M. A. WASSIF

## CONTENTS

*Partially printed from: "*El-Nesr, M. N., 2006. Subsurface Drip Irrigation System Development and Modeling of Wetting Pattern Distribution. Unpublished PhD Thesis, Faculty of Agriculture, Alexandria University, Egypt* <www.alexu.edu.eg/index.php/en>." Academic and research guidance/support by S.M. Ismail, T.K. Zien El-Abedin, and M.A. Wassif; and by Faculty of Agriculture at Alexandria University is fully acknowledged.  In this chapter: Jerusalem artichokes is *tartoufa (Arabic)*, and Feddan, Fed., is an Egyptian measure of land area, 1 Fed. = 4200 m$^2$.

## 3.1 INTRODUCTION

The coarse textured soils have poor deep percolation. However, this problem exists in both, surface drip irrigation and subsurface drip irrigation (SDI). The deep percolation is worst in SDI. When water is applied at top of a coarse textured soil, it takes some time (depending on the infiltration rate) to move away from the root zone. This time will definitely decrease if the dripping source is closer to the end of the root zone.

The main avenue for water losses under SDI is deep percolation, which is highest during the seedling stage and declines with the increase of root system [2]. On the other hand, Phene et al. [5] showed that deep percolation losses and runoff can be reduced with properly designed and managed SDI systems.

Barth [1] used an impermeable polyethylene foil below the lateral pipes: 60 cm wide and 0.06 mm thick plastic sheet at 30 to 40 cm depth. He concluded that this physical barrier significantly increased the amount of water held in the root zone, either from dripper line or from rain, and limited the deep percolation. He also stated that the V-shape plastic foil increased the amount of water storage. In addition, he developed a special installation equipment to release the dripper line and the V-shaped plastic foil simultaneously into the soil without disturbing the natural soil profile.

Welsh et al. [8] developed a vector-flow™ technique to increase the horizontal flow of water under SDI. The technique involves placing an impermeable V-shaped line just below the dripper line, i.e., the dripper line is placed over the small V-shaped stripe, which is only 7.5 cm wide. A dripper size of 3.5 L/h was used in a sandy loam soil. They concluded that their technique allows 70% of the applied water and was able to wet up to 90 cm wide zone in the upper 15 cm of soil, while only 25% of the applied water was spread without the technique.

The method of barricading water percolation through physical means can be a success in its job. It raises the water content above its location and leads to minimized leakage of water. The major drawbacks of this method are technical and economic problems to trench a wide and deep furrow to lay the physical barrier. Also, the hazard of air lack may appear in the root zone due to the moist environment created by the physical barrier, in addition to root dwarfness hazard.

Therefore, a new method was investigated which was called the "Hydraulic Barrier system or the bilateral drip system." This method barricades water without these problems. The hydraulic barrier method can be described as 1. Burying a secondary pipe-line similar to the primary one but beneath it; and 2. Dividing the required water volume between the two pipe-lines. This formulates the wetting pattern so as to increase its width and hence to increase the available water in the shallow root zone. This technique requires no extra trenching width, and does not cause air lack or root dwarfness.

In this chapter, we will discuss the research to evaluate the hydraulic barrier technique and to compare this technique with the previous barricading technique under field conditions.

## 3.2 MATERIALS AND METHODS

A field experiment was conducted in the North Sinai research station of the Desert Research Center in "El-Shaikh Zowayed" city, 30 km east of "El-Arish," and 12 km west of "Rafah" on the Egyptian- Palestinians' borders. At the site, the soil texture through particle size distribution, soil moisture characteristic curve, electrical conductivity (EC), and pH were also determined according to method by FAO [4]. Soil hydraulic conductivity was determined using van-Beers [7] method, while the infiltration rate was established using the method described by Philip [6]. These values are listed in Tables 1 and 2. The site was initially prepared by shallow disking to remove surface herbs and clean the land surface. No tillage operation was performed due to the texture of the soil.

Water was supplied with the drip network irrigation system through the control head consisting of: Main pump station, sand media filter, a screen filter. The lateral lines were installed in the field as required in each treatment. The dripper lines were of GR type with built-in emitters of discharge 4 L/h. The emitter spacing was 30 cm spacing. In order to bury the dripper lines, soil was trenched manually to the desired depth and the dripper line was laid. Then the gutter was covered. This operation was performed line by line as the soil tumbles back rapidly after trenching.

Soil moisture content was measured using a neutron scattering probe. The access tube was 120 cm tall and 65 mm outer diameter. The access tube was installed up to 105 cm depth, thus allowing 15 cm of the tube above the soil surface to mount the neutron scattering device on it. Each access tube was isolated from the bottom by a plastic sheet similar to the one used in the physical barrier in order to prevent the ground water from entering the tube. Seventy-two access tubes were installed by digging a hole by the soil auger after moistening the sand to increase the ability of sand carrying by the auger. When installing the access tubes in the plastic sheet zone, the access tube was attached to a special piece of plastic sheet at the desired depth, then a 50*50 cm² trench was dig till the depth of plastic sheet (at 40 cm). Then the installed plastic sheet was punched to allow the soil auger to go through and the auger was able to complete the rest of 65 cm out of 105 cm depth. The experimental treatments are summarized in Table 3, and the field layout is shown in Fig. 1.

**TABLE 1** The values for soil moisture characteristic curve for the experimental site soil.

| Suction, bar | 0.10 | 0.50 | 1.00 | 5.00 | 10.00 | 15.00 |
|---|---|---|---|---|---|---|
| Soil moisture, fraction (cm³/cm³) | 0.23 | 0.21 | 0.20 | 0.17 | 0.10 | 0.03 |

**TABLE 2** Some soil properties of the experimental site soil.

| Soil depth | Texture | EC | pH | Infiltration rate | Hydraulic conductivity |
|---|---|---|---|---|---|
| cm | | dS/m | — | cm/h | m/day |
| 0 to 30 | medium to fine sand | 5.86 | 7.72 | 66.62 | 24.6 |
| 30 to 60 | medium to fine sand | 4.51 | 7.72 | — | — |

**TABLE 3**    Barrier treatments at the site.

| | |
|---|---|
| 1. No barrier | • On soil-surface |
| | • Buried at 10, 20, and 30 cm |
| 2. Hydraulic barrier | • Not exists (Single lateral) |
| | • Exists (Double laterals, the upper was variable depth while the lower was fixed at 40 cm) |
| 3. Physical barrier | • Exists (a plastic sheet installed at 40 cm depth) |
| | • Not exists |

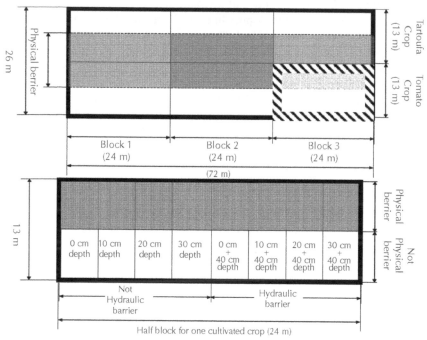

**FIGURE 1**    Experimental layout.

In each half of the site area, two crops were planted: Jerusalem Artichokes (Arabic "*Tartoufa*") and tomato. We compared two root types (tubers and normal roots) as affected by SDI system and its wetting pattern shape.

## 3.3   RESULTS AND DISCUSSION

### 3.3.1   SOIL WETTING PATTERNS UNDER FIELD CONDITIONS

Field measurement of soil moisture distribution for each experimental treatment was made using a field-calibrated neutron scattering probe. To obtain accurate and representatives observations,, measurements were taken after 6 to 8 h after irrigation was initiated. This ensured that water redistribution occurred. Moisture measurements were taken throughout the growing season.

It was noticed that the existence of the physical barrier pushes the water content isolines (contours) upward, above its location, in the upper 40 cm of soil profile. Figure 2 indicates the comparison among treatments: (P) to (N) and (B) to (H). The effect of hydraulic barrier on the wetting pattern was also observed by comparing treatments: (H) to (N) and (B) to (P). This increased the slope of the isolines in the space between dripper lines (it permit a narrower space between them).

However, the effect of hydraulic barrier can be observed clearly (as a barrier) when the space between the two-dripper lines is smaller. The narrower gap acts like a real barrier as shown in Fig. 3.

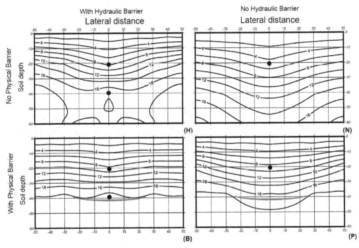

**FIGURE 2**   Moisture distribution for treatments in the tartoufa crop with physical and hydraulic barriers. Subsurface drip irrigation is at 20 cm soil depth. Note: **N** = No barriers, **H** = With hydraulic barrier, **P** = With physical barrier, and **B** = Both barriers.

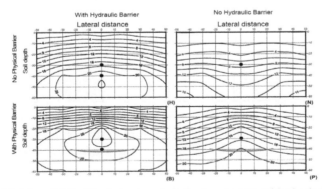

**FIGURE 3**   Moisture distribution for treatments in the tomato crop with physical and hydraulic barriers. Subsurface drip irrigation is at 20 cm soil depth. Note: **N** = No barriers, **H** = With hydraulic barrier, **P** = With physical barrier, and **B** = Both barriers.

The existence of the hydraulic barrier with 10 cm gap raises the soil water content above it as can be observed by comparing treatment (H) to (N). The reader will notice that the water isolines of pattern (P) are almost similar to pattern (H). This concludes that both physical and hydraulic barriers behave approximately in a similar way for a narrower gap.

### 3.3.2   CROP YIELD

Statistical analysis for the field data was done in split-split-plot design with hydraulic barrier existence {Hb} as a whole plots factor, physical barrier existence or plastic isolation {Pb} as a subplot factor. The remainder factor was burying depth of pipe-line depth {Dp}. The analysis of results indicated that only {Pb} parameter was significant by its own in most of the measured data. The means-comparison was also analyzed.

#### 3.3.2.1   TARTOUFA CROP

Total and marketable yield of tartoufa were measured. Total yield included the weight of all tubers regardless of its size or state, while marketable yield included tubers only that can be sold in the market (not so- small, not broken, and not suffering of any disease's syndromes).

The analysis of variance showed that when "total yield" was considered as dependent variable, only the factors {Pb}, {Dp}, and the interaction {Pb}x{Dp} were statistically significant. The hydraulic barrier {Hb} main effect and interactions were not significant in case of total tartoufa yield.

However, in case of "marketable yield," ANOVA showed that the same significant variables in "total yield" are significant too.

The interaction between burying depth and hydraulic barrier {Dp}x{Hb} was statistically significant. In other words, although the hydraulic barrier existence variable {Hb} was not significant, yet its interaction with depth {Dp} was significant.

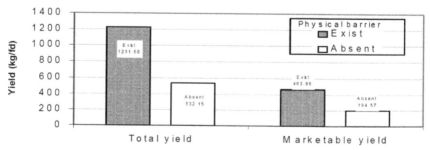

**FIGURE 4**   Tartoufa yield as affected by physical barrier existence. 1 fed = Egyptian measure of land area = 4200 m².

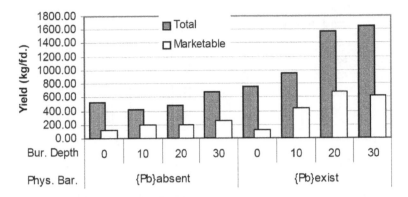

**FIGURE 5**   Tartoufa yield as affected by the physical barrier existence and burying depth. 1 fed = Egyptian measure of land area = 4200 m².

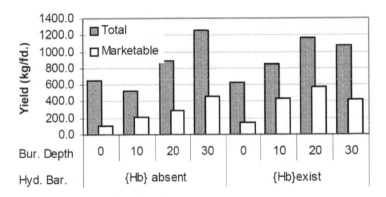

**FIGURE 6**   Tartoufa yield as affected by the hydraulic barrier existence and burying depth. 1 fed = Egyptian measure of land area = 4200 m².

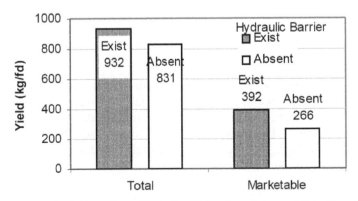

**FIGURE 7**   Tartoufa yield as affected by hydraulic barrier existence. 1 fed = Egyptian measure of land area = 4200 m².

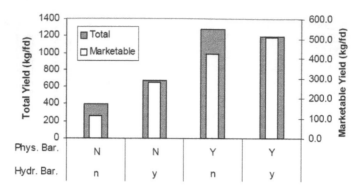

**FIGURE 8** Tartoufa yield as affected by hydraulic and physical barriers. Note: 1 fed = Egyptian measure of land area = 4200 m².

### 3.2.1.1 EFFECTS OF PHYSICAL, HYDRAULIC BARRIER AND BURRING DEPTH ON TARTOUFA YIELD

Regarding the main effect of {Pb}, Fig. 4 shows that the physical barrier existence almost doubles the yield of Tartoufa tubers. The physical barrier existence increases the yield by a factor of 2.31–2.38 compared to no physical barrier, in case of both total and marketable yield, respectively. Therefore, the physical barrier plays an important role in the tartoufa production.

As shown in Fig. 5, the {Pb}x{Dp} interaction showed the superiority of the physical barrier's treatments, with the yield increment with deeper burying depth. On the other hand, the {Hb}x{Dp} interaction indicated that deeper depths with the existence of the hydraulic barrier gave better results than the shallower depths (Fig. 6).

Although not statistically significant, the hydraulic barrier existence leads to 47% more marketable yield and 12% more total yield than in the absence of the hydraulic barrier (Fig. 7). The insignificance may be attributed to that the hydraulic barrier treatment has high error value as a main-plot factor.

In addition, the interaction between hydraulic and physical barriers is not statistically significant, however, the marketable yield increased in the existence of both barriers, while the total yield decreased in the existence of only the physical barrier. On the other hand, total and marketable yield increased in the absence of the physical barrier and the existence of the hydraulic barrier (Fig. 8).

### 3.3.2.1.2 EFFECT OF BURRING DEPTH

As shown in Fig. 9, tartoufa yield and burying depth were directly proportional. There was no significant difference neither between the pair, 30 cm and 20 cm, nor between the pair, 10 cm and 0 cm. However, there was significant difference between yields for these depth groups. This may be attributed to the Tartoufa tubers growing in the top 10 to 15 cm depth, thus preferring nonsaturated conditions.

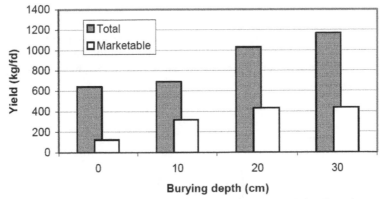

**FIGURE 9** Tartoufa yield as affected by burying depth. Note: 1 fed = Egyptian measure of land area = 4200 m².

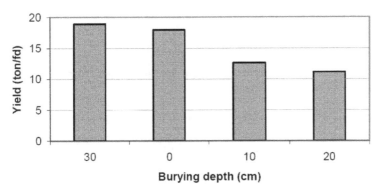

**FIGURE 10** Tomato yield as affected by burying depth. Note: 1 fed = Egyptian measure of land area = 4200 m².

### *3.3.2.2 TOMATO YIELD*

### *3.3.2.2.1 EFFECTS OF PHYSICAL, HYDRAULIC BARRIER AND BURRING DEPTHS ON TOMATO YIELD*

The tomato yield in the existence of physical barrier (about 20.78 ton/fed.) was significantly more than the yield for no physical barrier (which was about 9.49 ton/fed. However, unlike tartoufa, the ANOVA showed that only physical barrier treatment had significant effect on total yield. This concludes that neither the hydraulic barrier nor the burying depth had significant effect on the tomato yield. Although, the tomato yield was 12.98 ton/feddan (1 fed = Egyptian measure of land area = 4200 m²) in the existence of the hydraulic barrier compared to 17.29 ton /fed. in the absence of it (a reduction of 25%). This may be attributed to the shallow root zone of tomato and the deep burying depth of the hydraulic barrier (at 40 cm), which can take half the irrigation water away from the root zone.

### 3.3.2.2.2    EFFECT OF BURRING DEPTH

Unlike tartoufa, surface drip treatments gave better results than SDI results. Total tomato yield and burying depth were inversely proportional, except the 30 cm depth that gave the maximum yield (Fig. 10). The superiority of the 30 cm treatment (contrasting the trend) may be attributed to the closeness to the physical and hydraulic barriers both at 40 cm.

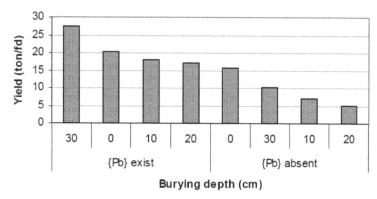

**FIGURE 11**    Tomato yield as affected by burying depth and physical barrier existence. 1 fed = Egyptian measure of land area = 4200 m$^2$.

In case of interaction {Pb}x{Dp}, Fig. 11 shows that yield increased inversely proportion to the depth except at the 30 cm depth as mentioned above. Detailed results of all treatments and measures are found at El-Nesr [3].

## 3.4    CONCLUSIONS

Both physical and hydraulic barriers act approximately the same manner for a narrow gap. The physical barrier existence almost doubled the yield of Tartoufa tubers. Actually, tuber total yield with a physical barrier was 2.31 times higher compared to a value with no physical barrier. Tuber marketable yield with a physical barrier was 2.38 times higher compared to a value with no physical barrier. The hydraulic barrier existence yielded 47% more marketable tubers and 12% more total tubers compared to the yield with no hydraulic barrier.

The tomato yield in the existence of physical barrier was about 20.78 ton /fed. In addition, it was significantly more than that with no physical barrier, which was about 9.49 ton/fed). The tomato yield was 12.98 ton/feddan with the hydraulic barrier compared to 17.29 ton /fed. With no hydraulic barrier (a reduction of 25%).

Tartoufa yield increased directly proportional to the burying depth. There was no significant difference neither between the pair, 30 cm and 20 cm, nor between the pair, 10 cm and 0 cm. However, there was significant difference between these depth groups.

Unlike Tartoufa, surface drip irrigation treatments gave better results than SDI treatments. Total tomato yield increased inversely proportional to the burying depth, except the 30 cm depth that gave the maximum yield.

Tartoufa tubers appear to prefer partial wetness to grow properly. However, better results were achieved with deeper laterals. Moreover, the yield of tartoufa gave better results mostly on double-lateral treatments. On the other hand for most tomato results, the single-tube treatments gave better results than double-tube treatments, because the latter emits water at two points. One variable is fixed at 0, 10, 20, 30 cm; and the other is fixed at 40 cm. However, tomato root system absorbs about 75% of its needs in the top soil layers. Therefore, if water exists on such layers, the yield may increase. It can be concluded that the second dripper line at 40 cm depth in the double-lateral treatment is not beneficial to the tomato crop; and it took half the water away from the tomato root zone. It can be inferred that if the second lateral was buried at 20 cm depth, the results might would been surprisingly different.

## 3.5 SUMMARY

"The hydraulic barrier technique" to barricade water deep percolation was investigated. The method involves burying a secondary dripper line under the primary one in order to increase water content in the deep layers, thus minimizing matric potential in these layers. It will also increase the lateral movement of soil moisture from the primary dripper line. The method was validated by a field experiment in Sinai sandy soil. The field experiment tested with four burying depths of primary dripper line, with and without the hydraulic barrier, compared with the physical barrier. Two crops were used: Jerusalem artichokes (Tartoufa) as example to tuber roots, and tomato as fibrous root.

The results indicated that the physical barrier extremely increased the crop yield for both crops, almost doubling the yield with no physical barrier value (2.35 for Tartoufa and 2.19 for Tomato) The hydraulic barrier existence increased the total crop yield of Tartoufa by about 12% and marketable yield by 47%. Reduction of 25% of Tomato yield was due to existence of hydraulic barrier. The Tartoufa yield increased directly proportional to the burying depth of the primary lateral while the situation is inverted for the tomato crop. The hydraulic barrier acted in a pattern similar to the physical barrier when the narrower gap between the two dripper-lines.

## KEYWORDS

- **Alexandria**
- **barrier**
- **barrier, hydraulic**
- **barrier, physical**
- **bilateral drip system**
- **crop yield**
- **deep percolation**

- drip line
- dripper line
- EC
- Egypt
- Feddan, Fed., (Egyptian measure of land area, 1 Fed. = 4200 m²)
- impermeable layer
- isolines
- Jerusalem artichokes, *tartoufa (Arabic)*
- lateral distance
- lateral line
- micro irrigation
- neutron scattering meter
- pH
- plastic foil
- root zone
- runoff
- sandy loam soil
- soil depth
- soil moisture
- soil profile
- subsurface drip irrigation, SDI
- tomato
- water percolation
- wetting pattern

## REFERENCES

1. Barth, H.K., 1995. Resource conservation and preservation through a new subsurface irrigation system. Proceedings of 5th International Microirrigation Congress by ASAE at Orlando – Florida. Pages, 168–174.
2. El-Berry, A. M., 1989. Design and utilization of subsurface drip irrigation system for fodder production in arid lands. Misr J. Agr. Eng., 6(2): 153–165
3. El-Nesr, M. N., 2006. Subsurface Drip Irrigation System Development and Modeling of Wetting Pattern Distribution. Unpublished PhD Thesis, Faculty of Agriculture, Alexandria University, Egypt.
4. FAO, 1970. *Methods of Soil and Water Analyzes.* Soil Bulletin # 10 by FAO, Rome, Italy.
5. Phene, C. J., Hutmacher, R.B., Ayars, J.E., Davis, K.R., Mead, R.M., and Schoneman, R.A, 1992. Maximizing water use efficiency with subsurface drip irrigation. ASAE International Meeting, Paper No. 92-2090.

6. Philip, J. R., 1957. The theory of infiltration – Part IV: Sorptivity and algebraic infiltration equations. Soil Science, 84: 257–264.
7. Van Beers, W. F. J., 1976. The auger hole method. International Institute for Land Reclamation (ILRI), Netherlands.
8. Welsh, D.F., U.P. Kreuter, and J.D. Byles, 1995. Enhancing subsurface drip irrigation through vector flow. Proceedings of 5th International Microirrigation Congress by ASAE at Orlando, Florida. 688–693.

# CHAPTER 4

# MICRO IRRIGATION DESIGN USING MICROCAD: EGYPTIAN DESERT

M. N. EL-NESR, S. M. ISMAIL, E. R. EL-ASHRY, and G. A. SHARAF

## CONTENTS

*Partially printed from: "*El-Nesr, M.N. (1999). Computer Aided Design and Planning of Drip Irrigation Systems. Unpublished MSc Thesis, Faculty of Agriculture, Alexandria University, Egypt* <www.alexu.edu. eg/index.php/en>." Academic and research guidance/support by S.M. Ismail, T.K. Zien El-Abedin, and M.A. Wassif; and by Faculty of Agriculture at Alexandria University is fully acknowledged. In this chapter: *faddan* is an Egyptian unit of area and 1 feddan = 1.038 acres (0.42 ha = 4200 m²).

## 4.1 INTRODUCTION

Micro irrigation system is a solid set system for supplying filtered water and fertilizers in a small quantity and under low pressure, directly near the root zone. Water through a trickle network flows in predetermined direction, starts from the water source, passing through control unit/conveying lines (main and submain), distribution pipes (manifolds and laterals) and finally to "emitters." Micro irrigation system design is a decision-making process to match the operational, physical and economic settings. The design is based on crop water requirements, hydraulic principles, emitter characteristics; field properties, economics of the system, and criteria of water application uniformity [5].

Fortunately, computers have come in general use, and any complicated calculations can easily be performed with a suitable program. The use of computers removes much tedious work associated with repetitive complex calculations and the manipulation of large pools of data. This results in more detailed analysis when compared to noncomputer aided design.

In general, irrigation computer models can be categorized by function into models for: Crop Water Requirements - IRPSYS [6]; Irrigation Scheduling - CWR-VB [8]; System Selection by Ponce et. al. [21]; Network Planning by Awady and Ahmed [3]; System Design – REDES by Gomes [11]; System Evaluation by Meshekt and Warner [19]; and Hydraulic Design/Cost Analysis by Sharaf [22]. However, Ismail [13] developed a computer simulation model to analyze and design a micro irrigation system that included parameters, such as: Emitter pressures and flow rates along laterals, emission uniformity and economical pipe size.

In this chapter, authors discuss a MicroCad interactive computer program to design a micro irrigation system for an Egyptian Desert. We developed this model for planning, hydraulic design and cost estimation of micro irrigation systems. MicroCad model can be used for the general field and research purposes.

## 4.2 MODEL DEVELOPMENT

For the development of MicroCad model to design a micro irrigation system, we will discuss briefly the topics, such as: 1. Planning rules, 2. Soil-Water-Plant relationships, 3. System hydraulics, and 4. Cost analysis.

### 4.2.1 PLANNING RULES

Planning a micro irrigation network involves a suitable layout of the pipes depending on the land shape and topography. This operation has no predefined steps and depends mainly on the designer's experience and knowledge, but some economic and hydraulic considerations must be observed. The entire irrigation project in the field is considered for an optimum planning. Some of the researchers consider "uniformity" as the most important factor in the planning process. Others consider the "hydraulic balance of the network," and "economic excellence."

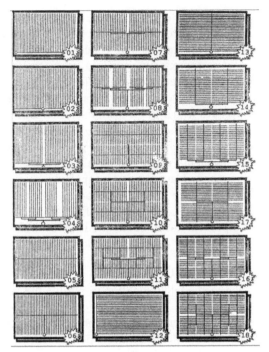

**FIGURE 1**   Planning categories for MicroCad [10].

The MicroCad technique of planning depends on categorizing the irrigation layouts into 18 general cases shown in Fig. 1. Assuming lateral line not to exceed "L" meters, manifold not to exceed "M" meters and considering laterals direction either in the direction of the short edge or in the long edge of the field. For example, case one in Fig. 1 can be applied if all of the following conditions are met:

   (a)  The direction of the laterals is by the field short edge;
   (b)  Long edge of the field "is less than or is equal to 'M'; and
   (c)  The short edge of the field is less than or is equal to 'L.'

**TABLE 1**   Planning rules for MicroCad technique for planning a trickle irrigation systems, 18 cases that are shown in Figs. 1 and 6a.

| Direction of laterals | | Long edge | | | |
|---|---|---|---|---|---|
| **Short** | ⎸⎸⎸⎸ | $\leq 1$ m | $\leq 2$ m | $\leq 4$ m | $\leq 8$ m |
| **edge** | | | | | |
| | | $\leq 1$ L | 1 | 2 | 3 | 4 |
| | | $\leq 2$L | 5 | 6 | 7 | 8 |
| | | $\leq 4$ L | — | 9 | 10 | 11 |
| | ≡ | | $\leq 1$L | $\leq 2$L | $\leq 3$L | $\leq 4$L |
| | | $\leq 1$ M | 12 | 13 | 14 | 15 |
| | | $\leq 2$ M | — | — | 16 | 17 |
| | | $\leq 4$ M | — | — | — | 18 |

On the other hand for the same direction of lateral lines, if the short edge of field is greater than 'L' and less than or equals '2L,' case 5 in Fig. 1 can be used, etc. Other cases can be applied according to the planning conditions given in Table 1.

By default, MicroCad sets maximum manifold length to 100 meters, maximum lateral length to 50 meters. However, MicroCad offers the option to automatically or manually modify these values if necessary. In general, it is recommended not to design a field greater than 16 hectares with a single main line. Alternatively, bigger fields can be planning by partitioning each to smaller fields.

MicroCad has the advantage of selecting any planning category (Fig. 1) by its number. In addition, any layout that differs from the suggested one can be applied, and any modification to the suggested layout could be applied too.

In Egypt, most land slopes are less than 0.1% especially in the new reclaimed lands. Therefore, the slope effect is neglected in the planning stage. Otherwise, the point of connection of pipes will not be at the middle. However, it will be shifted toward up-slope direction, in order to maximize the down-slope gain.

### 4.2.2   SOIL- WATER-PLANT RELATIONSHIPS

This section presents the methodology applied to resolve the initial design criteria including:

1.  Plant consumptive use under micro irrigation by applying SCS model [23] that includes a certain ratio depending on the percent of shaded area of the total land.

2.  Net irrigation depth which is equivalent to the soil water in the root zone depth, reduced by the percent of wetted area, and depleted by certain ratio so that the plant is free from water stress. It was computed according to Benami and Ofen [4].

3.  The Irrigation interval that is the integer number of days required by the soil-water to reach the desired critical point.

4.  Emission uniformity that is an indicator of the percentage of the average irrigation depth that is received by the least-watered area. It was calculated by the equation developed by Nakayama et al. [20].

5.  Irrigation system efficiency is the product of the application efficiency and the emission uniformity as shown by Wu and Gitlin [25].

6.  Leaching requirements that must be met to control the soil salinity. It was calculated according to Doorenbos and Pruitt [9].

7.  "Transpiration to application ratio" which is the ratio of irrigation water transpired to the total irrigation depth applied to the least watered areas. It is mainly affected by management, atmospheric conditions, amount of leaching water, and deep percolation. It was estimated according to Keller and Bliesner [17].

8.  The gross irrigation depth delivered to the plant including plant requirements, deep percolation, and leaching requirements. It was calculated according to Meshkat and Warner [19].

9.  Flow rate required per plant that is delivered from the pump to the root zone of a plant for a specific area in a certain time.

## 4.2.3 SYSTEM HYDRAULICS

### 4.2.3.1 MINOR LOSSES

The emitter flow is simulated by an emitter flow function described by Keller and Karmeli [18]. The connector barb of an emitter projects into the flow in lateral line pipe causing additional turbulence over the normal pipe friction. The method used to represent additional minor losses contributed by the barbs was the equivalent pipe length method [24].

### 4.2.3.2 DESIGN CRITERIA

The design criteria of lateral and manifold is based on partitioning the allowable flow variation or pressure variation for both the lateral and manifold. The default value for partitioning this variation is 55% on lateral and 45% on manifold. To perform acceptable uniformity in a micro irrigation field, the maximum discharge variation should not exceed 10% as recommended by ASAE standards [2].

### 4.2.3.3 ESTIMATION OF LOSSES DUE TO FRICTION IN PIPES

The friction loss along conveying lines (laterals and manifolds) or distribution lines (submain and main lines) can be computed by one of the following formulas: Hazen-Williams, Scobey-Blasius or any other formula suggested by the user.

### 4.2.3.4 DESIGN OF LATERALS

In general, there are two ways for a proper design of a lateral. The first is to select the size of a lateral line for a given length, the other is to determine the maximum length of lateral line for given flow conditions and ground slope when the lateral line is limited to a specific size. MicroCad supports the former method.

In literature, several techniques have been developed to select the lateral size. The most famous techniques are: 1. F-technique [7]; 2. Segment by segment technique [12]; 3. Statistical approach technique [1]; and selected graphical methods [17]. MicroCad supports the first and the second techniques. Also there are some new design techniques, such as: The energy gradient line approach [26], and the uniformity based technique [15].

### 4.2.3.4.1 F-TECHNIQUE

The F-Technique is based on calculating the friction loss in the lateral line with no outlets. Then the results are multiplied by a coefficient called "F factor or reduction factor," which depends on the number of outlets along the lateral line.

### 4.2.3.4.2 SEGMENT BY SEGMENT TECHNIQUE (SBS)

"Segment by Segment Technique" is based on calculating the total pressure head loss due to friction and minor losses between two adjacent emitters. Total pressure loss in the line is the summation of pressure losses in these segments.

### 4.2.3.5 MANIFOLD DESIGN

Manifold is the water delivery pipeline with multiple outlets that supplies water from the main lines to the laterals on one or both of sides of the pipeline. With MicroCad, it can be designed by techniques for design of laterals.

### 4.2.3.6   MAINLINE DESIGN

MicroCad supports two methods for designing the main line: Velocity limit (1.5 m/s) method; and the economical method [18].

### 4.2.3.7   THE CONTROL HEAD

The control head consists of devices to ensure the proper operation of the irrigation system. By default, MicroCad considers the components of pump, filtering system and valves as major items. Extra accessories such as water meter, pressure gages, pressure regulators, chemical injection equipment and automatic control device can be added as an option by the user.

### 4.2.3.8   TOTAL DYNAMIC HEAD

The Total Dynamic Head (TDH) is estimated by MicroCad as a sum of the dynamic lift, supply system losses, control head losses, friction losses in pipelines. In addition, a safety factor represents minor losses of around 10% of the sum of the friction head losses, as recommended by Keller and Bliesner [17].

### 4.2.3.9   PUMPING POWER

Micro irrigation systems require energy to move water through the pipe distribution network, and discharge it through emitters. This energy is provided by pumping unit. The pump is operated by an electric motor or an internal combustion engine. Micro-Cad computes parameters for pump selection including TDH, water horsepower, and brake horsepower. It also computes operating costs for different types of power unit.

### 4.2.4   COSTS OF MICRO IRRIGATION SYSTEM

Determining the expected annual cost of owning and operating any irrigation system is an important part of irrigation system design. The cost of each component is given by an analytical equation or set of equations, to be combined to from an analytical model for the cost of the entire system. The fixed costs include annual depreciation, interest costs and yearly expenditures for taxes and insurance. The annual operating costs include the cost of energy, maintenance and repair, and labor cost. The method used by MicroCad for cost estimation is discussed in details by James [14].

### 4.3   RESULTS AND DISCUSSION

Once the model has been developed, it was evaluated by field experiments for its validation. Ten *faddans* (an Egyptian unit of area equivalent to 1.038 acres (0.42 ha = 4200 m$^2$)) were selected for the validation field test at the Desert Development Center (DDC) of the American University in the south Tahreer region. The experimental site consisted of five feddans of orchard and five feddans of vegetable crops. For both fields, pressure, discharge and statistical uniformity were determined. The layout and dimensions of these fields are presented in Fig. 2. Emitter used in both fields was "Katif" pressure compensating type. Under field conditions, the emitter function for Katif emitter is defined by:

$$q = 4.47 \, [h]^{0.04} \tag{1}$$

where, $q$ = Emitter discharge; $h$ = pressure head; and 4.47 and 0.14 are emitter characteristic constants for a particular emitter type. The coefficient of flow variation (CV) was assumed as 11.4%. The distance between emitters along the lateral line was one meter. Distance between laterals was five meters along the manifold for orchards, while it was one meter in the vegetable field. Topography in both fields was almost flat.

## 4.3.1 ANALYSIS OF FIELD MEASUREMENTS

The measured discharge profile was compared with the predicted profile by the MicroCad as shown in Fig. 3. The predicted profile under-estimated the actual discharge profile, especially at the lateral far end. This occurs due to the un-controllable parameters, which affect flow rate such as: Undulating of field topography, emitter wear, temperature, emitter clogging, and velocity head.

Figure 4 shows the comparison between measured and predicted mean flow rates for emitter at different locations on lateral. Figure 5 shows the comparison between measured and predicted mean flow rates for lateral at different locations on manifold. The statistical analysis showed no significant differences between predicted and measured mean flow rates of emitters along the laterals, and mean laterals flow rates along the manifold.

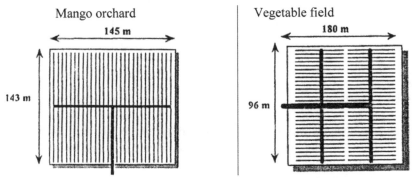

**FIGURE 2** Layout and dimensions of the field: Left – mango orchard field; Right – vegetable field.

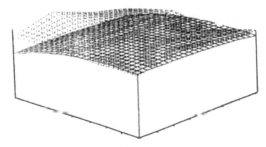

**FIGURE 3** Effects of length of laterals and manifold on the emitter flow profile for the orchard field. Dotted lines = Measured flow profile; and Solid lines = Predicted flow profile

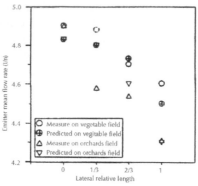

**FIGURE 4**    Actual versus mean emitter flow rates on lateral for vegetable and orchard fields.

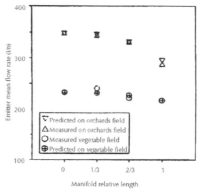

**FIGURE 5**    Actual versus mean emitter flow rates on manifold for vegetable and orchard fields.

Additional analysis was also performed to develop the relationships between the predicted pressure at the inlet and far-end of laterals and manifold to the actual measurements of the pressure at the same locations in field. The results showed no significant differences between the predicted and measured pressures. Table 2 presents the statistical analysis of the results, which correlate predicted flow and pressure by the model with the field measurements of flow and pressure.

**TABLE 2**    Statistical analysis for the predicted and measured flow rates and pressure for vegetable and orchard fields.

| Parameter | Item | LSD | |
|---|---|---|---|
| | | Vegetable field | Orchard field |
| Mean emitter flow rate, Lph | **Lateral** | 0.254* | 0.714 |
| Mean lateral flow rate, Lph | **manifold** | 16.1* | 45.02 |
| Pressure head, meters | **Both lateral and manifold** | 1.21* | 0.233 |

*Not significant at 95% confidence limit.

For the vegetable field, the measured field statistical uniformity was 88.7% compared to a predicted value of 91%. In case of orchard field, the measured field statistical uniformity was 88.3% compared to the predicted value of 90%. This small deviation between 88.7% and 88.3% can be attributed to the experimental error.

Some sample screen photos of the MicroCad module are shown in Figs. 6–15.

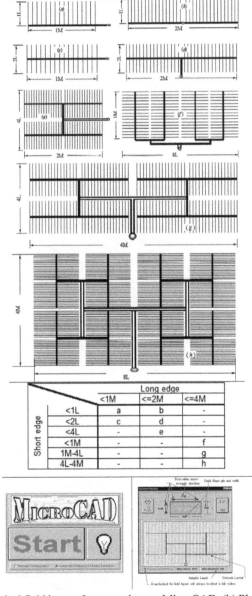

**FIGURE 6** (a) A typical field layout for example on a MicroCAD. (b) Planning stage: To insert field dimensions and the direction of laterals.

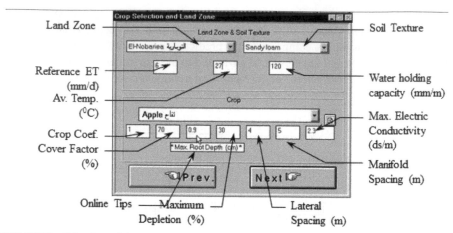

**FIGURE 7**   Selection of crop and soil properties from a dropdown menu.

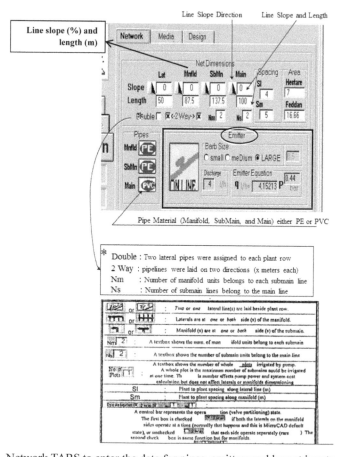

**FIGURE 8**   Network TABS to enter the data for pipes, emitters and layout inputs.

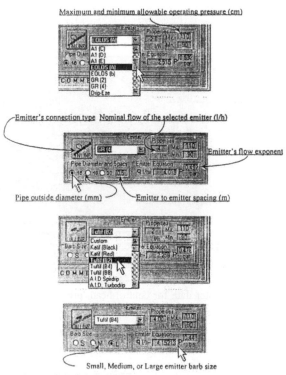

**FIGURE 9**   Selection of an emitter type and model. Data for other emitters can be entered through the "Custom" option in the dropdown menu.

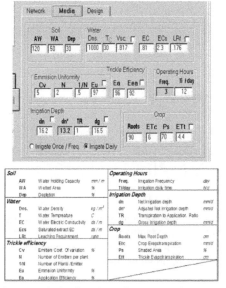

**FIGURE 10**   Media Tabs to enter the data for soil, water and crops.

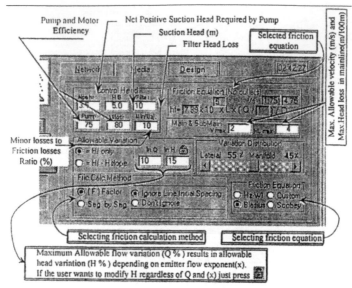

**FIGURE 11**    Design Tabs to enter data for control head, pressure distribution, and pressure limits. There is also a tab for design options.

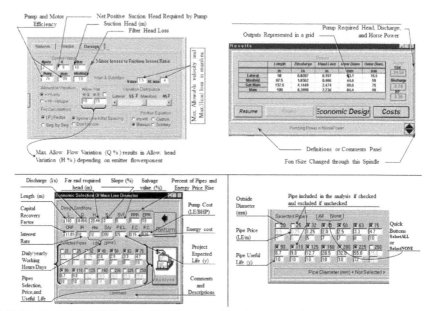

**FIGURE 12**    (a) LEFT: Design TAB for control head, pressure distribution and pressure limits (LEFT). Results for Network design: Mainline "Economic Design" can be accessed by pressing the button; Pump specifications (RIGHT). (b) Inputs for mainline economic design (optional method: LEFT); magnified view of previous form (RIGHT).

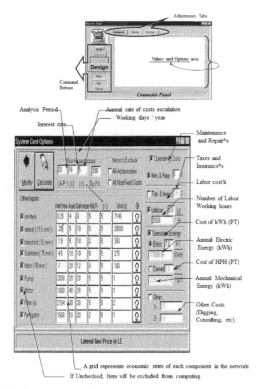

**FIGURE 13** Module for cost: main form.

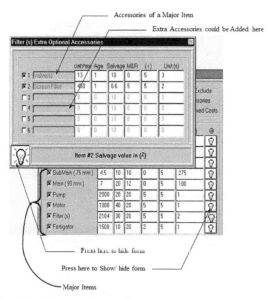

**FIGURE 14** Cost module for secondary system components.

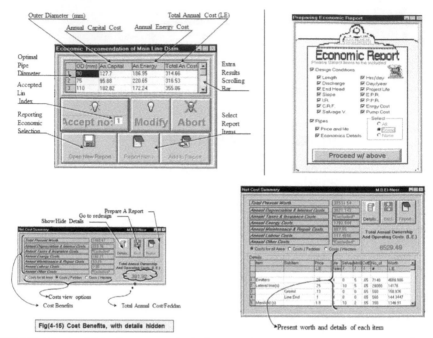

**FIGURE 15**   (a) Main line economic design (results) and report for economic design. (b) Module for cost benefit analysis.

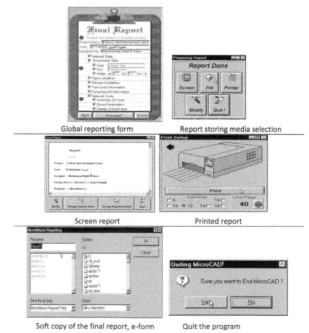

**FIGURE 16**   Reporting the data and results.

## 4.4  CONCLUSIONS

An interactive computer model has been developed to be used as a micro irrigation system design aid as well as a research tool. It is named as MicroCad. The functions of MicroCad model include those for planning the network layout, hydraulic design, and system costs. These three operations can be done by MicroCad either consequently or individually. However, MicroCad model is flexible in use so that designer can use it in testing and analyzing any alternative design: hydraulically and economically. The model can be used to calculate the annual costs of any system even if it is not designed by MicroCad. Nevertheless, the model has been constructed to design the system for all of the common design techniques. MicroCad can support databases, such as: The common emitters database, crops properties database, and land zones climatic and soil physical properties database. Field tests (discharge, pressure, and uniformity) of the model have been performed on about ten Faddans at Desert Development Center (DDC) of the American University in the south Tahreer region. The model correlated well with the measured data. The MicroCad requires that the user should be proficient in design concepts and the subject matter of micro irrigation design technology. Because MicroCad is not structured in a tutorial manner.

## 4.5  SUMMARY

To simplify the design and planning of micro irrigation systems, an interactive computer aided model called MicroCad has been developed. The MicroCad approaches the design in four stages: (1) Planning, (2) Soil-water relations, (3) Hydraulic design, and (4) Cost analysis. These four stages can be conducted either consequently or individually. The model can be used to calculate the annual costs of any micro irrigation system, even it is not designed using MicroCad. The unique feature of the model is its ability to design the system using all of common design techniques and have the ability to customize any of them. The MicroCad can support databases like: The common emitters database, crops properties database, and some Egyptian land zone climatic database, and soil physical properties database. For validity, field tests of flow, pressure and statistical uniformity were performed at the Desert Development Center (DDC) of American University - Cairo, Egypt. The model results correlated well with the field-measured data. The program was written in VISUAL BASIC for the IBM-PC.

## KEYWORDS

- **American University, Egypt**
- **Blasius**
- **Computer Aided Design, CAD**
- **cost analysis**
- **crop water requirements**
- **Desert Development Center, Egypt**
- **economic analysis**

- efficiency
- emission uniformity
- emitter
- energy gradient
- filtration system
- Hazen – Williams
- injection system
- irrigation depth
- irrigation depth
- irrigation scheduling
- lateral
- mainline
- micro irrigation
- microcad
- pressure losses
- pump
- statistical analysis
- submain
- surface irrigation
- total dynamic head
- Visual Basic
- water horsepower
- wetting pattern
- wetting zone

## REFERENCES

1. Anyjoi, H. and J.P. Wu, 1987. Statistical approach for drip lateral design. Trans. of ASAE, 30(1): 187–192.
2. ASAE Standards, I 992. EP405.1, S4358, and EP458.
3. Awady, M.N., and O.M. Ahmed, 1996. Planning micro irrigation systems by computer aid. 4th Conference of Misr J. Agric. Eng., 13(4): 87–104.
4. Benami, A. and A. Oren, 1984. *Irrigation Engineering - Sprinkler, Trickle, and Surface Irrigation: Principles, Design and Agricultural Practices*. Irrigation Engineering Scientific Publications (JESP). Haifa, Israel. 257 pages.
5. Bucks, D. A., F. S. Nakayama, and A. W. Warrick, 1982. Principles, practices and potentialities of trickle (drip) irrigation. In: "*Advances in Irrigation*, Editor D. Hillel." 220-290 pages.
6. Chavez-Morales, J., M. A. Marino, and E. A. Holzapfel, 1996. Computer packages for irrigation planning. 6th Conference on Computers in Agriculture, Cancun, Mexico.
7. Christiansen, J. E., 1942. Irrigation by sprinkling. Cal. Agr. Exp. Sta. Bull. 670. 49 pages.

8. Clarke, D., and K. M. S. El-Aslulri, 1996, Irrigation scheduling, a windows equivalent to the FAO CROPWAT program. 6th Conference on Computers in Agriculture, Cancun - Mexico.

9. Doorenbos, J., and W. 0. Pruit, 1984. *Crop Water Requirements*. FAO Irrigation and Drainage Paper 24.

10. El-Nesr, M. N. B., 1999. Computer aided design and planning of micro irrigation systems. MSc. Thesis, Agric. Eng., Faculty of Agriculture, Alexandria University, Egypt.

11. Gomes, H.P., 1996. REDES: A software for optimum design of micro irrigation pipeline systems. 6th Conference on Computers in Agriculture, Cancun, Mexico.

12. Howell, T. A., and E. A. Hiler, 1974. Micro irrigation lateral design. Trans. ASAE 17(5): 902–908.

13. Ismail, S. M., 1993. Hydraulic Simulation of drip irrigation system. Alex. J. Agric. Rec., 38(2): 25–43.

14. James, L. G., 1989. *Principles of Farm Irrigation System Design*. NY: John Wiley & Sons.

15. Kang, Y., and S. Nishiyama, 1996. A simplified method for design of micro irrigation laterals. Trans. ASAE, 39(5): 1681–1687.

16. Keller, J. 1965. Selection of economical pipe sizes for sprinkler irrigation systems. Trans ASAE, 8(2): 186–190, 193.

17. Keller. J., and R. D. Bliesner, 1990. *Sprinkler and Micro irrigation*. Chapman and Hall Pub.

18. Keller, J., and D. Karmeli, 1975. Micro irrigation design. Rain Bird Sprink. Manuf. Corp., 133 pages.

19. Meshket, M. and R. C. Warner, 1985. A user friendly interactive micro irrigation model. Proceedings on Drip/Micro irrigation in Action at the third International Drip Irrigation Congress by ASAE, Volume 1: 339–451.

20. Nakayama, F. S., D. A. Ducks, A. J. Clemmens, 1979. Assessing trickle emitter application uniformity. Trans. ASAE, 22(4): 816–821.

21. Ponce, .J. C. H., E. P. Penn, J. M. G. Camncho, and J., E. Medena, 1996. Selection and design of trickle and microsprinkler systems. 6th Conference on Computers in Agriculture, Cancun, Mexico.

22. Sharaf, G. A., 1996. Optimal design of micro irrigation submain-unit. Misr J. Ag. Eng., 13(3): 501–SIS.

23. Soil Conservation Service, 1983. *National Engineering Handbook*, Section 15 – Irrigation, Chapter 7: Micro irrigation. United States Department of Agriculture.

24. Watters, G, Z. and J. Keller, 1978. Micro irrigation tubing hydraulics. ASAE meeting paper No. 78, 2015.

25. Wu, I. P., and H M. Gitlin, 1975. Irrigation efficiencies for surface, sprinkle, and drip irrigation, Proc. of the 2nd World Congress, Water Resources Association, New Delhi, Vol. (1): 191–199.

26. Wu I. P. and Yue, R., 1993. Drip lateral design using energy gradient line approach. Trans. ASAE 36(2): 389–394.

# PART II
# MICRO IRRIGATION RESEARCH ADVANCES AND APPLICATIONS

# CHAPTER 5

# SUSTAINABLE SUBSURFACE MICRO IRRIGATION IN AUSTRALIA: VEGETABLES[1,2]

VIOLA DEVASIRVATHAM

## CONTENTS

[1]The author appreciates the assistance and support received from the following professionals and institutions: University of Western Sydney (UWS) and the Cooperative Research Center for Irrigation Futures (CRCIF) for funding this research; Professors Peter S. Cornish, Basant Maheshwari, and Geoff Creswell, University of Western Sydney for the valuable guidance and academic support throughout the research. This chapter is based on the M.Sc. (Agriculture, 2008) thesis by the author, University of Western Sydney, School of Natural Sciences.

[2]The numbers in parentheses refer to the literature cited.

## 5.1   INTRODUCTION

Efficient use of water is a key factor for irrigation management globally, with widespread efforts being made to increase water productivity and reduce the environmental impacts of irrigation. With future water scarcity and climate change, management of water will become an increasingly important issue in intensive vegetable production.

With peri-urban irrigated agriculture such as in the Sydney region, NSW, Australia, competition for water resources is acute, and the need for improved irrigation management is most important [190]. Intensive vegetable production is an important and expanding industry in peri-urban Sydney [138]. Many vegetable growers in this area use potable water as the major water source [63]. Competition for water with urban users has led to uncertainty about the security of future water supplies.

The Sydney region enjoys a subhumid climate with an average rainfall of 800–1000 mm spread over most months [38]. The rainfall and mild temperatures enable vegetables to be grown year-round, but access to irrigation is required in most months to provide water security [121]. Soils vary greatly, but the three dominant soil types used for vegetable production in Western Sydney are the alluvial soils along the Hawkesbury-Nepean river system and the texture-contrast soils based on Hawkesbury sandstone, both of which have sandy surface soils, and the soils derived from Wianamatta shale which have clay-loam surface soils [217].

The Sydney region is ideal for lettuce (*Lactuca sativa*), which is grown year-round [172] by direct seeding or by using nursery-raised 'transplants' [261]. Lettuce is irrigated mostly by overhead sprinklers [249]. Drip irrigation is not widely used for lettuce in NSW, apparently because of higher costs (low durability), interference with normal cultural practices necessitating removal between crops, and the lack of local guidelines to adapt system design and management to the diverse soils and annually varying climate [58, 122]. In other respects, drip irrigation should be suitable [261]. The irrigation water requirement of drip irrigated plants can be less than half that of sprinkler irrigated plants [249].

Subsurface drip irrigation (SDI) is an alternative to conventional drip irrigation, which could become an attractive option to lettuce growers in the Sydney region as the cost over the life of the product can be less than with surface tape, and because reduced tillage using semipermanent beds [230] has removed the need for deep cultivation between every crop.

The advantages of SDI compared to surface drip irrigation include direct application of water to the root zone, less evaporation from soil, potentially greater water use efficiency and fewer weed and disease problems [201]. SDI has been found to increase yield over surface drip irrigation [222]; furrow irrigation [100]; and sprinkler irrigation [69], providing the SDI system receives good irrigation scheduling [92].

SDI also has several important potential disadvantages, including 'tunneling,' variable soil surface water and risky establishment [149, 152, 180], poor germination and/or crop establishment [213]. Crop establishment is often poor [147] due to insufficient surface soil moisture to meet the demands of seedlings or seeds [278].

One approach to improve crop establishment with SDI has been to modify the buried drip tape by adding an impermeable plastic barrier below the tape, as in the Capillary Root Zone Irrigation system (CRZI). CRZI reduced variability in soil surface wetting but did not improve establishment [49, 67], mainly due to the hydraulic properties of the particular soils. CRZI has undergone extensive development, in particular the width of the impermeable layer (now 100 mm), and it is now sold under the trade name KISSS™ (hereafter called "modified SDI"). It has not been thoroughly evaluated. In addition to modifying tape design, surface soil moisture can be improved by shallow tape installation and increased irrigation frequency [42], although these approaches have met with limited success. Harris [102, 103] concluded that possibly crops cannot be established this way at all, without an additional source of water.

In this chapter, author discusses the adoption of SDI technology in vegetables by improving surface soil water conditions and crop establishment. Lettuce crop is used as an example because of its importance in the Sydney region. This chapter also answers the research questions for SDI technology in Australia: Does an impermeable layer beneath the drip tape (modified SDI) improve surface soil water conditions and crop establishment, compared with conventional SDI? Does the modified SDI offer any advantage over using conventional SDI with greater irrigation amount or frequency? Author also evaluated if the irrigation management with SDI should take into account soil type and evaporative demand. This research was based on the hypothesis that the strip of impermeable material below the buried drip tape created a temporary water table during each irrigation, thus increasing the upward flux of water to the soil surface.

## 5.2 REVIEW OF SUBSURFACE DRIP IRRIGATION IN VEGETABLE PRODUCTION

Irrigation in the Australian vegetable industry has traditionally been dominated by surface irrigation. However, increasing pressures on water availability, the potential yield increase through improved control of soil and plant water relationships, and the benefits of reduced labor/fertilizer/pesticide cost, have increased the interest of vegetable growers in alternative irrigation methods, including drip irrigation systems.

Drip irrigation has the potential to use scarce water resources most efficiently to produce vegetables [161]. The modern development of drip irrigation started in Great Britain during World War II and continued in Israel and other countries [43]. The major benefits of drip irrigation are the ability to apply low volumes of water to plant roots, reduce evaporation losses, and improve irrigation uniformity [225].

Subsurface drip irrigation (SDI) applies water below the soil surface, using buried drip tapes [8]. It has many benefits over conventional drip irrigation [241]. The biophysical advantages are the lower canopy humidity and fewer diseases and weeds [44]. The yield and quality of vegetable crops can improve with a buried drip system compared with a surface drip system [18, 201, 224]. Environmental benefits include the ability to manage nutrient and pesticide leaching and the threat to groundwater [147]. However, SDI has also potential challenges [103, 147, 149].

The review of SDI in this chapter considers: the vegetable production industry in Australia, with a particular focus on issues for production in the Sydney region; basic concepts of water and its management in irrigation; drip irrigation design and management and its adaptation to SDI, including identification of problems with SDI; and soil factors.

## 5.2.1  VEGETABLE PRODUCTION IN AUSTRALIA

Vegetable production in Australia is dominated by Queensland, New South Wales and Victoria, with more than 4,000 farms producing vegetables [2]. The largest area is in Queensland. Over the four years (2000–2004), the number of farms fell by 19%, but the industry continued to be dominated by small farms. The value of output from the typical vegetable farm rose from $281,000 in 2000–2001 to $387,000 in 2003–2004. Fifty years ago, the average Australian consumed around 130 kg of vegetables annually. Today, per capita consumption is 162 kg. If vegetable consumption increases at the same rate, the per capita consumption should reach around 188 kg in 2050 [10]. Consumer demand for vegetables is rizing over the long-range, so there is need for continued expansion in vegetable production. Most vegetable farming is characterized by intensive management including irrigation. A major problem faced by farmers is the availability and cost of irrigation water [114]. However, improvements in the productivity of irrigation water are being made. For example, the average return from vegetable production per ML rose from $1,762/ML in 1996–1997 to $3,207/ML in 2000–2001 [1]. An industry report attributes this to increased use of water-efficient delivery systems such as drip irrigation, irrigation scheduling and soil moisture monitoring, which help achieve a good quality product resulting in higher market prices [115].

## 5.2.2  THE NSW AND SYDNEY BASIN INDUSTRY

New South Wales's vegetable production districts are the Sunraysia, Riverina – Murray Irrigation Area (MIA)/slopes/tablelands, and Sydney Basin [173]. The Sydney Basin including Greater Western Sydney is a focus of this chapter, and these supply full range of fresh vegetables to the local market. The vegetable industry in NSW contributes approximately $300 million to the national economy. Nearly 26% of the total value of this industry is produced by the Sydney Basin [113], where the major vegetable crops are lettuce (*Lactuca sativa* L.), cabbage (*Brassica oleracea* var. Capitata), and cauliflower (*Brassica oleracea* var. Botrytis). Most 'Asian vegetables' in NSW are produced on 340 small farms with 5–20 acres in Western Sydney, equally contributed by Chinese, Cambodian and Vietnamese growers [187].

Lettuce is a common salad vegetable in Australia. Several types of lettuce are available: crisp head, butter head, romaine (cos) and leaf varieties [261]. Lettuce has a short growing season, commonly reaching maturity in about 6 to 10 weeks from sowing, depending upon the type. All commercial lettuce production uses 'transplants' or nursery-raised seedlings. The transplanting requires less time in the field [143, 268] thus allowing more intensive cropping, also overcomes crop establishment problems and the cost of thinning [261].

FAO [75] defined the area of farm units surrounding towns as 'peri-urban,' supplying fresh vegetables, fruit. In all countries, rural to urban migration is placing pressure on the peri-urban area where housing and industrial development interact with food production [31]. Sydney's peri-urban zone is characterized by an inner zone of market gardens, an intermediate zone of poultry-horticulture and an outer zone of dairy or mixed farming [138].

According to the Agricultural Land Classification System, the Western Sydney Peri-Urban Horticultural region is arable land Class-1 with high to very high productivity [127].

### 5.2.3  IRRIGATION WATER SOURCES AND ISSUES ARISING

Many peri-urban vegetable growers in the Sydney region use potable water from the Sydney water supply as their main water source [63], although there is also a significant industry based on irrigation from the Hawkesbury-Nepean River and its major tributary (South Creek) as well as farm dams. Out of 3,000 irrigators, approximately 1,500 are river pumpers, 750 draw from farm dams and the remaining 750 irrigators, mainly vegetable growers, use town water [114].

Charges for town water used by vegetable growers are based on the commercial Tier-1 rate of $1.20/kilo-liter. Peri-urban vegetable growers pay annually ~A$10,000 to A$20,000 for water [114]. The production is also affected by water restrictions during periods when water levels in the Sydney Water Reservoirs are low [250], and it will compete increasingly with urban and industrial users [169].

Farmers in this area practice intensive horticulture. Irrigation is excessive and not uniform, overhead sprinklers are most common, and mostly farmers do not use any form of irrigation scheduling method or soil water monitoring [168, 231]. Drip irrigation is rarely used in the Sydney region, apparently because surface drip systems are costly, they are said to interfere with normal cultural practices, and there are no guidelines for designing and managing drip systems across the diverse soils and climates of the Sydney region [58]. Excessive irrigation on the farms investigated by Cornish and Hollinger [59] was associated with high storm water runoff and nutrient loss from farms, although the magnitude of loss depended on soil type [122]. In two on-farm trials, it was observed [122] that SDI greatly reduced irrigation requirement. It also reduced storm water runoff because the soil profile was generally drier and accepted more rainfall before runoff occurred.

Subsurface drip irrigation can overcome two main drawbacks of surface drip irrigation. One is the high cost associated with frequent removal and replacement, provided the SDI system lasts long enough to offset the high initial set-up cost. The other is interference with cultural practices. Reduced tillage based on semipermanent beds [230] requires only shallow cultivation, potentially allowing SDI tube to remain undisturbed for many years, without impeding cultural practices.

These findings demonstrate the significant need in the Sydney region to improve irrigation efficiency and help address the problem of increasing cost of water and supply restrictions. Given the irrigation systems and management practices currently being used, there is scope to meet this need with drip or particularly subsurface drip irrigation, although installation and management will need to be adapted for the wide range of soils and the seasonal climate variation in the region.

## 5.2.4   SOIL WATER FOR IRRIGATION MANAGEMENT

### 5.2.4.1   SOIL AND PLANT WATER CONCEPTS

**Soil water potential** is expressed in energy terms (bars or MPa). The difference in energy between pure water and that of soil water at standard pressure and temperature is called the soil water potential. The total water potential can be expressed:

$$\psi_t = \psi_g + \psi_m + \psi_p + \psi_o \tag{1}$$

where, $\psi_t$ = the total soil water potential energy, $\psi_g$ = the gravitational potential energy, $\psi_m$ = the matric potential due to capillary pressure, $\psi_p$ = the pressure potential, $\psi_o$ = the osmotic potential due to salts [71]. To determine the potential energy status of soil water, piezometers, tensiometers and psychrometers are commonly used [85].

**Soil water content** is expressed as the mass of water in unit mass of soil (gravimetric) or as volume of water in unit volume of soil (volumetric) [133]. Gravimetric water content ($\theta_g$) is measured by weighing the soil when wet ($m_{wet}$) and again after drying at 105°C ($m_{dry}$).

$$\theta_g = [m_{wet} - m_{dry}] / [m_{dry}] \tag{2}$$

**Volumetric water content** ($\theta_v$) is the volume of liquid water per volume of soil, and can be calculated from $\theta_g$ using bulk density ($\rho$) in Eq. (3), where: $\rho_{water}$ = 1.0 g/cm³. Relationships between water content and potential are important for understanding water flow in soil [189].

$$\theta_v = volume_{water} / volume_{soil} = (m_{water} / \rho_{water})/ m_{soil} / \rho_{soil} = [\theta_g * \rho_{soil}] / [\rho_{water}] \tag{3}$$

### 5.2.4.2   THE SOIL MOISTURE CHARACTERISTICS AND CONCEPTS OF AVAILABLE SOIL WATER

**The energy of soil water** and soil water content are related by the moisture characteristic curve [210]. In saturated soil, all pores are filled with water and the water potential is zero. As suction is increased, progressively smaller pores drain so the soil water content decreases and the water potential becomes more negative. At very high suctions, only the very small pores retain water. In light to medium textured soils (sands, sandy loams, loams and clay loams), soil structure can evidently affect the soil moisture characteristic, while in heavy textured soils the influence of structure is less distinct [275].

**Field capacity** is defined as the water content of the soil (at a soil moisture tension of 0.33 bars) following drainage of a saturated soil profile underlain by dry soil for about 24 – 48 h depending on the soil type [104]. The soil water potential at field capacity is variously defined as around −0.1 bar to −0.3 bar (−0.01 to −0.03 MPa) depending on soil texture and whether the soils have been homogenized or they are structured under the field conditions [185]. The permanent wilting point (PWP) is the soil water content at which plants are unable to absorb soil water (at a soil moisture tension of 15 bars), and wilt permanently [158]. The soil water potential at this point is usually considered to be 15 bars [223], although the actual value will depend on plant type and the demand for water. The available water in a soil is the amount of water that can be used by plants for their growth and development. It is a difference between the soil water contents at field capacity and the permanent wilting point.

### 5.2.4.3   SOIL-PLANT-ATMOSPHERE CONTINUUM

Soil-plant-atmosphere relationship recognizes that all components of the field environment (the soil, the plant, the atmosphere), when taken collectively form a physically integrated and dynamic system. The water movement inside the system is known as soil-plant-atmosphere continuum (SPAC) [117]. Although water generally moves from soil to the plant and then into the atmosphere, when the soil is dry and the atmosphere is near saturation, yet water may move in small quantities from plants into soil [223]. The flow path of water through SPAC is a complex process with a series of resistances offered by the different components of the system. Plants offer little resistance when the soil has sufficient moisture and the atmospheric conditions are moderate [71]. When soil dries, water deficit develops in plants and stomata close partially or completely. Under this condition, plants offer greater resistance to water movement [16].

### 5.2.4.4   SOIL WATER MOVEMENT AND HYDRAULIC CONDUCTIVITY

The sum of the suction and gravitational potentials is defined as the hydraulic head [116]. The hydraulic head determines the direction and rate of water movement. Water moves from soil with lower to higher potential. In this chapter, author considers upward flux of water, soil matric potential and evaporative demand. The scientific principle underpinning evaluation of the modified SDI in this research is that the water required for crop establishment is met by upward flux from the subsurface drip.

Hydraulic conductivity is a measurement of the ability of the soil to conduct water and depends upon the permeability of the soil to water [71]. Knowledge of the hydraulic conductivity of soil is important to the understanding of soil-water behavior including the movement of water and solutes within the soil profile and studies of water uptake by plant roots. Hydraulic conductivity depends greatly on soil water content [179], so it is often determined for both the saturated and unsaturated conditions [146]. Saturated hydraulic conductivity pertains to the conductivity of soil when all pores are filled with water, whereas conductivity is unsaturated when pores are partially filled. The soil factors affecting hydraulic conductivity include the pore geometry, soil structure and presence of entrapped air in the soil pores [133].

### 5.2.4.5   PLANT WATER RELATIONS

Total plant water potential ($\psi$) includes three components (ignoring gravitational):

$$\Psi = P + \pi + \tau \qquad (4)$$

Where: $\psi$ = water potential in the plant, P = pressure or turgor, $\pi$ = osmotic potential, $\tau$ = soil matric potential [263]. Stomatal closure starts if plant water stress occurs, following decreasing soil water potential, indicated by a fall in $\Psi$ below a threshold value. The decrease of 0.5–1.0 MPa in soil water potential normally takes place over days and weeks. As the soil continues to dry, a plant can be considered under water stress, although there may be little change in the midday water potential of exposed leaves [273]. The $\Psi$ at which the stomata close will depend on the osmotic potential in the leaves and rate of drying.

Plants suffering from water deficits have a reduced leaf area and reduced root and shoot development [139]. Leaf area or leaf area index (LAI) is an important growth parameter for irrigation management [129]. During early crop growth, LAI is low and influenced by row spacing. Although transpiration is low at this stage, significant evaporation can take place when the topsoil is wet. In dry soil, evaporation decreases [215].

## YIELD THRESHOLD DEPLETION

Yield threshold depletion (YTD) is the amount of water that can be depleted from the soil before there is an effect on yield or quality of crop. If the YTD is known, the soil water balance can also show the maximum time allowable between irrigation. Commonly, a crop should be irrigated before reaching the YTD level. YTD depends upon soil, plant and climatic factors. Crops differ in their sensitivity to water stress. Yield threshold depletions are often less for vegetable crops than field crops [87], because shallow rooted plants exploit less soil and therefore, are less well buffered against changes in soil water.

## SOIL WATER BALANCE

For irrigation research, the soil water balance is defined in Eq. (5), where: ASW is available soil water at times 1 and 2; $(ASW_1 - ASW_2)$ is the change in soil water during the interval $t_1$ to $t_2$, and P = precipitation, I = irrigation, ET = evapotranspiration, $R_o$ = surface runoff and D = deep percolation beyond the root zone in an interval $t_1$ to $t_2$ [223]. If $ASW_1$ is the desired state and $ASW_2$ is the present state, then irrigation required to return the soil water to the desired state (the replenishment of water use in the period), $(ASW_1 - ASW_2)$ can be estimated by assuming $R_o$ and D as zero, irrigation requirement = ET – (I+P). On-site measurements of both $R_o$ and D are difficult and are commonly ignored in practice. Irrigation above the irrigation requirement is presumed drainage [206]. The same applies to rainfall after irrigation.

$$ASW_1 - ASW_2 = P + I - (ET + R_o + D) \tag{5}$$

In budgeting approaches to irrigation scheduling, ET is estimated from potential evaporation combined with the use of a crop coefficient [108]. A simplified water balance Eq. [39, 223] is defined below to calculate the components of the water balance when water was applied to a bare soil surface:

$$E = I - D \tag{6}$$

where: E = Evaporation, I = Irrigation and D = Drainage. In this chapter, the Eq. (6) is used to calculate E.

## 5.2.4.6   MONITORING SOIL AND PLANT WATER IN IRRIGATION SCHEDULING

Successful operation and management of an irrigation system requires a proactive monitoring approach for managing soil water. There are three different approaches to monitor and to schedule the irrigation [85]:

i.   Soil-based methods estimate soil water status by its appearance, feel or, more objectively, by water content or suction.

ii.  Plant-based methods include visible symptoms such as wilting, that reflect leaf turgor and thus indirectly leaf water potential, the Scholander or 'pressure bomb' that measures plant water potential, and noncontact thermometry with an infrared thermometer (a water stressed plant transpires less and is cooled less by evaporation).

iii. The water budget approach, which estimates crop water use from weather data and, from this, the irrigation requirement.

Measurements of soil water can be used to indicate when to irrigate, thus avoiding over and under irrigation. Soil water sensors measures either soil water potential (SWP) or volumetric soil water content (VSWC). Water potential can be measured by tensiometer, gypsum blocks and granular matrix sensor [238]. A variety of FDR (frequency domain reflectometer) [247], TDR (time domain reflectometer) [50] and capacitance probes [77] are available for measuring volumetric soil water content.

## TENSIOMETER

Tensiometer measures only soil water potential. It does not provide direct information on the amount of water held in the soil [274]. The use of tensiometers for irrigation scheduling has been widely reported for over 30 years [86, 109, 207], although they have been infrequently used in practice in peri urban area of Australia [168].

Enough research has been conducted on the appropriate depth of placement and water potential guidelines. Recommendations vary with soil type and crop. As an example, the tensiometer should be placed about 15 cm deep in the soil for shallow rooted crops (e.g., lettuce, [194]) and at 30 cm for deep-rooted crops (e.g., tomatoes, melons, [120]).

The main limitation with tensiometers is that they operate only in water potential up to $-75$ kPa. Further drying leads to breaks in the water column and a high degree of maintenance [84]. Also farmers will often want to deplete soil water beyond the range of the tensiometer, meaning that some interpretation needs to be made, for example from soil water tension deeper than the zone of greatest root proliferation.

## GRANULAR MATRIX SENSOR/GYPSUM BLOCK

The granular matrix sensor is similar to the gypsum block, although apparently more durable. It operates on the principle that resistivity of the block depends on its moisture content, which in turn depends on soil water potential. Like the gypsum block, the granular matrix sensor has been reported to have slow response times in some circumstances and each sensor needs calibration [236]. However, both sensors are inexpensive. Granular matrix sensors operate in the range 0–0.2 MPa, and therefore, have a wider range of applications than the tensiometer. The granular matrix sensors (and tensiometers) are most suitable for automatic drip irrigation [184]. There appear to have been advances in design and performance over time, and it has been a very effective irrigation scheduling aid for drip irrigated mint and onions on silt loam soils [237].

## WETTING FRONT DETECTOR: CAPACITANCE PROBE/FREQUENCY DOMAIN REFLECTOMETER

The wetting front detector, which originated from Australia, is a soil moisture-monitoring device, which can be used to detect wetting fronts. The 'FullStop' wetting front detector is simplest and is comprised of specially shaped funnel, a filter and a float mechanism. The funnel of the detector is buried in the soil within the root zone of the crop [247]. If sufficient water or rain falls on the soil to move to the funnel, it passes through a filter. This water activates a float mechanism, which operates an indicator flag above the soil surface. The wetting front detectors and a capacitance-type device (the 'Diviner' – frequency domain reflectometry) were the best tools (together) to monitor soil water with beans and melons under SDI in the Cowra district of NSW [247]. The Diviner was useful at identifying deficits and the wetting front detectors was suitable to identify over irrigation and also useful for nitrate monitoring in leachate.

The EnviroSCAN capacitance probe is widely used to measure soil water content throughout the soil profile and schedule irrigation for orchard crops in Australia. Data are downloaded to monitor water content through the soil profile and to schedule irrigation [77]. EnviroSCAN enables low cost continuous logging of soil water, which is important to detect infiltration or the flux of water between soil layers over relatively short time periods. However, it measures only a very small volume of soil, and there is a relatively large interface between the soil and access tube, so errors can be high. The probes are not useful in cracking or crusted soils.

## TIME DOMAIN REFLECTOMETER (TDR)
A TDR emits a pulse charge of electromagnetic energy, using sensors or 'wave guides' buried in the soil. The pulse signal reaches the end of the sensor and is reflected back to the TDR control unit. The time taken for the signal to return is related to the water content of the soil surrounding to the probe [50, 274]. The use of multiwire probes in the TDR provided rapid determination of soil profile water content and offers the capability of monitoring the dynamics of the soil water volume around a point source to differentiate soil water conditions at different vertical and horizontal soil volumes [245]. It is, however, expensive.

## NEUTRON PROBE
The neutron scattering method (neutron probe) measures volumetric water content of soil indirectly using high-energy neutrons emitted from the probe. Neutron probe method is suitable for coarse or medium textured soils but not suitable for measurements near soil surface and in shallow soils without special calibration [48]. The neutron probe has been in use in some sectors of the irrigation industry for many years, and has proven to be suitable for a range of applications from row crops like cotton [134] to trickle irrigated vegetable crops [61]. Disadvantages of the neutron probe are the high initial cost, high regulatory requirements (training and licensing), and the need for careful calibration.

In this chapter, tensiometers and gypsum blocks were used to determine soil water potential, and the theta probe was used to measure volumetric water content.

## 5.2.4.7    WATER BALANCE APPROACHES IN IRRIGATION SCHEDULING
The soil water balance represents the integrated amount of water in the soil at a particular time. The water balance method is an indirect way of monitoring water status, using simplifications of the soil water balance equation. It is used to estimate crop water use [85] from climatic data [6]. Climatic parameters including solar radiation, temperature, relative humidity and wind have either direct or indirect effects on crop

water use through their influence on evaporation and transpiration [123]. Various methods of estimating crop water use from meteorological information are used [123]. The combination of soil evaporation (E) and transpiration (T) make up the total water use, which is commonly referred to as evapotranspiration (ET). Estimation of evapotranspiration generally uses four factors: reference evapotranspiration ($ET_r$) based on a specific type of crop, a crop factor ($K_{cb}$) that describes both the dynamic seasonal and developmental change in the crop evapotranspiration in relation to $ET_r$ a soil factor ($K_{cs}$) which describes the effect of low soil water content on transpiration and having close relationship with crop growth parameters such as rooting and a soil factor ($K_{so}$), which describes the evapotranspiration amount from either rainfall or irrigation. The crop water use is represented by the following Eq. (6):

$$ET_c = ET_r [(K_{cb} . K_{cs}) + K_{so}] \tag{7}$$

Reference evapotranspiration ($ET_r$), expressed in mm/day, can be estimated by different methods such as modified Blaney-Criddle method, the modified Jensen-Haise method, the Penman-Monteith combination equation, or directly by pan evaporation. Evaporation pans of various designs have been widely used throughout the world as an index of reference evapotranspiration ($ET_r$). To calculate the particular crop water use or crop evapotranspiration, crop coefficient values are used [211]. The crop coefficient ($K_c$) value varies between crops and growth stages. Crop evapotranspiration ($ET_c$) is calculated by multiplying crop coefficient ($K_c$) and reference evapotranspiration ($ET_r$).

The water balance approach has been developed for irrigation management to estimate ET from large areas. Its application is difficult in micro irrigation because of the multidimensional water application pattern [155].

### 5.2.4.8   WATER USE EFFICIENCY

Generally, plant growth is directly related to transpiration (T), although under field conditions changes in soil moisture result from both T and soil evaporation (E) [117]. E and T are commonly summed to give evapotranspiration (ET), which can either be measured as change in soil water or estimated using well known equations. Both farmers and scientists are concerned with water use efficiency. In irrigated crops, efficiency of water use can be affected by the method, amount, and timing of irrigation.

Water use efficiency (WUE) has been defined in various ways and it is important to understand the differences. WUE is a ratio of dry matter produced (Y) per unit of water transpired by a crop (T), and is expressed as kg/mm or kg/ha/mm [164].

$$WUE = Y / T. \tag{8}$$

The Eq. (8) relates the biomass production relative to the water actually used by the plant, and should more correctly be termed the 'transpiration efficiency' (TE). The TE of different crops may vary with differences in photosynthetic mechanism ($C_3$, $C_4$, and CAM) and vapor pressure deficit [163, 265].

$$WUE = Y_e / ET \tag{9}$$

The term $Y_e$ / ET shows the agronomic yield of the system relative to total water use, and is a more correct use of the term 'water use efficiency' or agronomic water use efficiency [164].

Soil surface modifications such as tillage and retaining surface residue may influence WUE [110] by reducing soil evaporation (E) and increasing crop transpiration (T). One potential advantage of SDI is the reduced soil evaporation [242]. Water use efficiency is an amount of water transpired relative to the amount of irrigation applied (tons of yield per ML water), which is also called irrigation efficiency. The poor soil structure, profile salinity, and irrigation management restrict the expansion and efficiency of the plant root system that will reduce water use efficiency [162]. Overall agronomic efficiency of water use ($F_{ag}$) in irrigated systems is defined by FAO [76] using an adaptation of the soil water balance:

$$F_{ag} = P/U, \qquad (9a)$$

where: P is crop production (total dry matter or the marketable yield) and U is the volume of water applied. The components of U are expressed by the following equation:

$$U = R + D + E_p + E_s + T_w + T_c, \qquad (9b)$$

where: R is the volume of water lost by runoff from the field, D is the volume drained below the root zone (deep percolation), $E_p$ is the volume lost by evaporation during the conveyance and application to the field, $E_s$ is the volume evaporated from the soil surface, $T_w$ is the volume transpired by weeds and $T_c$ is the volume transpired by the crop. Overall irrigation efficiency (Eo) is calculated by multiplying the efficiencies of the components. For a system, it includes reservoir storage, water conveyance, and water application:

$$E_o = (E_s) \times (E_c) \times (E_a), \qquad (9c)$$

where: $E_s$ = reservoir storage efficiency, $E_c$ = water conveyance efficiency, $E_a$ = irrigation application efficiency. In all agricultural systems, low water use efficiency can occur: when soil evaporation is high in relation to crop transpiration, early growth rate is slow (e.g., crop establishment stage), water application does not correspond to crop demand, and when shallow roots are unable to use deep water in the profile. This has been demonstrated during the early growth phase of potato [195]. These factors are especially pronounced in intensive vegetable production [80]. Irrigation control may increase water use efficiency (yield/water used). "Water use" here is a sum of ET and deep percolation [264].

*5.2.4.9   IRRIGATION SCHEDULING TO IMPROVE WATER USE EFFICIENCY*

Irrigation scheduling is an application water at intervals based on the crop needs, with the primary objective of managing soil water within defined limits. It is the process by which an irrigator determines the timing, amount and quality of water to be applied to the crop [23, 211]. Scheduling is intended to maximize irrigation application efficiency by minimizing runoff and percolation (drainage) losses [262].

Several tools for measuring soil water in irrigation scheduling have been considered before in this chapter. Whether measured directly, or predicted indirectly using climatic data and crop water use models, soil water status is of primary importance for irrigation scheduling. The use of indirect and direct measurement has often been compared, but it appears that the benefits of each approach are situation-specific and not clearly defined. As an example, using direct measurement of soil water to schedule subsurface drip irrigation of tomatoes was no better than using indirect prediction, at least in terms of total fruit yield [160]. However, the direct measurement of soil moisture required significantly less water than indirect prediction. Thus direct measurement of soil water gave higher irrigation efficiency.

The irrigation scheduling using crop evapotranspiration ($ET_c$) has been compared with volumetric soil water content measured by TDR, for tomato in a silty clay loam [266]. The surface drip irrigation had drainage during crop establishment when water was applied at a higher rate than crop evapotranspiration.

Sensors must be placed in the active root zone in proximity to the emitter. Sensor placement in SDI systems varies, but is mostly located midway between emitters [124].

### 5.2.5  MICRO IRRIGATION AND ITS ADAPTATION IN SDI MANAGEMENT

Drip irrigation allows water to be applied uniformly and slowly at the plant location so that essentially all the water is placed in the root zone [137]. Drip systems are categorized according to their placement in the field:

- Surface drip irrigation: Water is applied directly to the soil surface.
- Subsurface drip irrigation: Water is applied to below the soil surface through perforated pipes.

Subsurface drip irrigation (SDI) has been used in Australia and elsewhere for crops including citrus, cotton, sugarcane, some vegetables, sweet corn, ornamentals, lucerne and potato [5, 22, 150, 214, 237, 259]. SDI is an efficient irrigation method with potential advantages of high water use efficiency, fewer weed and disease problems, less soil erosion, efficient fertilizer application, maintenance of dry areas for tractor movement at any time, flexibility in design, and lower labor costs compared to a conventional drip irrigation system. However, there are also potential disadvantages with SDI, which mainly relate to poor or uneven surface wetting and risky crop establishment [47, 147, 213].

*5.2.5.1  DESIGN AND INSTALLATION OF SUBSURFACE DRIP IRRIGATION*

The SDI systems comprise of a filter leading to the main supply tube, submain, laterals that convey water to the emitters [104].

*LATERAL DRIP LINE*

Tapes and tubes are available for use as laterals. Tape products are thinner than tubes [186]. Commonly, tube wall thickness ranges from 0.4 mm to 1.5 mm [98]. Two classes of tape wall thickness have been identified [47]. Flexible thin-walled (0.15 mm to 0.30 mm) tapes are typically used for shallow installation, while thicker-walled (0.38 mm to 0.50 mm) tapes are installed deeper or where the soil does not provide sufficient support to prevent collapse by equipment or soil weight. The 0.38 mm thickness of tape has been used for potato (*Solanum tuberosum* L.), corn (*Zea mays* L.), alfalfa (*Madicago sativa*) and pinto bean (*Phaseolus vulgaris* L.) production in sandy loam soils. Successful production of lucerne with SDI was recorded [252] in Victoria, using 0.38 mm tape.

## TAPE INSTALLATION DEPTH

The use of surface drip irrigation *versus* subsurface drip irrigation varies by region and by crop, and is often based on perceived constraints on the vertical placement of the drip tape/tube or laterals [53]. With SDI, the choice of drip tape depth is influenced by crop, soil, climate characteristics and anticipated cultural practices, but it generally ranges from 0.02 to 0.7 m [43]. It is often in the range of 0.05 to 0.2 m for shallow rooted horticultural crops. From the literature, a depth of 0.15 m for lettuce is appropriate on the sandy soils at UWS, in this chapter.

Although installation depth is generally decided for horticultural reasons, another consideration for determining depth is that deeper placement (0.45 m) will be required if the primary aim is to reduce soil evaporation and capture the potential benefit of improved water use efficiency (yield and quality) that is possible with SDI [35].

With the shallow systems, relatively deeper installation will reduce soil evaporation and also allow a wider range of cultural practices. However, deeper installation may limit the effectiveness of the SDI system for seed germination/crop establishment. Deeply placed drip lines may require an excessive amount of irrigation for germination/crop establishment. This practice can result in off-site environmental effects [43], and can reduce water-use efficiency. Deeper placement may restrict the availability of surface applied nutrients and other chemicals [44].

Relatively shallow tape placement has been tried for many years to assist germination [42]. Recent examples include broccoli on sandy loam soil [216] and corn on a silt loam soil [150]. Germination of tomato (*Lycopersicon esculentum* Mill.) under SDI was better with drip line depth of 0.15 and 0.23 m than at 0.3 m on clay loam soil [226]. It can be assumed that shallow placement is especially important for establishment if there is no supplementary source of surface irrigation.

Shallow placement of drip tape is generally required also for satisfactory growth of shallow rooted crops in sandy soils, which have limited capillary water movement [30], although this is not always the case. For example, higher zucchini (*Cucurbita pepo*) yield at 0.15 m depth was observed than 0.04 m depth on a coarse loam soil [219].

In Australia, tape depth of 0.25 to 0.30 m is used in the Queensland cotton (*Gossypium spages*) industry on cracking clay soils [214]. There are regional differences in the tape placement, with growers in NSW generally installing more deeply than in Queensland [214].

## LATERAL SPACING

An overview of published studies show that lateral spacing ranges from 0.25 to 5 m for SDI, as determined by crop behavior, cultural practices soil and properties. Wider lateral spacing is practiced in heavy textured soil [43]. Closer spacing is recommended for sandy soil [198]. Lateral spacing is generally one drip line per row/bed or an alternative row/ bed with one drip line per bed or between two rows [149]. With row crops such as tomatoes, laterals are often spaced 1 to 2 m apart. Lateral spacing of 1.5 m in subsurface drip-irrigated corn was successful in a silt loam soil [65]. Lateral placement of 0.3 m is recommended for SDI in the loamy sandy soil of South Carolina for vegetable crops: cowpea (*Vigina unguiculata*), green bean (*Phaseolus vulgaris*), yellow squash (*Cucurbita pepo*), muskmelon (*Cucumis melo*) and broccoli (*Brassica oleracea*) [46]. Lateral spacing of 2 m intervals on a "1:2 drip tape:crop row" has been successful in Queensland for cotton [214].

The above discussion indicates that closer drip line spacing (0.3 m) and two drip lines per three rows of crop is appropriate for lettuce on sandy soils in this chapter.

## INSTALLATION
Lateral lines are laid along the contour of the land as closely as practicable to avoid pressure variations within the line due to elevation change [93]. The first step in installing a successful SDI system is maintaining proper hydraulic design. This allows the system to deal with constraints related to soil characteristics, field size, shape, topography, and water supply. Lateral diameter and length influence water application uniformity [141]. In vegetable crops in the USA, it has been observed that a tape diameter of 125–200 mm has been the industry standard and common for SDI where rows range from 90 m to 180 m [152]. In Greece, 17 mm polyethylene pipe was used at the shorter row length of 30 m for sugar beet (*Beta vulgaris* L.) research using SDI [222].

## EMITTERS
Emitters are plastic devices, which precisely deliver small amounts of water [119]. Two types of emitters are available. Point-source emitters discharge water from individual or multiple outlets. Line-source emitters have perforations, holes, porous walls, or emitters extruded into the plastic lateral lines [11]. Line-source emitters are generally used for widely spaced crops such as vines, ornamentals, shrubs and trees. Point source emitters are used for small fruits, vegetables and closely spaced row crops [36]. The emitters used for SDI are much the same as those used for surface drip irrigation, however, the emitter is fixed internally in the drip line [104].

## EMITTER SPACING
Soil characteristics and plant spacing determine emitter spacing. Emitter spacings in Queensland are mostly between 0.3 m to 0.75 m for row crops [105]. In cotton a 0.3 m emitter spacing was used for in sandy loam soil in the USA [140]. Similarly, an emitter spacing of 0.3 m was suitable for corn production for deep silt loam soils under SDI [148]. In a semiarid environment, 0.45 m emitter spacing was used in clay loam soils for drip-irrigated corn [125]. In general, emitter spacing should normally be less than the drip lateral spacing and closely related to crop plant spacing [149].

## FLUSHING CAPACITY
A critical area of design that impacts on system performance is the flushing capacity. Many SDI systems appear to have been installed with inadequate flushing capacity, resulting in sediment deposition, reduction in flow volumes and blockages [204]. This also produces higher back-pressures in the mains, which may also affect the system performance [149]. Retrofitting large valves or increasing the number of valves may solve some of the flushing problems [214].

### 5.2.5.2   WATER APPLICATION UNIFORMITY
Water application uniformity in micro irrigation depends on system uniformity and spatial uniformity in the field [277]. The **system uniformity** is affected by system

design factors such as lateral diameter and emitter spacing [276], and manufacturing variation [26]. It is also considered to include emitter clogging [27]. The parameters used to evaluate micro irrigation system application uniformity are: the Uniformity Coefficient (UC); emitter flow variation ($q_{var}$); and Coefficient of Variation (CV) of emitter flow [25, 276]. Using these parameters, the values of these uniformity parameters were determined for various drip tape products [13]. System uniformity values predicted by design or evaluation models were found to be similar for both surface drip irrigation and SDI [45].

The spatial uniformity in the field refers to variation in soil water. In addition to system design factors noted above [277], variation due to field topography and soil hydraulic properties are also observed [41, 42].

The causes of nonuniformity include unequal drainage and unequal application rates [40]. Even where system uniformity is high, variation in soil properties, such as hydraulic conductivity, can affect drainage and lead to variation in soil water content. Application uniformity may be directly related to yield [156, 243]. Non-uniformity in one field (45%) was estimated to be mainly due to pressure differences, with only 1% due to unequal drainage and 2% due to unequal application rate [40]. The typical manufacturing coefficient of variation in tube today is only 0.02 to 0.06, which will be negligible [40]. Soil 'excavating' by subsurface emitters was shown to increase flow rate by 2.8% to 4%, but not sufficiently to affect uniformity calculations [221].

One consequence of nonuniform application is increased drainage [21, 200], assuming irrigation for uniformly good crop growth. Drainage may also occur if the application is uniform but the soil water holding capacity or hydraulic properties are not uniform.

Obtaining sufficiently moist soil for germination and crop establishment by applying uniform irrigation to soils, which are inherently variable, is a challenging issue for SDI [195]. It has been observed that to provide adequate irrigation water for potato plants in the early growth period, the field was overirrigated, leading to more downward movement of water on sandy loam soil than upward capillary movement of water [195].

Overall, minimizing nonuniformity of the drip system requires: a design, which considers the topography of the field [277], periodic checking of the system [55], and irrigation scheduling (volume and frequency) [41]. Greater irrigation uniformity can be achieved by using pressure-compensating emitters in surface and subsurface drip [227]. Flow meters are widely recommended to check the system performance in sub surface drip irrigation [4]. They are used to determine the rate and volume of water applied in an automated irrigation control system [12].

## 5.2.5.3  COMPARISON OF UNIFORMITY IN SURFACE DRIP IRRIGATION AND SUBSURFACE DRIP IRRIGATION

In SDI, emitter clogging and accumulation of salt caused by evaporation is less than in surface drip irrigation [118]. More uniform water content was observed in the root zone with SDI than surface drip [83]. In an SDI system, more uniform water content in root zone was observed than surface drip, and thus drainage was less with SDI [21, 200].

### 5.2.5.4   MANAGEMENT OF SDI

**Discharge rate and irrigation frequency in relation to crop and soil type**: SDI systems generally consist of emitters that have discharge rates less than 8 L/h [8]. A discharge rate of 0.25 L/h gave high yield of corn in sandy loam soils of Israel [9], although the difference in yields between discharge rates was not statistically significant. In a silt loam soil, a discharge rate of 0.5 L/h gave the highest onion (*Allium cepa*) yield [238]. In a drip irrigation system, frequency and emitter discharge rate determine the soil water availability and plant water uptake pattern [56, 57] and consequently yield [37, 74].

Illustrating the importance of matching irrigation frequency to soil type, a coarse textured sandy soil required drip lines with higher flow rates and shorter irrigation cycles than clay soil [220]. Similarly, shallow rooted vegetable crops on fine sandy soils in Florida required frequent (once or more per day) water application [92]. Conversely, in a clay loam soil, drip irrigation applied every second day achieved maximum tomato yield [62]. High frequency irrigation seems to be especially important for coarser-textured soils. High frequency SDI gave best yields of processing tomato in a sandy loam soil [13] and of potato in loamy soils in China [269]. High frequency water application under drip enables maintenance of salts at reasonable levels within the rooting zone [182].

The main reported benefit of increased irrigation frequency with SDI is the increased yield. A less commonly reported benefit of increased irrigation frequency is improved crop establishment [198]. As crop establishment is a common problem in SDI, it is surprising that there seem to be relatively few studies of irrigation frequency in relation to crop establishment. More frequent or pulsing irrigation, which involves applying small increments of water multiple times per day rather than applying large amount for long duration, has been advocated to improve surface and near surface soil moisture wetting for crop establishment [149]. However, there is a lack of operational guidelines for SDI [149]. In Australia, a comparison of pulsed and continuous irrigation on a Hanwood loam soil in NSW revealed very little difference between treatments, leading the author to conclude that responses depended on tape depth and soil type [178].

Other potential benefits of high frequency SDI are reduced deep drainage of water [13], although for this it will require uniform water application, uniform soil and crop growth. High frequency SDI may have lower water requirement, as shown by Wendt et al. [272].

The flow rate of the drip line has to match with the particular soil type. When soil hydraulic conductivity decreases, the pressure head of the soil next to the emitter will increase, which reduces the flow rate of emitters [270]. On the other hand, emitter discharge decreases due to backpressure, which depends on the soil type, possible cavities near the dripper outlet, and the drip system hydraulic properties [232]. When the pressure in the emitter increases this may significantly reduce the source discharge rate [154].

It was noted earlier in this chapter that soil types on which intensive horticulture is practiced in the Sydney basin vary from uniform sandy alluviums to loam overlying heavy, poorly drained clay. This variation presents a challenge to farmers to match discharge rate to soil type and select appropriate irrigation frequencies, especially when a wide range of crops is grown. Crop type also influences optimum irrigation frequency, even among vegetable crops. For example, on loam soil, cantaloupe (*Cucumis melo*) yield was higher with weekly irrigations compared to daily irrigations, while onion yield was higher for daily irrigation compared with weekly irrigation [37].

In most cases, supplementary irrigation has been used in establishment [126, 228]. Of the many studies dealing with irrigation management with SDI, few appear to have independently varied management for the crop establishment and growth periods other than adjust the crop factor. It appears that crops are often overwatered in the crop establishment period [73, 195] to ensure establishment. This has been reported to increase drainage [126].

The research has been scarce on the need to vary irrigation frequency during the crop growth to meet different water requirements. Frequent irrigation may be needed for good crop establishment, and frequent irrigation subsequently should reduce deep drainage, and increase water use efficiency. This approach is analogous to securing establishment by increasing irrigation rate above the crop requirement determined by $K_c$ and $ET_r$ [124], but with less risk of increased drainage.

### 5.2.5.5   FERTIGATION VIA DRIP IRRIGATION

Although this chapter is not concerned directly with **fertigation**, yet the application of nutrients via irrigation water have some considerations directly relevant to SDI. Therefore, the fertigation will be briefly reviewed here. Fertigation is a sophisticated and efficient method of applying fertilizers with irrigation water [167]. It contributes to higher yields and better quality by increasing fertilizer efficiency [111, 130], regardless of which type of drip irrigation method is being used. In addition, minimization of leaching below the root zone may be achieved by fertigation [90, 96].

Although fertigation can be used with any type of drip irrigation system, a major potential advantage of SDI is that water and nutrients are potentially used more efficiently when compared to surface installation [201]. Frequency of fertilizer injection can range from once a week to daily for drip irrigated vegetable crops [170]. Combined SDI and nutrient management technologies have been developed for several vegetable crops, including collard, mustard, spinach, and romaine lettuce [254, 255, 256] and corn [151]. SDI combined with fertilizer management has been found to increase the marketable yield of tomato, sweet corn and cantaloupe [13], sweet corn [18], cabbage and zucchini [219].

SDI provides incremental application of nitrogen and water. With good management, this has been reported to reduce $NO_3^-$ leaching and contamination of groundwater in lettuce production [256]. For crops such as broccoli, celery and lettuce, N-uptake is low in the first half of the season and higher before harvest. Fruiting crops such as tomatoes, pepper and melons require little N until flowering, then increase N-uptake, reaching peak uptake during fruit set. These factors need consideration for drip irrigation with fertigation [106].

Water and fertigation requirements need to be established for each crop, as significant differences occur among crop and soil types. For example, watermelon yield may be increased by maximizing the interactive effects of water and nitrogen applied through SDI on sandy loam soil [203], whereas for broccoli production with SDI on sandy loam soils, fertigation frequency had no effect on yield [257].

The substantial drainage activity was observed during the crop establishment period of tomato under drip irrigation [266], when the roots explore only a small volume of soil and water absorption capacity is small [132]. The excessive irrigation and associated drainage of tomatoes during establishment caused large N-losses [267]. So, if extra irrigation is required to ensure crop establishment and this creates a risk of drainage, the fertigation regime needs to be varied to minimize the risk of N-leaching.

SDI may also manage the placement and availability of immobile nutrients (e.g., P). The restricted mobility of the phosphate ions implies that preirrigation mixing of P in both clay and sandy soils is necessary, supplemented by addition to the irrigation solution, to obtain a uniform P concentration in the soil volume [19]. Immobile nutrients are delivered at the center of the soil root volume rather than on top of the soil in SDI [171]. Fertigation with P in SDI has improved yield, root growth and environmental performance of tomato [13] and sweet corn [202].

Potassium is easily soluble in water and applied through drip irrigation. Daily low rate application of nitrogen and potassium with a high frequency drip irrigation system improved nutrient uptake efficiency of sweet corn in sandy soils and reduced leaching loss [198].

## 5.2.5.6    GROWTH AND YIELD OF VEGETABLES IN SURFACE DRIP IRRIGATION AND SUBSURFACE DRIP IRRIGATION

As a general guide, crops that are suitable for surface drip irrigation are also suited to SDI [149]. With good agronomic practices, increased yields have been reported for a wide range of crops. These include lettuce [100]; sugarbeet [222, 233]; soluble solid content in transplanted muskmelon (*Cucumis melo* L.) [107]; onion [97, 237]; and green bean (*Phaselous vulgaris* L.) [177].

The crop response to SDI differs with crop growth characteristics and rooting pattern [149]. In lettuce, little yield difference was found between SDI and furrow irrigation in a sandy loam soil [100]. Potato yield was increased 27% with SDI over sprinkler irrigation, while reducing irrigation needs by 29%, provided there were drip lines in each crop row [69]. SDI had greater yield and higher water use efficiency than surface drip, furrow and sprinkler irrigation with cantaloupe, zucchini and oranges when irrigation depth was close to the consumptive use [66].

Information on root distribution is useful to understand crop responses to irrigation and fertigation, especially with the limited wetted soil volume that develops under SDI [202]. The root length and rooted soil volume of sweet corn was improved by frequent irrigation with shallow SDI [198]. The frequent irrigation maintained a portion of the root zone within the optimal matric potential range. In high-frequency irrigated corn, root length density and water uptake patterns are determined primarily by the soil water distribution under the drippers, whether the drippers are placed on, or beneath the crop row [57]. Most of the root system is concentrated in the top 40 cm of the soil profile in drip-irrigated tomatoes [166].

Unfavorable results obtained with drip irrigation have often resulted from inadequate root growth and distribution [34] in heavy textured soil [174]. Supply of aerated water with SDI can maintain aeration of the root zone in heavy clay soils and significantly increased yield of soya-bean and zucchini [22].

SDI can minimize the period between crops, especially with reduced tillage, and facilitate more intensive cropping. Multiple cropping with SDI has several practical advantages. The SDI does not require staking of the drip tubing during initial plant development, does not interfere with machine or manual thinning, weeding, spraying and harvesting of crops as does surface drip irrigation of vegetable crops [37]. A continuous cropping system of head lettuce and cabbage by using no tillage is a potential advantage with SDI [51]. Minimal tillage on semipermanent beds has been widely adopted in the Sydney region, although not with SDI [230]. Multiple cropping of vegetables such as cowpea, green bean, squash, and muskmelon in the spring season and broccoli in the autumn season were possible without yield reduction in a humid area [46].

### 5.2.5.7 PROBLEMS ENCOUNTERED WITH SDI

There are potential disadvantages with SDI, including high initial investment cost, clogging of emitters by various means, 'tunneling' of soil, and difficulties with uneven wetting and poor plant establishment [50, 152, 180]. The specific benefits and disadvantages of SDI in Australia [212, 213] are:

1. *Crop establishment*: In the absence of supplementary irrigation, germination and crop establishment with subsurface drip irrigation depends on unsaturated water movement (i.e., upwards or laterally from the buried emitter). Therefore, important determinants of uniform germination/establishment include the distance from the emitter to the seed/transplant, soil properties (structure, texture, hydraulic conductivity) and preceding water content [49].

2. *Soil and water interaction*: Emitter discharge rate can exceed the ability of some soils to distribute the water in the soil [144]. The water pressure in the region around the outside of the emitter may exceed atmospheric pressure thus altering emitter flow. This leads to the "tunneling" of emitter flow to the soil surface causing undesirable wetting spots in the field. Small soil particles may be carried with the water, causing a 'chimney effect' that leads a preferential flow path. The 'chimney' may be difficult to permanently remove.

The rest of this section deals with the crop establishment issue, especially in relation to wetting pattern, which varies with soil type [32]. This is a particular issue for developing SDI for the Sydney Basin because of the wide variation in soil types. Where soil types vary greatly between farms, it is both costly and challenging to undertake the research and develop extension recommendations for irrigation design and management that are not clear and unambiguous. In fields with heterogeneous soils, there can be uneven wetting with its inherent problems.

It has been shown in this chapter that SDI is commonly placed relatively deeply in the soil, even for shallow-rooted horticultural crops, to reduce soil evaporation or to facilitate tillage operations. Consequently, the variable wetting pattern and inadequate surface wetting of SDI often provides insufficient surface soil moisture to meet the demands of seeds [278] or seedlings.

Several reviews have concluded that crop establishment can be difficult with SDI [47, 147, 213], at least for germination of shallow-planted seeds. In most situations, a crop cannot be established using subsurface drip irrigation alone [103]. Therefore, a parallel surface drip irrigation system represents an added cost to SDI, while it will also reduce water use efficiency during the period of surface irrigation, and increase the risk of deep drainage.

For cotton, germination remains one of the greatest challenges for SDI [213], although the problem extends beyond germination to include the whole establishment period, including establishment from transplanted seedlings. Problems arizing from the poor wetting pattern may persist through the crop growth, unless efforts are made to have an adequate the wetting pattern that will match to the crop root zone [18].

As discussed previously, wetting patterns can be managed by varying dripper discharge rate and spacing [165], influencing the dripper interface [176], increasing irrigation frequency [198] or amount [124], and reducing the depth of installation [195]. It may also be approached through modifying the SDI tape design [271]. Accordingly, research has been undertaken to improve crop establishment under SDI following a range of approaches. However, from the literature discussed previously, none of the solutions involving shallow tape installation or higher discharge rates will be satisfactory under all circumstances.

The modification of a drip tape is the most likely approach to achieve satisfactory performance under a wide range of soil and climatic conditions. Even with this, to achieve adequate surface wetting and remove the risk of poor crop establishment [278] under all circumstances, it is likely that situation-specific guidelines will be needed for irrigation rate and frequency. Thus for SDI to be adopted in the Sydney region, and to enhance its adoption elsewhere, further research is needed on the modification of drip tape to improve surface wetting, and into development of appropriate guidelines for irrigation rate and frequency.

The modification in SDI design by adding an impermeable membrane has the potential advantages of changing the wetting pattern [178] and inhibiting the downward percolation of water [271]. To counter problems of poor germination, a new technique was suggested for manipulating the wetting pattern of SDI using an impermeable membrane to transform the point source of water in drip lines to a broad band source from which a capillary force operates to draw water upward and outward [271]. Another new subsoil irrigation system consisted of a V-shaped device, which released foil and pipe simultaneously into the soil [17]. Although the impervious layer is intended to reduce downward percolation [271], it is hypothesized in this chapter that any benefit may arise because the layer creates a temporary water table, from which the upward flux of water is increased.

Modifying the drip tape to include the impermeable layer was commercialized in the Capillary Root Zone Irrigation (CRZI) product. It was evaluated in loam and sandy loam soils [49]. The results indicated that CRZI provided a more uniform wetting pattern but failed to improve crop establishment in English spinach. In this case, however, crop establishment was considered to be good (~50%) with standard subsurface drip because of the particular soil properties that gave rise to adequate surface water. So, despite the improved wetting pattern, germination was not better. The results did show that an impermeable barrier can be beneficial for surface wetting. Similar results have been obtained with lettuce germination [67].

It appears that more research is needed to define the conditions under which the crop establishment problems arise and to reduce the technical barriers to SDI. Barriers to the adoption of SDI include the need to adapt the system design and management to local soil and climatic conditions and constraints. CRZI has undergone extensive development and is now sold under the trade name Kapillary Irrigation Subsurface System (KISSS™). The advantage of this product over conventional SDI for vegetable seedling establishment has not been evaluated.

## 5.2.6  SOIL PROPERTIES AND SDI PERFORMANCE: VEGETABLES

### 5.2.6.1  ROLE OF SOIL TEXTURE AND STRUCTURE

The furrow, drip and subsurface drip irrigation systems have been compared for lettuce on sandy loam soils [100]. There was more sand and less silt under furrow irrigated plots in the top layer of soil (0–0.3 m) due to greater infiltration than the drip irrigated plots. Sand, silt and clay contents of the 0–0.3 m depth interval were quite constant with distance in SDI. Changes in clay content, cation levels and the pore space around emitters were observed in long-term SDI with tomato, rock-melons and onions [15]. These changes could have inhibited the movement of water by altering soil hydraulic properties and reducing the spread of the irrigation wetting-front in clay soils. In one study in heavy textured soil in a region where secondary salinity is a problem, SDI increased the rate of salinization compared to furrow irrigation because of improved soil structure and reduced slaking and dispersion in subsoil which led to increased solute movement through the soil profile [128].

Slaking and dispersion are used to measure the structural stability of soil [64]. Gypsum improves soil structural stability and economic use of gypsum depends on soil properties and seasonal condition [78, 88]. Soil conditioners applied by drip irrigation have also increased water stable aggregation in the wetting zone around the drippers [235].

Drip irrigation can improve plant water availability in medium and low permeability fine-textured soil, and in highly permeable coarse-textured soil in which water and nutrients move quickly downward from the emitter [60]. Continuous irrigation at a rate equal to evapotranspiration was optimal for medium textured soils while greater application rate was required for coarse textured soils to minimize deep percolation losses [83]. Many experiments have been conducted in both modeling and field research to investigate plant water availability and root uptake pattern in different soil types [181, 182, 192, 193, 259].

### 5.2.6.2  ROLE OF SOIL HYDRAULIC PROPERTIES

Knowledge of soil hydraulic properties assists in the design of irrigation systems [175]. Non-uniformities in hydraulic properties and infiltration rates are considered to be major reasons for inefficient drip irrigation system and may cause nonuniformities in soil water content and can potentially affect plant growth. Soil hydraulic conductivity is a limiting factor for water uptake by plants under drip irrigation, particularly in sandy soils [159]. However, in clay loam soils, subsurface drip irrigation resulted in very nonuniform soil water contents above the depth of emitters [7], which may be corrected by using a membrane under the drip tube.

### 5.2.6.3  SOIL CHEMICAL RESPONSES TO SURFACE DRIP IRRIGATION AND SUBSURFACE DRIP IRRIGATION

For row crops, the emitters are often placed at the center of row beds, below which most salt loading or leaching would probably occur. In one study, soil electrical conductivity, pH and soluble cations were lower under SDI than surface drip [188], suggesting increased leaching. The conversion of fertigated ammonium sulfate and urea into nitrate-N caused acidification in the wetted soil volume to the surface (0–20 cm) of silt loam soils, also suggesting an increase in leaching [112]. Similarly, acidification throughout the soil profile was observed in vegetable beds in tomato crops [248], again suggesting leaching of $NO_3$. This hypothesis finds support in an investigation of commercial production of processing tomato where SDI, combined with excessive fertilizer application, was thought to cause the leaching of nitrogen (and phosphate) to ground water depths [248]. Under drip irrigation of tomato crops on sandy loam soils, greater drainage occurred during the crop establishment period, which increased the leaching of nitrates previously stored in the soil profile [266].

From these research studies, it seems possible that vegetables crops may be over-irrigated using both SDI and surface drip. Therefore, there is a the need to irrigate above crop water requirement in order to maintain acceptable soil moisture in the soil surface, especially in the case of SDI.

### 5.2.6.4  SOIL WETTING PATTERN

A basic need for better drip irrigation systems is the knowledge about the moisture distribution pattern, shape and volume of soil wetted by an emitter [157]. The volume of wetted soil represents the amount of water stored in the root zone. Its depth should coincide with rooting depth while its width should be related to the spacing between emitters. One possibility for controlling the wetted volume of a soil is to regulate the emitter discharge rate based on the soil hydraulic properties [29, 165].

The wetting front is an important factor in drip infiltration, indicating the boundaries of the wetted soil volume [29]. A simple technique known as the pit method was developed for design and management of drip systems [20].

Soil texture is an unreliable predictor of wetting and for adopting different emitter spacings. For different soil texture, site-specific information on soil wetting is required [259]. Under given climatic conditions, the effect of soil type on the depth-width-discharge combination is influenced by water holding capacity and hydraulic conductivity of the soil [279].

The wetting pattern with SDI can be affected not only by irrigation management, but also SDI design aspects such as emitter spacing and drip line depth. Dripper function can also be modified after installation. In one study, heterogeneity of the soil in the neighborhood of a subsurface emitter that had been disturbed by farm equipment resulted in low emitter flow, leading the authors to suggest using soil conditioners to improve and stabilize soil structure around the dripper [234].

The wetting pattern has also been enhanced by the addition of plastic barriers beneath the drip line [33, 49].

### 5.2.6.5  POTENTIAL OF SDI IN AUSTRALIA

The irrigation industry is under pressure to improve water use efficiency and reduce environmental impacts. In the Sydney region, drip irrigation is not widely used for vegetable production, although it has the potential to improve irrigation performance. From the literature review in this chapter, it can be concluded that SDI might improve water use efficiency, and reduce environmental impacts more than surface drip irrigation. It will overcome two important objections to drip irrigation: The high ongoing cost and the disruption to normal cultural practices.

However, SDI may have significant problems with poor or uneven surface wetting, leading to problems with crop germination and establishment. The research reported in this chapter was undertaken with the broad aim of providing a foundation for the adoption of SDI in the Sydney region, by addressing the problem of risky plant establishment.

Increased irrigation frequency and irrigation amount may improve surface wetting, although SDI can increase drainage during the establishment period, which appears to be related to the increased irrigation amount. Shallow tape placement is also helpful, but this has practical limitations due to farm cultural practices.

A promizing innovation is the inclusion of a narrow impermeable plastic barrier below the drip line and geotextile layer above the drip line, designed to improve surface wetting. The most recent version of this product has not been evaluated for its effects on surface wetting and crop establishment. Thus there is a need to test whether surface wetting is improved, and also whether this leads to improved establishment.

The research question was: "Does an impermeable layer beneath the drip tape improve surface soil water conditions and crop establishment compared with conventional SDI?" As the impermeable layer adds to the cost of SDI, it is also important to know if the modified tape has any benefit that cannot be achieved by varying irrigation rate or frequency, both of which are known to affect wetting patterns but have received little attention in relation to crop establishment. So, a subsidiary question was: "Does the modified SDI offer any advantage over using conventional SDI with greater irrigation amount or frequency?

## 5.3  LETTUCE CROP: MODIFIED SDI AND CONVENTIONAL SDI

The need for improved irrigation systems and practices in the Western Sydney horticultural area was identified in the literature review in this chapter. SDI offers significant potential benefits over conventional drip irrigation, but adoption has been slow because of problems with crop establishment. Lettuce is grown commercially from seedlings purchased from nurseries [70] in Australia. Poor establishment of seedlings with SDI is due to insufficient surface soil moisture [278] and subsequent 'transplant shock' [260]. This is a result of limited upward water movement from the drip line to the soil surface or near the soil surface [149], resulting in an uneven supply of water to seeds or seedlings.

Assured establishment with any irrigation system requires consistent management of water and the system infrastructure [196]. With SDI, shallow tape installation combined with frequent water application is one approach to improve crop establishment on sandy soil [42], although there is no general guideline to irrigation frequency. Irrigation application volume during establishment may be increased to more than the requirement estimated by using the crop factor [124], but this may result in increased deep drainage and reduced water-use efficiency.

An alternative approach is to modify the subsurface drip tape to improve surface soil moisture. The 'Vector flow' drip line with an impermeable membrane reduces drainage and encourages longitudinal movement of water [271]. This transformed the wetting pattern from a point source to a broad band source of water and encouraged greater upward movement of water. This concept was later modified by adding geotextile material above the impermeable membrane (capillary root zone irrigation - CRZI), which facilitated mass flow along the line. This improvement in SDI provided a more uniform wetting pattern than conventional SDI, but it failed to provide an advantage in germination of English spinach [49, 67]. The CRZI has undergone extensive development and is now sold under a new trade name as KISSS™.

This Chapter evaluates improvements in soil water and crop establishment with the modified SDI (KISSS™) compared with conventional SDI. Field experiments were undertaken in autumn and spring, to determine if an impermeable layer beneath the drip line, and geotextile material above the drip line, will improve surface soil moisture relations and crop establishment with SDI in a sandy soil. It was also evaluated if the modified SDI was any better for establishment than increasing irrigation frequency or amount with conventional SDI.

### 5.3.1  METHODS AND MATERIALS

The lettuce crop was established at the University of Western Sydney - Hawkesbury Campus (UWSH) at Richmond, 64 km west of Sydney – NSW (33.62°S latitude and 150.75°E longitude at an elevation of 20 m above mean sea level). The mean annual rainfall at the experimental site was 800 mm [38]. Table 1 indicates physical properties of the soil [3]. The soil type was a Clarendon sand, a freely draining coarse sand, brownish-gray in color to a depth of 75 cm; a light gray and yellowish brown sandy clay for 75–210 cm depth. Parent material of this soil type is coarse sandy alluvium of the Nepean River (Pleistocene).

**TABLE 1**   Soil profile details of the experimental site [3].

| Soil depth (cm) | Texture description | Color description | Horizon |
|---|---|---|---|
| 0–22.5 | Coarse sand | Brownish gray | A1 |
| 22.5–55 | Coarse sand | Light gray | A2 |
| 55–75 | Coarse sandy loam | Light gray and pink mottled | B1 |
| 75–210 | Clayey sand | Grey and yellow brown | B2 |

The field was plowed and beds were formed in the north to south direction. SDI tapes were installed manually in each bed at a nominal depth of 15 cm in each row with lateral spacing of 30 cm. Both drip tape types comprised of 1.6 Lph emitters at 50 cm spacing, which is equivalent to application rate of 3.2 L.m$^{-1}$.h$^{-1}$. An automatic battery operated irrigation controller was installed in the main line of the SDI system.

### 5.3.1.1   FIELD LAYOUT AND GENERAL AGRONOMICAL PRACTICES FOR LETTUCE CROP

The experiments were randomized complete block designs with 8 replications in Experiment 1, in autumn 2007 (Fig. 2) and 4 replications in Experiment 2, in spring 2007 (Fig. 3). Beds were 6 m long and 1 m wide. The lettuce (*Lactuca sativa*) seedlings were transplanted at 30 cm spacing between plants and rows. There were three rows per bed and two drip lines located between plant rows. Compound fertilizer @ 100 kg/ha (12% N, 5.2% P and 14.1% K) was incorporated three days prior to transplanting. Insecticide (Entrust$^R$ @ 60 g/ha) was applied as necessary during plant establishment.

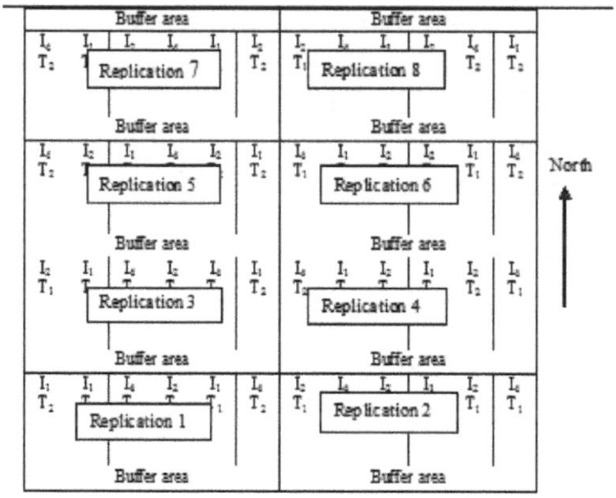

**FIGURE 1**   Field experiment 1 layout.

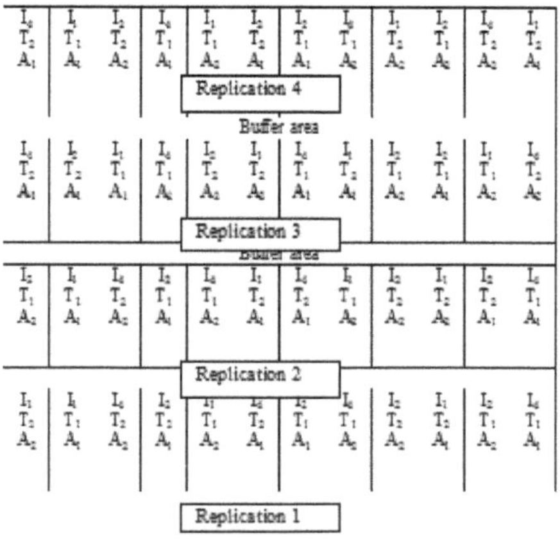

**FIGURE 2**   Field experiment 2 layout.

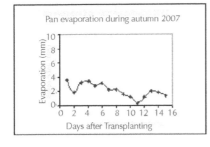

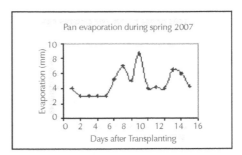

**FIGURE 3** Pan evaporation for both seasons.

### 5.3.1.1.1 EXPERIMENT 1 – AUTUMN 2007 (FIG. 1)

Crop establishment was evaluated for two drip tape types with three irrigation frequencies using equal amounts of water applied to all treatments. The treatments were two drip tape types ($T_1$-M.SDI, $T_2$-C.SDI) and three irrigation frequencies (1, 2 and 4 times per day designated $I_1$, $I_2$ and $I_4$). The amount of water applied was determined by multiplying the Class A pan evaporation value of the previous day by a crop factor of 0.4.

Plant numbers were recorded in all plots at 14 days after transplanting (DAT) and percentage plant establishment was calculated. Leaf number, leaf length and width, plant fresh weight and dry weight were assessed in each treatment. Four plants were randomly selected immediately after transplanting, tagged and observed for leaf growth parameters during the crop establishment period. The number of leaves was counted daily from 4 to 11 days after transplanting. Lettuce leaf length and width were measured at 7 and 14 days after transplanting. Fresh weight of lettuce was estimated at 14 days after transplanting by harvesting four plants randomly from each bed. The fresh weight samples were dried in a fan-forced dehydrator at 80° C for a minimum of 72 h to determine dry weight.

Volumetric soil moisture content was measured daily with a Theta probe (model of ML-2x a Delta T-Device, Measurement Engineering Australia). The measurements were taken twice randomly from the bed at 5–10 cm depth, within 10 min after irrigation.

### 5.3.1.1.2 EXPERIMENT 2 – SPRING 2007 (FIG. 2)

The spring trial evaluated the modified SDI under a higher evaporative demand. Treatments were the same as in Experiment 1, except that crop factors of 0.4 ($A_1$) and 0.8 ($A_2$) were also compared to determine if, under a higher evaporative demand, increased irrigation amount (higher crop factor) provides a better wetting soil surface and improved plant establishment.

Crop establishment was determined 15 DAT by counting all plants. Six plants per plot were randomly selected after transplanting, tagged and observed for leaf number every three days from 3 to15 DAT. Leaf length and width were measured at 7 and 14 DAT. Plant fresh weight was measured in plants from three locations: the center of the drip line and on both sides of the drip line. Six plants per location per plot were collected at random and weighed. Volumetric soil water content at 5–10 cm depth was measured during establishment, in the center and both sides of the drip line, before irrigation.

### 5.3.1.2 WEATHER DATA

Pan evaporation and rainfall were recorded within 100 m of the experimental site. Evaporation data are given in Fig. 3. Evaporation was higher in the spring. Total rainfall of only 4.6 and 1 mm was received during the autumn and spring experiments, respectively.

### 5.3.1.3 STATISTICAL ANALYSIS OF DATA

Crop establishment data were calculated as percentages and transformed by square root transformation [246]. Analysis of variance (ANOVA) was conducted on both experiments using MiniTab *ver.* 14. Data for leaf appearance in each plot were analyzed to calculate the regression coefficients. The leaf appearance rate was subjected to an ANOVA.

### 5.3.2 RESULTS

#### 5.3.2.1 CROP ESTABLISHMENT

In both autumn and spring, the mean establishment for M.SDI (100% and 99%) was significantly ($p < 0.001$) greater than for the C. SDI (93% and 97%) as shown in Table 2. In both seasons there was no significant difference between irrigation frequencies. There was also no significant difference due to crop factor in the spring experiment.

#### 5.3.2.2 NUMBER OF LEAVES

In autumn, M.SDI had significantly more leaves at each time during crop establishment from 4 to 11 days after transplanting (Table 3). However, irrigation frequency did not show any significant difference in number of leaves at any time. As no treatments had been imposed before day 4, the significant difference between tape types must reflect a chance difference in the size of seedlings when they were transplanted although the seedlings were selected at random. Therefore, subsequent differences cannot be attributed with certainty to treatment effects. Leaf appearance rate (LAR) was calculated with regression analysis in each treatment over time. An ANOVA was performed to find the regression coefficient, which is the LAR (Table 3a). The LAR over the establishment period did not show any significant difference due to tape type or irrigation frequency.

**TABLE 2**   The effects of tape type, irrigation frequency (IF) and crop factor (CF) on survival (%) of lettuce (14 DAT). DAT = Days after transplanting.

| Irrigation frequency (IF) No. per day | Autumn 2007 | | | Spring 2007 | | | | |
|---|---|---|---|---|---|---|---|---|
| | M.SDI C.SDI IF | | | M.SDI C.SDI IF | | | | |
| | | | | 0.4 | 0.8 | 0.4 | 0.8 | |
| One | 10.0 | 9.6 | 9.8 | 9.9 | 9.9 | 9.9 | 9.9 | 9.9 |
| | (99.8) | (92.6) | (96.2) | (98.0) | (98.8) | (97.2) | (98.0) | (98.0) |

**TABLE 2** *(Continued)*

| Irrigation frequency (IF) No. per day | Autumn 2007 | | | Spring 2007 | | | | |
|---|---|---|---|---|---|---|---|---|
| | M.SDI C.SDI IF | | | M.SDI C.SDI IF | | | | |
| | | | | 0.4 | 0.8 | 0.4 | 0.8 | |
| Two | 10.0 | 9.6 | 9.8 | 9.9 | 9.9 | 9.8 | 9.8 | 9.9 |
| | (99.4) | (91.8) | (95.5) | (98.4) | (98.4) | (95.9) | (95.5) | (97.0) |
| Four | 10.0 | 9.8 | 9.9 | 10.0 | 10.0 | 9.8 | 9.8 | 9.9 |
| | (100) | (95.9) | (97.8) NS | (99.2) | (99.6) | (96.2) | (96.6) | (98.0) NS |
| Mean tape | 10.0 (99.8)* | 9.7 (93.3) | — | 9.9 (98.8)* | | 9.8 (96.6) | | — |
| Mean Crop factor | — | — | — | 0.4 = 9.9 (97.4) | | 0.8 = 9.9 (97.8) NS | | — |

* Significant at p<0.001, within experiment. Values are square root transformed means. Values in parenthesis have been retransformed to the original scale. NS = Not significant.

In the spring experiment, leaf numbers in the M.SDI were significantly greater than in the C.SDI during the crop establishment period from 3 to day 15 DAT (Tables 4a to 4b). Irrigation frequency initially had no effect on leaf number, but by the end of the crop establishment period, leaf numbers were significantly higher in the treatment with most frequent irrigation. Leaf numbers were consistently higher with the higher crop factor. The LAR over the establishment period in spring was significantly greater (p<0.05) with the M.SDI, higher crop factor, and more frequent irrigation (Table 5). The interaction between tape type and irrigation frequency, tape type and crop factor were also significant (p<0.05). The LAR with M.SDI was 0.80 leaves/day, compared with conventional SDI at 0.74 leaves/day (p<0.001). The greatest LAR (0.84 leaves/day) was with the M.SDI and highest frequency of irrigation. The main effects of tape type and IF, and the interaction between Tape type x IF, CF x Tape type were significant.

**TABLE 3** Leaf number during lettuce establishment, for two tape types and three irrigation frequencies in autumn 2007.

| Tape type | Day 4 | | | | Day 5 | | | | Day 6 | | | | Day 7 | | | |
|---|---|---|---|---|---|---|---|---|---|---|---|---|---|---|---|---|
| | I. F. | | | | I.F. | | | | I.F. | | | | I.F. | | | |
| | 1 | 2 | 4 | Mean | 1 | 2 | 4 | Mean | 1 | 2 | 4 | Mean | 1 | 2 | 4 | Mean |
| M.SDI | 6.3 | 6.3 | 6.5 | 6.3* | 6.8 | 6.6 | 7.1 | 6.9* | 6.9 | 6.9 | 7.2 | 7.0* | 7.4 | 7.4 | 7.8 | 7.5* |
| C.SDI | 6.0 | 6.2 | 6.0 | 6.1 | 6.6 | 6.6 | 6.4 | 6.5 | 6.6 | 6.4 | 6.4 | 6.5 | 7.0 | 7.0 | 7.0 | 7.0 |
| Mean | 6.1 | 6.2 | 6.3 | | 6.7 | 6.6 | 6.6 | | 6.8 | 6.7 | 6.8 | | 7.2 | 7.2 | 7.4 | |

**TABLE 3**   *(Continued)*

| Tape type | Day 4 | | | | Day 5 | | | | Day 6 | | | | Day 7 | | | |
|---|---|---|---|---|---|---|---|---|---|---|---|---|---|---|---|---|
| | I. F. | | | | I.F. | | | | I.F. | | | | I.F. | | | |
| | 1 | 2 | 4 | Mean | 1 | 2 | 4 | Mean | 1 | 2 | 4 | Mean | 1 | 2 | 4 | Mean |
| Tape type | Day 8 | | | | Day 10 | | | | Day 11 | | | | — | | | |
| | I. F. | | | | I. F. | | | | I. F. | | | | — | | | |
| | 1 | 2 | 4 | Mean | 1 | 2 | 4 | Mean | 1 | 2 | 4 | Mean | | | | |
| M.SDI | 7.4 | 7.4 | 7.8 | 7.5* | 7.8 | 7.9 | 8.4 | 8.1* | 8.7 | 8.3 | 9.1 | 8.8* | — | | | |
| C.SDI | 7.0 | 7.0 | 7.0 | 7.0 | 7.5 | 7.4 | 7.3 | 7.4 | 8.3 | 8.3 | 8.3 | 8.3 | | | | |
| Mean | 7.2 | 7.2 | 7.4 | | 7.7 | 7.7 | 7.7 | | 8.6 | 8.3 | 8.7 | | | | | |

**TABLE 3A**   Lettuce leaf appearance rate (LAR) during crop establishment (leaves per day) for two tape types and three irrigation frequencies in autumn 2007.

| Tape type | Irrigation frequency, IF Numbers/day | | | |
|---|---|---|---|---|
| | 1 | 2 | 4 | Mean |
| M.SDI | 0.44 | 0.36 | 0.45 | 0.41 **NS** |
| C.SDI | 0.38 | 0.39 | 0.40 | 0.39 |
| **Mean** | **0.41** | **0.38** | **0.42 NS** | |

NS – Not Significant.

**TABLE 4A**   Number of leaves on day 3 (Spring 2007).

| IF Irrigation per day | M.SDI | | C.SDI | | Mean | Mean | | Mean | |
|---|---|---|---|---|---|---|---|---|---|
| | | | | | IF | C.F. x I.F. | | Tape x IF | |
| | 0.4 | 0.8 | 0.4 | 0.8 | | 0.4 | 0.8 | M.SDI | C.SDI |
| One | 6.5 | 6.8 | 5.8 | 6.5 | 6.4 | 6.2 | 6.6 | 6.6 | 6.2 |
| Two | 6.1 | 6.4 | 6.1 | 5.9 | 6.2 | 6.1 | 6.2 | 6.3 | 6.0 |
| Four | 6.1 | 6.8 | 6.3 | 6.3 | 6.4NS | 6.2 | 6.5 | 6.4 | 6.3 |
| **Mean** | **6.4*** | | **6.2** | | — | **6.2 *** | **6.5** | **NS** | — |
| Mean | 6.2 | 6.7 | 6.1 | 6.2 | — | — | — | — | — |
| C.F. x tape | | NS | | | | | | | |

The main effect of tape type and the CF x IF interaction were significant.
* Significant at p<0.05, NS – not significant. DAT – days after transplanting.

**TABLE 4B** Number of leaves on day 6 (spring 2007).

| IF Irrigation per day | M.SDI | | C.SDI | | Mean IF | Mean C.F. x I.F. | | Mean Tape x IF | |
|---|---|---|---|---|---|---|---|---|---|
| | 0.4 | 0.8 | 0.4 | 0.8 | | 0.4 | 0.8 | M.SDI | C.SDI |
| One | 8.1 | 8.3 | 6.8 | 8.1 | 7.8 | 7.5 | 8.1 | 8.2 | 7..4 |
| Two | 7.9 | 8.0 | 7.5 | 7.5 | 7.7 | 7.7 | 7.7 | 7.9 | 7.5 |
| Four | 7.8 | 8.7 | 8.1 | 7.4 | 8.0NS | 7.9 | 8.1 | 8.3 | 7.8 |
| **Mean** | **8.1 \*\*** | | 7.5 | | | 7.7NS | 8.0 | NS | 8.1 \*\* |
| Mean C.F.x tape | 8.0 | 8.3 NS | 7.4 | 7.7 | — | — | — | — | — |

The main effect of tape type was significant.
\*\* Significant at p<0.01, NS – not significant.

**TABLE 5** Leaf appearance rate (LAR, leaves per day) during crop establishment for tape type, irrigation frequency and crop factor (Spring 2007).

| IF (Irri./day) | M.SDI 0.4 0.8 | | C.SDI 0.4 0.8 | | I.F. Mean | Mean Tape x I.F. M.SDI C.SDI | |
|---|---|---|---|---|---|---|---|
| One | 0.76 | 0.80 | 0.72 | 0.74 | 0.76 | 0.78 | 0.73 |
| Two | 0.74 | 0.80 | 0.76 | 0.75 | 0.76 | 0.77 | 0.75 |
| Four | 0.81 | 0.87 | 0.72 | 0.77 | 0.79* | 0.84 | 0.74 |
| **Mean C.F. x Tape** | 0.77 * | 0.82 | 0.74 | 0.75 | | * | |
| **Mean Tape** | **0.80 \*\*\*** | | 0.74 | | | | |
| **Mean C.F.** | 0.75** for 0.4 CF | | 0.79 for 0.8 CF | | | | |

\*, \*\*, \*\*\* Significant at p<0.05, 0.01, 0.001, respectively.

### 5.3.2.3   LEAF SIZE

*LEAF LENGTH*

Comparing tape types, the M.SDI had greater leaf length at both 7 and 14 DAT (p<0.001) in the autumn experiment (Table 6). The effect of irrigation frequency (p<0.001) and interaction between tape type and irrigation frequency (p<0.01) were significant at 7 DAT. Four irrigations per day in M.SDI recorded the greatest leaf length at 7 DAT (7.5 cm). At 14 DAT, the effect of irrigation frequency was again significant (p<0.001). The interaction between tape type and irrigation frequency was not significant (p<0.05).

**TABLE 6**   The effect of tape type and irrigation frequency (IF) on leaf length (cm) at 7 and 14 days after transplanting (autumn 2007).

| Tape type | 7 DAT | | | | | 14 DAT | | | |
|---|---|---|---|---|---|---|---|---|---|
| | Irrigation frequency/day | | | | | Irrigation frequency/day | | | |
| | 1 | 2 | 4 | Tape-Mean | | 1 | 2 | 4 | Tape-Mean |
| M.SDI | 6.8 | | 6.9 | 7.5 | 7.1 | 7.1 | 7.1 | 7.8 | 7.3 |
| C.SDI | 6.4 | | 6.5 | 6.6 | 6.5*** | 6.6 | 6.7 | 7.1 | 6.8*** |
| Mean IF | 6.6 | | 6.7 *** | 7.0 | | 6.9 | 6.9 | 7.4 | |
| | | | | | | | *** | | |
| Mean | | | | | | | | | |
| Tape x IF | | | * | | | | NS | | |

The main effects of both tape type and IF were significant at 7 and 14 DAT, and the tape x IF interaction at 7 DAT.

*, *** Significant at p<0.05, p<0.001

**TABLE 7**   The effects of tape type, irrigation frequency (IF) and crop factor (CF) on leaf length 7 days after transplanting (spring 2007).

| IF | M.SDI | | C.SDI | | Mean I.F. | Mean Tape x I.F. | |
|---|---|---|---|---|---|---|---|
| (Irri./day) | 0.4 | 0.8 | 0.4 | 0.8 | | M.SDI | C.SDI |
| One | 8.3 | 8.4 | 8.4 | 8.4 | 8.4 | 8.3 | 8.4 |
| Two | 8.5 | 8.4 | 8.1 | 8.2 | 8.3 | 8.5 | 8.1 |
| Four | 8.8 | 8.9 | 8.6 | 8.6 | 8.7 | 8.8 | 8.6 |
| | | | | | *** | | |
| Mean | 8.5 NS | 8.4 | 8.5 | 8.4 | | * | |
| C.F. x Tape | | | | | | | |
| **Mean Tape** | **8.5 **** | | **8.4** | | | | |
| Mean C.F. | 8.4 NS for 0.4 CF | | 8.5 for 0.8 CF | | | | |

The main effects of both tape type and IF were significant.

*, **, *** Significant at p<0.05, 0.01, 0.001 respectively

## LEAF WIDTH

Values for leaf width are shown using interval plots to reveal the greater variability of the C.SDI (Figs. 4 and 5). With respect to treatment responses, in autumn, leaf width of 3.0 cm at 7 DAT for M.SDI was significantly greater (p<0.001) than in conventional SDI with 2.8 cm. Four irrigations per day recorded 3.0 cm leaf width, which was greater than in less frequent irrigations (Fig. 4). The interaction between tape type and irrigation frequency was significant (p<0.001) at 7 DAT. The same trends were observed at 14 DAT.

In the spring experiment, leaf width was greater than in autumn, at both 7 and 14 DAT. In both observations, the mean leaf width for M.SDI (3.4 cm and 4.0 cm) was significantly greater than conventional SDI (3.3 cm and 3.9 cm), respectively. Compared to three irrigations per day, plants receiving four irrigations per day recorded 3.5 cm and 4.0 cm width at 7 and 14 DAT, respectively. Crop factor did not show any significant difference at 7 DAT, while at 14 DAT it was significant (p<0.001). The interaction between tape type and irrigation frequency was statistically significant (p<0.05) at 7 DAT, but not at 14 DAT.

**TABLE 8**   The effects of tape type, irrigation frequency (IF) and crop factor (CF) on leaf length 14 days after transplanting (spring 2007).

| IF (Irri./day) | M.SDI 0.4 | 0.8 | C.SDI 0.4 | 0.8 | Mean I.F. | Mean Tape x I.F. M.SDI | C.SDI |
|---|---|---|---|---|---|---|---|
| One | 9.3 | 9.6 | 9.3 | 9.3 | 9.3 | 9.4 | 9.3 |
| Two | 9.5 | 9.5 | 9.0 | 9.1 | 9.2 | 9.5 | 9.1 |
| Four | 9.7 | 10.1 | 9.4 | 9.5 | 9.7*** | 9.9 | 9.5 |
| Mean C.F. x Tape | 9.5 | 9.7** | 9.2 | 9.3 | | NS | |
| Mean Tape | 9.6** | | 9.3 | | | | |
| Mean C.F. | 9.4*** for CF = 0.4 | | 9.5 for CF = 0.8 | | | | |

All main effects and the interaction between tape type and CF were significant.

**, *** Significant at p<0.01, 0.001, respectively

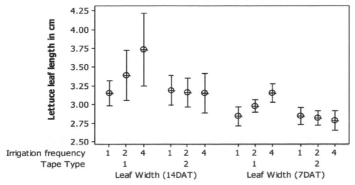

**Interval plot of leaf width at 7 and 14 DAT during autumn 2007**
95% CI for the Mean

**FIGURE 4**   Effects of tape type and irrigation frequency on lettuce leaf width during autumn 2007.

## 5.3.2.4   LETTUCE FRESH WEIGHT

In the autumn experiment, plants in the M.SDI treatment had significantly (p<0.001) higher fresh weight (19.2 g/plant) than C.SDI (16.6 g/plant) (Table 9). More frequent irrigation (four/day) gave greater fresh weight of 20.4 g/plant (p<0.001) than less frequent irrigation. The interaction between tape type and irrigation frequency was not significant.

**TABLE 9**   The effects of tape type and irrigation frequency (IF) on fresh weight (g/plant) at 14 DAT (autumn 2007).

| I.F. | Tape type | | Mean |
|------|-----------|------|------|
| (Irri. /day) | **M.SDI C.SDI** | | **I.F.** |
| One | 17.4 | 14.2 | 15.8 |
| Two | 18.6 | 16.4 | 17.5 |
| Four | 21.6 | 19.1 | 20.4 *** |
| **Mean** | **19.2 ***** | **16.6** | |

*** Significant at p<0.001.

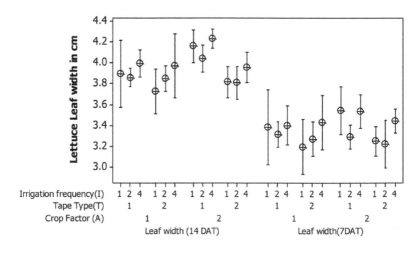

**FIGURE 5**   Effects of tape type and irrigation frequency on lettuce leaf width during spring 2007.

**TABLE 10** The effects of tape type, irrigation frequency (IF) and crop factor (CF) on fresh weight (g/plant) at 14 DAT (spring 2007).

| IF (Irri./day) | M.SDI 0.4 0.8 | | C.SDI 0.4 0.8 | | Mean I.F. | Mean Tape x I.F. M.SDI C.SDI | |
|---|---|---|---|---|---|---|---|
| One | 17.3 | 21.4 | 12.9 | 18.6 | 17.6 | 19.4 | 15.8 |
| Two | 18.2 | 23.6 | 14.9 | 19.5 | 19.0 | 20.9 | 17.2 |
| Four | 22.4 | 27.6 | 17.5 | 20.5 | 22.0 | 25.0 | 19.0 |
| Mean C.F. x Tape | 19.3 | 24.2** | 15.1 | 19.5 | *** | ** | |
| **Mean Tape** | **21.7 ***** | | **17.3** | | | | |
| Mean C.F. | 17.2*** for 0.4 C.F | | 21.9 for 0.8 C.F | | | | |
| **, *** Significant at p<0.01, 0.001 respectively. | | | | | | | |

In the spring experiment, the overall plant growth was greater in M.SDI than C.SDI (p<0.001) (Table 10). The crop factor of 0.8 gave significantly higher fresh weight of 21.9 g/plant (p<0.001) compared with the 0.4 crop factor (17.2 g/plant). Significant increases in lettuce fresh weight in response to irrigation frequency were evident. The interaction between tape type and irrigation frequency was significant (p<0.01), with four irrigations per day and the M.SDI giving the highest fresh weight of 25.0 g/plant.

Plant fresh weight responded to the plant's position in relation to the drip tape. Fresh weight in the conventional SDI varied significantly (p<0.05) between positions, with weight being greater in the center of the bed between two drip lines (19.2 g/plant) compared with either side of the drip line (16.2 and 16.6 g/plant) (Table 11). The M.SDI had uniform growth of plants across all positions.

*5.3.2.5 LETTUCE DRY WEIGHT*

In the autumn experiment, plant dry weight increased at the greatest irrigation frequency (4/day), with mean values increasing from 1.6 g/plant to 2.0 g/plant (p<0.001) from one irrigation per day to four irrigations per day (Table 12). The mean weight in the modified SDI treatment was 1.8 g/plant, which was significantly greater than the 1.6 g/plant in the conventional SDI (p<0.001). The interaction between tape type and irrigation frequency was not significant (p>0.05).

**TABLE 11** The effect of plant position, in relation to the drip line, on lettuce fresh weight (g/plant) at 14 DAT (spring 2007).

| Plant position | M.SDI | C.SDI |
|---|---|---|
| One side of drip line | 21.7 | 16.2 |
| Center of 2 drip lines | 22.0 | 19.2 |
| Other side of drip line | 21.6 | 16.6 |
| **Mean** | **NS** | ***** |

*** Significant at p<0.001.

Plant dry weight in the spring experiment responded similarly to the autumn experiment, with the main effects of tape type and irrigation frequency both being highly significant (Table 13). However, in the spring experiment, the interaction between tape type and irrigation frequency was also significant ($p < 0.001$). In addition, the main effect of irrigation amount (crop factor) was significant ($p < 0.001$). Plant dry weight responded to the plant's position in relation to the drip tape in the same way as fresh weight (Table 14).

**TABLE 12**   The effect of tape type and irrigation frequency on dry weight (g/plant) at 15 DAT (autumn 2007).

| Tape type | Irrigation frequency/day | | | Mean |
|---|---|---|---|---|
| | 1 2 4 | | | |
| M.SDI | 1.7 | 1.7 | 2.0 | 1.8*** |
| C.SDI | 1.4 | 1.5 | 1.9 | 1.6 |
| **Mean** | **1.6** | **1.6** | **2.0*** | |

*** Significant at $p < 0.001$.

**TABLE 13**   The effects of tape type, irrigation frequency (IF) and crop factor (CF) on dry weight (g/plant) during spring 2007.

| IF | M.SDI | | C.SDI | | Mean I.F. | Mean Tape x I.F. | |
|---|---|---|---|---|---|---|---|
| (Irri./day) | 0.4 0.8 | | 0.4 0.8 | | | M.SDI C.SDI | |
| One | 1.5 | 1.9 | 1.2 | 1.6 | 1.7 | 1.7 | 1.4 |
| Two | 1.6 | 1.7 | 1.4 | 1.5 | 1.6 | 1.7 | 1.4 |
| Four | 1.9 | 2.1 | 1.4 | 1.4 | 1.7** | 2.0 *** | 1.4 |
| Mean Tape | 1.8 *** | | 1.4 | | | | |
| Mean C.F. | 1.5*** for 0.4 C.F | | 1.7 for 0.8 C.F | | | | |

All main effects and the tape x IF interaction were significant.

**, *** Significant at $p < 0.01$, $0.001$, respectively.

**TABLE 14**   The effect of plant position, in relation to the drip line, on lettuce dry weight (g/plant) at 15 DAT (spring 2007).

| Plant Position | M.SDI | C.SDI |
|---|---|---|
| One side of drip line | 1.8 | 1.2 |
| Center between 2 drip lines | 1.8 | 1.6 |
| Other side of drip line | 1.8 | 1.4 |
| **Mean** | **NS** | * |

* Significant at $p < 0.05$.

**TABLE 15** The effects of tape type and irrigation frequency on volumetric soil water content (v/v) during establishment (autumn 2007) after irrigation.

| Tape type | First week after treatment commenced, Irrigation frequency/day | | | Second week after treatment commenced, Irrigation frequency/day | | |
| | 1 | 4 | Mean | 1 | 4 | Mean |
|---|---|---|---|---|---|---|
| M.SDI | 0.192 | 0.246 | 0.219* | 0.180 | 0.251 | 0.215* |
| C.SDI | 0.175 | 0.201 | 0.188 | 0.164 | 0.204 | 0.184 |
| Mean | 0.183 | 0.223** | | 0.172 | 0.228*** | |

*, **, *** Significant at p<0.05, 0.01, 0.001, respectively.

### 5.3.2.6 VOLUMETRIC SOIL WATER CONTENT

Volumetric soil water content, measured *after* irrigation in the autumn trial, responded significantly to tape type and irrigation frequency (Table 15). Soil in the modified SDI treatment was wetter than in conventional SDI at both times of measurement. The effect of irrigation frequency was also significant.

In the spring trial, the volumetric soil water content *before* irrigation responded to crop factor, tape type and irrigation frequency (Table 16). The M.SDI had higher water content (0.05 v/v) than conventional SDI (0.04 v/v). The highest soil water content (0.09 v/v) was recorded in the M.SDI with 0.8 C.F. and four irrigations per day.

**TABLE 16** The effect of tape type, irrigation frequency (IF) and crop factor (CF) on soil water content (v/v) during establishment (spring 2007), before irrigation.

| IF (Irri./day) | M.SDI 0.4 | 0.8 | C.SDI 0.4 | 0.8 | Mean I.F. | Mean C.F. x I.F. 0.4 | 0.8 | Mean Tape x IF M.SDI | C.SDI |
|---|---|---|---|---|---|---|---|---|---|
| One | 0.016 | 0.026 | 0.013 | 0.038 | 0.023 | 0.015 | 0.032 | 0.021 | 0.025 |
| Four | 0.049 | 0.088 | 0.038 | 0.067 | 0.060 | 0.045 | 0.077 | 0.068 | 0.053 |
| | | | | | | *** | | | |
| Mean Tape | 0.045 *** | | 0.039 | | *** | | | *** | |
| Mean C.F. x Tape | 0.032 * | 0.057 | 0.026 | 0.053 | | | | | |
| Mean C.F. | 0.03*** for 0.4 CF | | 0.06 for 0.8 CF | | | | | | |

All main effects and interactions were significant.

*, *** Significant at p<0.05, 0.01 respectively.

**TABLE 17**    The effect of emitter position on soil water content (v/v) (spring, 2007), before irrigation.

| Plant position | M.SDI | C.SDI |
|---|---|---|
| One side of drip line | 0.045 | 0.038 |
| Center between two drip lines | 0.044 | 0.041 |
| Other side of drip line | 0.045 | 0.038 |
| Mean | NS | ** |

** Significant at $p<0.01$, NS – Not Significant.

There was a significant effect of emitter position on soil water content, but only in the C. SDI ($p<0.01$) (Table 17). Higher water content was observed between the drip lines (0.041 v/v) than to either side of them (0.038 v/v).

### 5.3.3    DISCUSSION

#### 5.3.3.1    OVERALL PERFORMANCE

Subsurface drip irrigation (SDI) offers many potential advantages to vegetable growers, but the adoption of SDI has been slow, partly because of increased risk of poor establishment which is related to poor or uneven surface wetting [68]. The two field experiments reported here suggest that modifying SDI with an impermeable layer beneath the drip tape and geotextile above the drip has the potential to improve surface soil water, plant establishment and early growth.

In general, lettuce establishment presently requires surface irrigation to supplement SDI [102]. The shallow rooted lettuce has shown uneven plant establishment during conditions of environmental stress, and it is highly influenced by irrigation management. Seedling establishment is one of the most critical stages of lettuce growth [260]. These considerations suggest that SDI would not be suitable for lettuce. However, the results obtained from the M.SDI (KISSS™) improved soil water, crop establishment and growth compared to C.SDI.

These results were achieved under both low and relatively high evaporative demand (in autumn and spring) and on light-textured soil, which has traditionally presented the greatest difficulties for SDI [159]. On the basis of these results, the modifications to SDI evaluated here do address the main concern with using SDI for vegetables such as lettuce, that is, the poor establishment.

Although the differences in mortality were relatively small, they appear to reflect a lower level of stress in the plants following transplanting, and this resulted in improved growth. Plants in the modified SDI had higher leaf appearance rates (at least in spring), wider and longer leaves, and ultimately had higher fresh and dry weights at the end of the establishment period (15 DAT). The measurements of soil water revealed wetter conditions at the soil surface with the modified SDI.

## 5.3.3.2   COMPARISON BETWEEN THE TYPES OF SDI

In both experiments, the modified SDI system resulted in better survival of the transplanted seedlings. Whilst the difference was numerically small at an average of 99% for M.SDI and 95% for the conventional SDI, it indicated a superior environment for establishment in the M.SDI system.

This superior environment was reflected in higher leaf appearance rates in the spring experiment. In both experiments, leaves were longer and wider. Leaf appearance rates were higher in the spring experiment than in autumn, presumably reflecting the higher temperatures. In direct seeded lettuce, [256] reported a leaf appearance rate of 1–2 leaves/day at 19 days after sowing, which is higher than in either of the two field experiments reported here, but this may reflect differences in temperature. For transplanted seedlings in Australia, the leaf numbers in this study were comparable to the crops reported by other authors [183].

Plant fresh and dry weights were substantially greater with the M.SDI with increases in fresh weight over the conventional SDI of 16% and 25% in the autumn and spring experiments, respectively.

A further advantage of the modified SDI system was that plants were more uniform. If the differences present at 15 DAT were carried through to maturity, this would mean more uniform sizes and harvest dates, which would be a significant advantage for marketing [24]. With plants harvested in the vegetative stage, it is usual for differences established early in development to continue through until harvest [142]. Whilst delaying harvest of smaller plants may result in more comparable plant weights, this is undesirable as it increases the duration of harvest and harvest costs. In some plants, a delay will mean plants progress into reproductive development and are not harvestable.

The improved plant establishment and growth in the M.SDI system was very likely due to improved plant water status rather than any other factor, although only soil moisture was measured. In the autumn experiment, soil water content in the surface soil after irrigation averaged 21.7% in the modified SDI compared with 18.6% in the conventional SDI. The difference between the tape types was greatest with the high irrigation frequency (4/day), in which soil water content was 5% higher than with the conventional SDI.

In the spring experiment, the soil water was measured just before irrigation began as a measure of the driest conditions, or the maximum stress, that newly transplanted seedlings might have encountered. The difference between the tape types overall was small, although statistically significant. Soil in the modified SDI was consistently wetter than with the conventional SDI at comparable irrigation frequencies and crop factors (irrigation amounts).

The best evidence that the improved growth of the M.SDI was related to improved surface water lies in the close relationship between plant weight and soil water, regardless of the source of variation in soil water: tape type, crop factor, or irrigation frequency, or location within the plot.

The finding that an impermeable layer under the SDI improved surface water confirms the earlier findings [49, 67] using an early version of M.SDI ('CRZI') in lighter-textured soil. Also, the researchers found no horticultural benefit from the improved surface water with the modified SDI. This was because, even with the poor wetting of the C.SDI, crop establishment and growth in their experiments was satisfactory given the hydraulic properties of the particular soil.

Whilst the present experiments establish that the new modified SDI (KISSS™) is superior to C.SDI, it does not establish whether product development since the original modified SDI product (CRZI) is responsible for the improved performance. However, other investigators have shown that an impermeable layer improve surface wetting of SDI. Any additional advantage of M.SDI (KISSS™) may lie in the geotextile fabric, which has now been included [17, 33, 271]. This will presumably hold more water against drainage and prolong any upward flux to the soil surface.

### 5.3.3.3   RESPONSES TO CROP FACTOR AND IRRIGATION FREQUENCY

Plants responded to both increased crop factor (CF), increased irrigation depth, and irrigation frequency (IF). However, for every combination of CF and IF, the growth of plants with the modified SDI (KISSS™) was greater than with conventional SDI. The modified SDI was as effective in improving surface soil water, compared with conventional SDI, as increasing irrigation amount and frequency. Under any combination of irrigation frequency and crop factor, the modified SDI (KISSS™) was better than the conventional SDI.

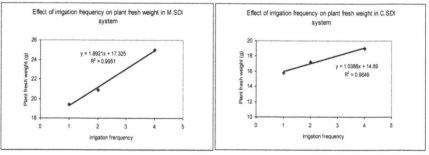

**FIGURE 6**   Plant fresh weight response to irrigation frequency in modified SDI (left) and conventional SDI (right)

The response to irrigation frequency is summarized in Fig. 6. Not only did the M.SDI gave greater plant weight overall, but the response to irrigation frequency was greater than with the C.SDI. The greater response to increased irrigation frequency in the M.SDI cannot be explained with certainty, but it is likely that with smaller volumes of water applied in each irrigation, there is increasing probability of all of the water being retained above the impermeable layer, within the geotextile.

Other studies have shown that short, frequent pulses of water are required to maintain optimal soil water regime on sandy soils [83]. Greatest water use efficiency was found at frequent intervals with drip irrigation [79]. Similarly, more frequent water application through SDI produced better results in vegetable production [74]. Thus more frequent irrigation has been shown to improve lettuce performance [240]. One guiding principle in micro irrigation on sandy soils is to water frequently for good plant establishment, possibly four irrigations per day [153].

Although the response to irrigation frequency reported here is consistent with studies by other authors, yet it appears that the M.SDI responded more to the greater irrigation frequency. It is significant that, despite the improved surface water status with the modified SDI (Tables 15 and 16, Fig. 6), further improvements were possible with increased irrigation frequency while providing the same amount of irrigation overall.

The data for soil water are consistent with the hypothesis that a fraction of the water applied at each irrigation is retained in a temporary water table which forms along the drip tape, above the impermeable layer of the modified SDI, at least in the KISSS™ product in which geo-textile has been included. A water table will increase upward flux of water towards the soil surface. This benefit will persist after irrigation until the water table has been depleted [116]. The greater the frequency of irrigation in a period of time, the more of the total irrigation water applied that is available for transfer from the water table to the soil surface for evapotranspiration, and less water is available for drainage.

In the spring experiment, when evaporative demand was relatively high (5.8 mm/day), both soil water and plant weight in the modified SDI system responded to increasing the amount of irrigation depth. The implication is that even with the modified SDI (KISSS™), soil water may limit crop establishment under high evaporative demand, unless the crop is overwatered. Frequent irrigation may not be adequate, despite the benefit under these conditions (Fig. 6). That is, there appears to be an additional advantage in scheduling or regulating irrigation by using a higher crop factor (0.8) than suggested by the low leaf area of transplants [124].

Where it is imperative to reduce drainage and improve irrigation efficiency, the modified SDI (KISSS™) offers distinct advantages over the conventional SDI, even if more frequent irrigation is required.

## 5.3.4   CONCLUSIONS

Generally the modified SDI performed better than the conventional SDI in both seasons. For shallow rooted crops like lettuce, the modified SDI had a positive effect on soil moisture and plant performance. The soil water response was consistent with the hypothesis that creating a localized water table encourages upward capillary water movement, maintaining more favorable conditions in the root zone. The crop establishment, leaf appearance rate, leaf size and plant fresh weights were all higher in the modified SDI than conventional SDI. Frequent irrigation in the modified SDI further improved lettuce establishment in sandy soil. Under more extreme evaporative demand, it may be necessary to increase the amount of irrigation water applied. Whether this is required or not, it is clear that the water requirement of modified SDI will be lower than conventional SDI.

## 5.4   EFFECT OF MODIFIED SDI AND IRRIGATION FREQUENCY ON SOIL MOISTURE AND COMPONENTS OF THE SOIL WATER BALANCE

### 5.4.1   THEORETICAL BASIS

To improve irrigation management under SDI, greater understanding of saturated and unsaturated hydraulic conductivity and upward flux or capillary rise of water is needed. Evaporation from soil depends on supply of energy and removal of vapor, which together determine evaporative demand of the atmosphere and on continual supply of water [116]. Where a water table occurs close to the soil surface, steady-state flow may take place from the saturated zone beneath, through the unsaturated layer to the surface and thus evaporation continues without changing soil water content. The evaporation rate with a water table present depends on soil properties and depth of water table and it increases with increase in the suction at the soil surface [81, 82]. When a shallow water table is present, the external environmental conditions may influence evaporative demands [116]. So the relationship between water content and matric potential is important to understand water flow in soil [189]. In this section, we will discuss the upward flux (evaporation from bare soil), soil water content and soil water potential.

## 5.4.2   CONSTRUCTION OF A GLASSHOUSE APPARATUS

In review of literature of this chapter, it is proposed that an impervious layer beneath the drip line in sub surface drip irrigation can increase the upward flux of water, to replace water lost by soil evaporation or uptake by newly transplanted seedlings. Therefore, if the irrigation amount remains unchanged, then drainage will be reduced, and surface soil water content is increased, by such a layer. It is further proposed that the rate of water movement will depend on soil physical properties and evaporative demand. Therefore, a glasshouse experiment was designed to quantify the water balance components and effects on soil water content of placing an impermeable layer beneath the drip line. This section describes the development and construction of the glasshouse apparatus.

Under field conditions, it is usual for soil to drain under gravity unless the soil profile is water-logged [117]. Therefore, when the soil water may be at zero suction at the point where water is applied, there will be a downward suction gradient [117, 208]. The apparatus described here provided a hydraulic gradient in pots, in which the effects of SDI design can be quantified under controlled conditions. To provide this hydraulic gradient, a 'tension table' [218] connected to a 'hanging water' column was made at the bottom of each pot.

**The experimental apparatus** consisted of a water inflow (irrigation) and outflow (drainage) system and hanging water column including burette (Fig. 7). A tension table made of silica flour was used to maintain negative water potential at the bottom of each pot. The pots were plastic boxes 50 x 35 x 30 cm, each containing 25 cm depth of soil above a tension table. The tension table was connected with tubing to hanging water column with a burette to indicate the suction in the tension table as the burette was raised or lowered. The tubing also served to collect drainage from the soil in the pot, as described here. The irrigation setup consisted of irrigation main line, sub mains, connectors to each pot, and drip tape within each pot. The water pressure was maintained at 20 psi (138 kPa) using a pressure gauge in the main line.

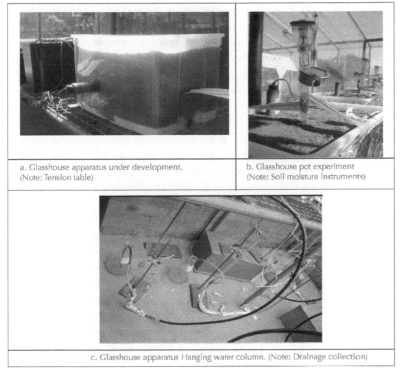

a. Glasshouse apparatus under development.
(Note: Tension table)

b. Glasshouse pot experiment
(Note: Soil moisture instruments)

c. Glasshouse apparatus Hanging water column. (Note: Drainage collection)

**FIGURE 7** Glass house apparatus. (a) Glasshouse apparatus under development. (Note: Tension table); (b) Glasshouse pot experiment. (Note: Soil moisture instruments); (c) Glasshouse apparatus – Hanging water column. (Note: Drainage collection).

As described in the materials and methods of this section, no plants were present. Irrigation amount was estimated from the duration of irrigation and the irrigation rate, which was measured at 138 kPa. Evaporation from the Soil (E) was estimated using: $E = [I - D]$, where: I is the amount of irrigation water applied and D is the drainage water collected.

Silica 'flour' was used to create the tension table. There are different grades of silica flour available: 60, 80, 100 and 120. All four grades were tested in a Haines apparatus, also referred as a Buchner funnel [91], including a hanging water column. The objective was to choose a grade of silica which (i) remained saturated, and therefore, held suctions, at up to about −100 kPa, and (ii) had high conductivity at this suction so that the tension table and soil above it responded quickly to changes in suction when the burette was raised or lowered.

Coarse silica (60G) was selected for the tension table. This porous material (silica 60G) was used to equilibrate the water in the soil sample. Using an external supply of water (hanging water column) at the desired suction, equilibrium was attained in the tension table. The tension table also allowed drainage from the pot.

**Hanging water column:** The tension table was connected to a hanging water column which includes a long rubber tube attached to a burette. The soil was slowly wetted through the burette until the pot was saturated at the surface. Then the water level in the burette was lowered to give a suction of 60 cm which was maintained finally by lowering the level of the water in the burette. Excess water from the soil flowed out of the burette until equilibrium was reached at −60 cm (~5.9 kPa). After reaching equilibrium, any drainage following irrigation the excess water in the tension table was collected through the burette. The drainage from the drip tube was collected in a separate bottle.

Maintenance of experimental setup: The experimental setup was kept in a controlled environment in a glasshouse. In the system, the sub mains, the hanging water column tube and drainage pipe were black plastic pipes to avoid algal growth inside the pipe. The pots were covered with aluminum foil to prevent algal growth. The suction pipe of the hanging water column was checked daily and maintained without air.

### 5.4.3   INTRODUCTION

Efficient and environmentally sound irrigation requires management of the water balance. Drainage and runoff losses and soil evaporation must be minimized, and transpiration maximized. The type of irrigation system, and its management, can both affect each of these components of the water balance. A review of literature in this chapter identified the soil water balance approach as one of the methods to schedule or manage irrigation to minimize undesirable deep drainage and runoff, while satisfying crop water demand [42, 47, 124, 147, 278]. In order to improve plant establishment with SDI, farmers may overirrigate [124] leading to increased drainage, or install the tape at shallow depth and water more frequently [42].

The field experiments in this chapter showed that modified SDI (M.SDI) had higher surface soil water, more even water distribution and better crop establishment than conventional SDI (C.SDI) with the same irrigation amount and frequency. Even when only half the amount of irrigation water was applied to the M SDI (0.4 CF, spring experiment), crop establishment was as good as with C.SDI irrigated at the higher CF (0.8). The results with M.SDI imply greater upward flux and less downward flux of water, although neither component of the water balance could be measured. Greater upward flux may increase soil evaporation, but it may also provide a more favorable soil moisture environment for establishing plants and lead to water savings.

Increased irrigation frequency in the field experiment had no effect on establishment, but it did result in wetter soil surfaces and improved seedling growth with both types of drip tape. For the spring experiment, the response was greater with the modified tape (the interaction was significant). A response to irrigation frequency has been reported previously [42], but the interaction found in the field experiment suggests that irrigation frequency may need to be further increased to obtain the full benefit of the modification to the SDI. This important observation requires further work for confirmation.

Water flowing out from a buried emitter moves vertically and laterally to wet a volume of soil [279]. Knowing the dynamics of water within the soil volume surrounding the emitter creates the opportunity to design improved irrigation systems as well as improve management of both water and chemicals [279]. For a given soil, knowing the dynamics of water in the wetted volume can help to determine emitter spacing and the duration of irrigation. The shape and the dimensions of the wetting pattern around a buried source depend on the soil type as well as on the applied volume of water [209] and the frequency of application [199].

This glass house experiment quantified the components of the water balance under irrigation with conventional and modified subsurface drip irrigation, and over a range of irrigation frequencies, in two soil types. The apparatus enabled irrigation amount to be controlled, drainage to be quantified, and evaporation to be calculated. The soil water content above the emitter and soil water potential near the soil surface were also measured.

In this section, the author tests the hypothesis that an impermeable layer under the drip tape reduces drainage and, given the same irrigation amount, increases the upward flux of water resulting in wetter surface soil and higher evaporation. The author also assessed the soil water response to irrigation frequency.

## 5.4.4  MATERIALS AND METHODS

### 5.4.4.1  SITE
The experiment was conducted in a controlled temperature glasshouse at UWS, Hawkesbury Campus. The experiment was carried out without plants, so the only upward flux of water was by soil evaporation.

### 5.4.4.2  EXPERIMENTAL DESIGN
The treatments consisted of factorial combinations of two drip tape types and two soil types with two replicates arranges in a randomized complete block design. The two tape types were modified subsurface drip ($T_1$) and conventional subsurface drip ($T_2$). The two soil types were sand ($S_1$) and sandy loam ($S_2$). Each treatment was subjected to a sequence of different irrigation frequencies, one per two days; and one, two and four per day ($I_{0.5}$, $I_1$, $I_2$, $I_4$). There was insufficient glasshouse space to run the irrigation frequencies concurrently.

There were three phases of experimentation. The first phase used a fixed irrigation rate (mm/day), which did not vary despite small variation in evaporation from day to day. The second phase used a fixed crop factor with varying irrigation rate, depending upon the previous day's evaporation. Phases 1 and 2 were both under a low evaporative demand. The third phase also used a fixed crop factor and varying irrigation rate, but with high evaporation.

### 5.4.4.3  SOIL TYPES
Soils were collected from two different locations on the UWS Hawkesbury campus. Soil particle size analysis was determined by the hydrometer method [136]. The soils were sieved and placed into the pots in the glass house apparatus (Table 18). After filling the pots with soil, they were brought to saturation by subsoil watering, and then drained to −60 cm suction. Weed seeds were encouraged to germinate by sprinkling water for a few days before applying glyphosate (360 g a.i./L).

**TABLE 18**   Particle sizes of the sand and sandy loam soils.

| Texture | Fraction | Composition (%) |
|---------|----------|-----------------|
| Coarse Sandy soil | Sand | 91.5 |
| Sandy Loam soil | Silt | 8.0 |
| | Clay | 0.5 |
| | Sand | 85.5 |
| | Silt | 6.0 |
| | Clay | 8.5 |

### 5.4.4.4   TAPE TYPES

Two tape types (M.SDI and C.SDI) were compared. Both tape types were installed at 15 cm depth as they were installed in the field experiments. Drip tape emitters had 1.6 Lph flow rate at 20 psi (138 kPa) pressure.

### 5.4.4.5   IRRIGATION RATE AND EVAPORATION ($E_{PAN}$) MEASUREMENTS

In Phase 1, irrigation rate was fixed at 5.3 mm/day (800 mL/pot/day), which was 0.8 of the pan evaporation of the previous four days in the glasshouse (6.0 mm/day). In Phases 2 and 3, the amount of water required for each irrigation was calculated by multiplying a 'crop factor' by the previous day's pan evaporation in the glasshouse, or over the previous two days with the $I_{0.5}$ treatment. For this phase, the crop factor of 0.8 was the crop factor used in field experiment 2, following the recommendation of Ref. [124] that a high crop factor be used for crop establishment using seedling transplants, even if transpiration is low. Having assessed the results of Phase 1 and the field experiments, Phases 2 and 3 used a crop factor of 0.4.

Evaporation in the glass house was measured by using a single pan equal in size to the treatment pots. It was an identical tub to that used in the pots, with the walls covered in aluminum foil. Depth of evaporation was measured daily with a ruler, the volume of evaporation calculated, and then replaced. The phases of the experiment were carried out under a 'low' (Phases 1 and 2) or 'high' (Phase 3) evaporative demand achieved by varying temperature from 25° C to 35° C (with diurnal variation of 5° C). Maximum/minimum and wet /dry bulb thermometers were used to measure temperature and relative humidity daily.

### 5.4.4.6   SOIL WATER MEASUREMENTS

**Theta probe**: Volumetric soil water content was measured by a single Theta probe (model of ML-2x a Delta T-Device from Measurements Engineering Australia) inserted vertically in each pot. The sensor probes were located to read at 5–10 cm depth, 10 cm away from the drip tape. The probes were connected to a data logger (Data logger Tbug, Measurement Engineering Australia) programed to take measurements at 15 min. intervals, which were generally before irrigation, at the time of irrigation and after irrigation.

**Gypsum blocks**: Soil water potential near the soil surface was measured daily, following irrigation and when drainage had ceased, by using gypsum blocks (GB Light, Measurement Engineering Australia) and expressed as kPa. The GB Light is used in all soil types and has a range from 0 to 200 kPa. The sensors were connected to GBug data logger (MEA), which was programd to take readings at two-hour intervals. The blocks were placed in all treatments horizontally at 3–5 cm depth and 10 cm away from drip tape.

***Tensiometer*:** These also measure soil water potential (kPa), but over a narrower range but with greater precision than the gypsum blocks. They were used in slightly deeper soil where only small variations in potential were expected. They were installed with the ceramic cup at 5–10 cm depth, above the drip line and 10 cm away from the emitter. Data were recorded when drainage had ceased.

### 5.4.4.7 WATER BALANCE COMPONENTS

Drainage was measured daily. Water was collected after irrigation when no further drainage occurred (up to 3 h) and weighed. Since the irrigation amount was known and there was no runoff, soil evaporation could be determined simply from the water balance equation: $E_s = [I - D]$. For estimation of $E_s$, a steady-state of soil water content on a day-to-day basis was considered. That is, water moved from the irrigation input to the atmosphere without contributing to changes in soil water. It is an approximation for steady state as soil water content varies diurnally. Daily drainage and soil evaporation were observed until they approached an apparent 'steady state.' Data for the last 3 days were averaged to provide estimates for D and $E_s$ pertaining to any treatment or phase, and were subsequently analyzed.

### 5.4.4.8 STATISTICAL ANALYSIS

Analysis of variance (ANOVA) was conducted on data using MiniTab *ver.* 15 statistical program. An ANOVA on soil water and drainage data was undertaken for each irrigation frequency within each evaporative demand (Phase). The estimates of soil evaporation were analyzed by linear regression to examine the effect of irrigation frequency.

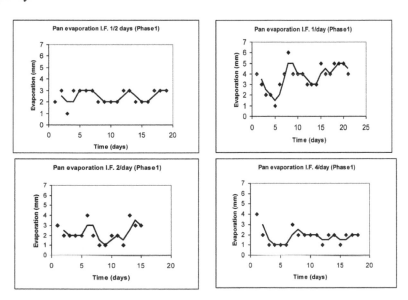

**FIGURE 8** *(Continued)*

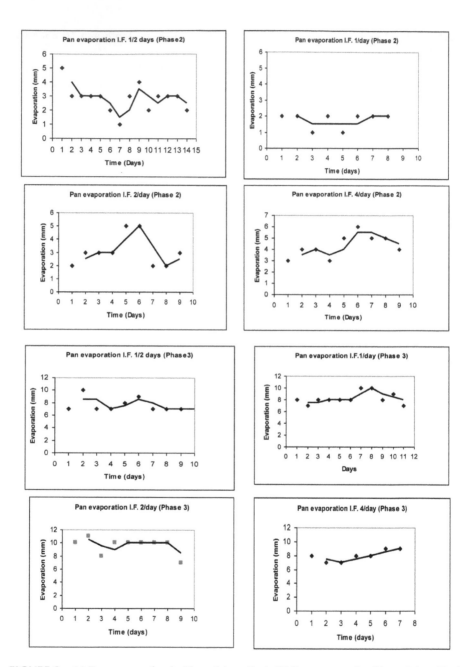

**FIGURE 8**    (a) Pan evaporation in Phase 1 (mm/day). (b) Pan evaporation Phase 2 (mm/day). (c) Pan evaporation in Phase 3 (mm/day).

## 5.4.5 RESULTS

### 5.4.5.1 PAN EVAPORATION IN THE GLASSHOUSE

Figure 8 shows the moving average for evaporation in all treatments. Evaporation in last few days of each 'run' coincides with estimates of soil evaporation or drainage at approximate 'steady state.' In the last three days, evaporation was highest in the 1/day and lowest in 4/day treatments. Evaporation tended to be lower in the 1/day and higher in the 2/day treatments.

### 5.4.5.2 DRAINAGE

In all three phases of the experiment, the mean drainage at 'steady state' for the modified SDI was less than for conventional SDI. Soil types generally did not show significant differences.

**Phase 1**: In this phase of the experiment, drainage from the different irrigation frequency treatments was compared with equal application of irrigation water to all treatments on every day. Within each irrigation frequency, the modified SDI gave significantly less drainage than in the conventional SDI treatment (Table 19). Within irrigation frequency, the effects of soil type and its interaction with tape type were not significant ($p > 0.05$). Over all irrigation frequencies, drainage was 20% less in the modified SDI (from Table 19). Higher irrigation frequency (four/day) in the modified SDI resulted in least drainage (194 mL/pot/day). Less frequently irrigated treatments overall resulted in the highest drainage, but the statistical significance of this difference could not be tested.

**Phase 2:** In this phase, irrigation amount varied according to daily pan evaporation, in contrast to Phase 1 in which irrigation amounts were fixed. Despite the change in irrigation scheduling strategy, the modified SDI again had significantly less drainage than conventional SDI within any of the irrigation frequencies. Overall, drainage was 29% less in the modified SDI than in conventional SDI (calculated from Table 20). Once again, drainage was least with the high irrigation frequency.

**Phase 3**: Drainage was again significantly less in the modified SDI than the conventional SDI at all irrigation frequencies. Averaged over all irrigation frequencies, drainage was 32% less under modified SDI than with the conventional SDI (from Table 21). Least drainage (230 mL/pot/day) was obtained in the modified SDI with frequent irrigation (four/day). Again, high irrigation frequency substantially reduced drainage.

**TABLE 19** The effects of tape type, irrigation frequency and soil type on drainage (mL/pot/day) (Phase 1) at steady state.

| I.F. (Irri./day) | Sand | | Sandy loam | | Mean M.SDI | Mean C.SDI |
|---|---|---|---|---|---|---|
| | M.SDI | C.SDI | M.SDI | C.SDI | | |
| One/2 days | 257 | 326 | 279 | 330 | 268** | 328 |
| One/day | 251 | 323 | 226 | 328 | 239*** | 326 |
| Two/day | 269 | 318 | 272 | 322 | 270 NS | 320 |
| Four/day | 194 | 256 | 193 | 248 | 194** | 252 |

**, *** Significant $p < 0.01$, $0.001$ respectively (Each frequency analyzed as a single experiment).

**TABLE 20** The effects of tape type, irrigation frequency and soil type on drainage (mL/pot/day) (Phase 2) at steady state.

| I.F. (Irri./day) | Sand | | Sandy loam | | Mean M.SDI | Mean C.SDI |
|---|---|---|---|---|---|---|
| | M.SDI | C.SDI | M.SDI | C.SDI | | |
| One/2 days | 365 | 456 | 352 | 455 | 358*** | 455 |
| One/day | 122 | 168 | 116 | 161 | 119** | 164 |
| Two/day | 116 | 153 | 112 | 154 | 114*** | 154 |
| Four/day | 196 | 333 | 203 | 342 | 199*** | 337 |

**, *** Significant at $p<0.01$, $0.001$ respectively (Each frequency analyzed as a single experiment).

**TABLE 21** The effects of tape type, irrigation frequency and soil type on drainage (mL/pot/day) (Phase 3) at steady state.

| I.F. (Irri./day) | Sand | | Sandy loam | | Mean M.SDI | Mean C.SDI |
|---|---|---|---|---|---|---|
| | M.SDI | C.SDI | M.SDI | C.SDI | | |
| One/2 days | 318 | 445 | 315 | 434 | 317*** | 439 |
| One/day | 278 | 453 | 261 | 454 | 270*** | 454 |
| Two/day | 372 | 490 | 370 | 489 | 371*** | 489 |
| Four/day | 233 | 350 | 227 | 348 | 230*** | 349 |

*** Significant at $p<0.001$ (Each frequency analyzed as a single experiment).

### 5.4.5.3 EVAPORATION FROM BARE SOIL

The results are shown in Table 22 for all phases of the experiment, together with the irrigation and drainage data from which they were derived. All data are reported as mm/day, after dividing the volumetric data by the surface area of the pot. As there was no apparent effect of soil type on drainage, soil evaporation is the pooled data for soil types.

Because irrigation amount is constant across treatments for any given period, the calculated $E_{soil}$ values must be the inverse of drainage. Evaporation was thus always higher for the M.SDI than C.SDI. It also tended to be higher at the higher irrigation frequencies. The response to irrigation frequency is shown in Fig. 9. In each phase of the experiment, $E_{soil}$ increased with irrigation frequency. All of the regressions were statistically significant. Whilst soil evaporation from both tape types responded to increasing frequency, it is notable that within any Phase, the greatest upward flux from C.SDI (at four irrigations/day) was no higher than with the M.SDI with two or fewer irrigations per day.

### 5.4.5.4 VOLUMETRIC SOIL WATER CONTENT

The soil water content was measured after irrigation at 10 cm above and 10 cm to the side of the emitter. The data reported in Tables 23–25 were measured after irrigation, for all three phases of the experiment.

The modified SDI generally had higher soil moisture content than conventional SDI, although the difference was not always significant, and there were exceptions in Phase 1. The most consistent response was an increase in water content with modified SDI at the highest irrigation frequency, where the difference between tape types was always significant. There was no significant effect of soil type.

***Phase 1***: The soil water content averaged over soil types was significantly higher (p<0.05) with modified SDI (31%) than with conventional SDI (27%) but only with frequent irrigation (four/day). At lower irrigation frequencies, the trend was for the modified SDI to have lower water content than conventional SDI, but the differences were not significant. Soil type did not show any significant difference between the irrigation frequencies.

***Phase 2***: The soil water content was consistently higher in the modified SDI compared with the conventional SDI, but the differences were statistically significant only with the most frequent irrigation (Table 24). Again, soil type did not show any significant difference in the soil water content.

***Phase 3***: Soil water content was again higher in the modified SDI than the conventional SDI in all irrigation frequencies (Table 25). The difference was statistically significant (p<0.05) in the once/two days and four times daily treatments.

**TABLE 22**  Water balance components at 'steady-state' for the three phases of the glasshouse experiment: pan evaporation ($E_{pan}$), irrigation (I), drainage (D) and soil evaporation (averaged over replicates and soil types).

| Irr./day | Av. $E_{pan}$ (mm/d) | Av. I (mm/d) | Av. D (mm/d) | | Av. $E_{soil}$ (mm/d) | |
|---|---|---|---|---|---|---|
| | | | **M.SDI** | **C.SDI** | **M.SDI** | **C.SDI** |
| | | | *Phase 1* | | | |
| 0.5 | 3.0 | 4.6 | 1.5 | 1.9 | 3.0 | 2.7 |
| 1 | 4.7 | 4.6 | 1.4 | 1.9 | 3.2 | 2.7 |
| 2 | 3.2 | 4.6 | 1.6 | 1.8 | 3.0 | 2.7 |
| 4 | 2.0 | 4.6 | 1.1 | 1.4 | 3.5 | 3.1 |
| | | | *Phase 2* | | | |
| 0.5 | 2.7 | 1.6 | 0.9 | 1.2 | 0.7 | 0.4 |
| 1 | 2.0 | 1.2 | 0.7 | 0.9 | 0.5 | 0.2 |
| 2 | 2.3 | 1.4 | 0.4 | 0.8 | 1.0 | 0.6 |
| 4 | 4.7 | 2.8 | 1.1 | 1.9 | 1.7 | 0.9 |
| | | | *Phase 3* | | | |
| 0.5 | 7.0 | 4.0 | 1.8 | 2.5 | 2.2 | 1.4 |
| 1 | 8.0 | 4.6 | 1.5 | 2.6 | 3.0 | 2.0 |
| 2 | 9.0 | 5.2 | 2.1 | 2.8 | 3.1 | 2.4 |
| 4 | 8.7 | 5.0 | 1.3 | 2.0 | 3.7 | 3.0 |

**TABLE 23**  The effects of tape type, irrigation frequency and soil type on soil water content (%) (Phase 1) at steady state.

| I.F. (Irri./day) | Sand | | S. Loam | | Mean M.SDI | Mean C.SDI |
|---|---|---|---|---|---|---|
| | M.SDI | C.SDI | M.SDI | C.SDI | | |
| One/2 days | 28.8 | 31.5 | 27.3 | 26.7 | 28.1NS | 29.1 |
| One/day | 25.1 | 30.8 | 27.2 | 27.3 | 26.1 NS | 29.1 |
| Two/day | 26.2 | 30.2 | 29.7 | 24.3 | 27.9 NS | 27.2 |
| Four/day | 31.9 | 27.8 | 30.3 | 25.9 | 31.1* | 26.9 |

* Significant at p<0.05, NS-Not Significant (Each frequency analyzed as a single experiment).

**TABLE 24**  The effects of tape type, irrigation frequency and soil type on soil water content (%) (Phase 2) at steady state.

| I.F. (Irri./day) | Sand | | Sandy loam | | Mean M.SDI | Mean C.SDI |
|---|---|---|---|---|---|---|
| | M.SDI | C.SDI | M.SDI | C.SDI | | |
| One/2 days | 27.4 | 26.4 | 30.2 | 26.8 | 28.8 NS | 26.6 |
| One/day | 29.2 | 28.3 | 28.9 | 28.7 | 29.1 NS | 28.5 |
| Two/day | 28.1 | 27.4 | 28.8 | 28.6 | 28.4 NS | 27.9 |
| Four/day | 29.7 | 27.5 | 30.6 | 27.1 | 30.1* | 27.3 |

* Significant at p<0.05, NS-Not Significant (Each frequency analyzed as a single experiment).

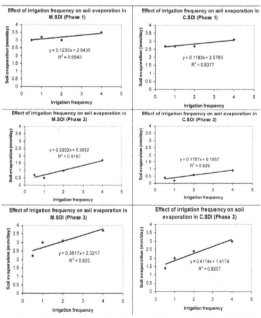

**FIGURE 9**  Effects of tape types and irrigation frequencies on mean soil evaporation (mm) from bare soil at 'steady sate.'

**TABLE 25** The effects of tape type, irrigation frequency and soil type on soil water content (%) (Phase 3) at steady state.

| I.F. (Irri./day) | Sand | | Sandy loam | | Mean M.SDI | Mean C.SDI |
|---|---|---|---|---|---|---|
| | M.SDI | C.SDI | M.SDI | C.SDI | | |
| One/2 days | 26.2 | 24.7 | 30.9 | 26.1 | 28.6* | 25.4 |
| One/day | 27.4 | 26.6 | 29.3 | 28.1 | 28.3 NS | 27.3 |
| Two/day | 27.1 | 25.4 | 33.9 | 30.1 | 30.5 NS | 27.7 |
| Four/day | 27.5 | 25.9 | 34.1 | 29.1 | 30.8* | 27.5 |

* Significant at p<0.05, NS-Not Significant (Each frequency analyzed as a single experiment).

### 5.4.5.5   SOIL WATER POTENTIAL AT 3–5 CM DEPTH

This was measured daily, following irrigation, using gypsum blocks (Tables 26–28). This near-surface measurement is the depth of transplanting for lettuce.

*Phase 1*: Responses in water potential to tape type were small but broadly mirrored those for volumetric water content. The effect of tape type was significant (p<0.01) in the high irrigation frequency (Table 26). Wetter surface soil (-7 kPa) was obtained in the modified SDI with four irrigations per day, compared with the conventional SDI (-9 kPa). The interaction between soil type and tape type was also significant (p<0.01) in the high irrigation frequency. As irrigation frequency decreased, the soil near the surface became drier in both tape types, and the effect of tape type was also not significant.

*Phase 2*: Significant (p<0.01) but contrasting responses were found between tape types (Table 27). In the treatment with four irrigations per day, modified SDI soil was wetter, whereas the converse was true with one irrigation/day.

*Phase 3*: Soil was slightly wetter in the modified SDI compared with conventional SDI in all irrigation frequencies, but the tape difference was significant only with high irrigation frequency (four/day) (Table 28). The effect of soil types was not significant.

**TABLE 26** The effects of tape type, irrigation frequency and soil type on soil water potential (-kPa) (Phase 1) at steady state (3–5 cm depth).

| I.F. (Irri./day) | Sand | | Sandy loam | | Mean M.SDI | Mean C.SDI |
|---|---|---|---|---|---|---|
| | M.SDI | C.SDI | M.SDI | C.SDI | | |
| One/2 days | 14 | 11 | 11 | 10 | 12 NS | 10 |
| One/day | 13 | 13 | 9 | 12 | 11 NS | 13 |
| Two/day | 10 | 10 | 9 | 8 | 10 NS | 9 |
| Four/day | 7 | 10 | 8 | 9 | 7 ** | 9 |

** Significant at p<0.01, NS-Not significant (Each frequency analyzed as a single experiment).

**TABLE 27** The effects of tape type, irrigation frequency and soil type on soil water potential (-kPa) (Phase 2) at steady state (3–5 cm depth).

| I.F. (Irri./day) | Sand | | Sandy loam | | Mean M.SDI | Mean C.SDI |
|---|---|---|---|---|---|---|
| | M.SDI | C.SDI | M.SDI | C.SDI | | |
| One/2 days | 8 | 11 | 11 | 11 | 9 NS | 11 |
| One/day | 15 | 11 | 10 | 10 | 12** | 10 |
| Two/day | 12 | 11 | 12 | 11 | 12 NS | 11 |
| Four/day | 8 | 11 | 11 | 10 | 9** | 11 |

** Significant at p<0.01, NS-Not significant (Each frequency analyzed as a single experiment).

**TABLE 28** The effects of tape type, irrigation frequency and soil type on soil water potential (-kPa) (Phase 3) at steady state (3–5 cm depth).

| I.F. (Irri./day) | Sand | | Sandy loam | | Mean M.SDI | Mean C.SDI |
|---|---|---|---|---|---|---|
| | M.SDI | C.SDI | M.SDI | C.SDI | | |
| One/2 days | 11 | 12 | 11 | 12 | 11 NS | 12 |
| One/day | 11 | 13 | 11 | 12 | 11 NS | 12 |
| Two/day | 11 | 13 | 12 | 14 | 12 NS | 14 |
| Four/day | 9 | 12 | 11 | 13 | 10** | 12 |

** Significant at p<0.01, NS –Not significant (Each frequency analyzed as a single experiment).

### 5.4.5.6  SOIL WATER POTENTIAL AT 5–10 CM DEPTH

The soil water potential at 5–10 cm depth measured with a tensiometer was less negative (higher soil moisture) with the modified SDI (Tables 29–31). The response was more consistent than with the other two measures of soil water, with no cases of conventional SDI being wetter than the modified SDI, and more of the differences being statistically significant. The effect of soil type on water potential was not significant (p>0.05). There was a trend for the soil to be wetter with the more frequent irrigation, but this could not be tested statistically.

**TABLE 29** The effects of tape type, irrigation frequency and soil type on soil water potential (-kPa) (Phase 1) at steady state (5–10 cm depth).

| I.F. (Irri./day) | Sand | | Sandy loam | | Mean M.SDI | Mean C.SDI |
|---|---|---|---|---|---|---|
| | M.SDI | C.SDI | M.SDI | C.SDI | | |
| One/2 days | 8 | 10 | 8 | 9 | 8** | 9 |
| One/day | 10 | 9 | 9 | 9 | 9 NS | 9 |
| Two/day | 8 | 11 | 7 | 9 | 7** | 10 |
| Four/day | 6 | 8 | 6 | 8 | 6* | 8 |

*, ** Significant at p<0.05, 0.01, NS –Not significant (Each frequency analyzed as a single experiment).

**TABLE 30**  The effects of tape type, irrigation frequency and soil type on soil water potential (-kPa) (Phase 2) at steady state (5–10 cm depth).

| I.F. (Irri./day) | Sand | | Sandy loam | | Mean M.SDI | Mean C.SDI |
|---|---|---|---|---|---|---|
| | M.SDI | C.SDI | M.SDI | C.SDI | | |
| One/2 days | 7 | 9 | 7 | 9 | 7* | 9 |
| One/day | 5 | 8 | 5 | 5 | 5* | 6 |
| Two/day | 5 | 5 | 5 | 5 | 5NS | 5 |
| Four/day | 5 | 8 | 4 | 5 | 4* | 6 |

* Significant at p<0.05, NS – Not Significant (Each frequency analyzed as a single experiment).

**TABLE 31**  The effects of tape type, irrigation frequency and soil type on soil water potential (–kPa) (Phase 3) at steady state (5–10 cm depth).

| I.F. (Irri./day) | Sand | | Sandy loam | | Mean M.SDI | Mean C.SDI |
|---|---|---|---|---|---|---|
| | M.SDI | C.SDI | M.SDI | C.SDI | | |
| One/2 days | 7 | 7 | 6 | 7 | 6 NS | 7 |
| One/day | 6 | 7 | 5 | 7 | 5 * | 7 |
| Two/day | 5 | 8 | 4 | 6 | 5 NS | 7 |
| Four/day | 5 | 7 | 5 | 6 | 5 * | 7 |

* Significant at p<0.05, NS – Not Significant (Each frequency analyzed as a single experiment).

## 5.4.6  DISCUSSION

The results of this experiment reveal the potential for the modified SDI to increase water-use efficiency and improve plant establishment, thus confirming the results of the field experiments. The greater upward flux of water, that is the calculated soil evaporation, was sufficient to keep the soil wetter while at the same time reducing drainage. The improved soil moisture is expected to improve crop growth during establishment, even in sandy soils with frequent irrigation (daily) under vegetable production [72].

### 5.4.6.1  DRAINAGE AND SOIL EVAPORATION

Reducing drainage can improve irrigation efficiency and improve the uniformity of applied water [95]. It empathy should also help to reduce overirrigation [135, 251] and the associated environmental risks. The modified SDI resulted in lower drainage than the conventional SDI with all irrigation frequencies under different evaporative demands, and with different irrigation rates. The average reduction was 20, 29, and 32% in the three Phases, a mean reduction in drainage of 27%. It appears that the reductions in drainage are greatest when evaporative demand is greatest, reflecting higher soil evaporation under these conditions. The results demonstrate that irrigation system design has the potential to manage drainage below the root zone in the way proposed for subsurface drip [13, 65, 202]. More importantly, it shows specifically that modifying the drip tape to include an impermeable layer beneath the tape substantially reduces the drainage found even with conventional SDI.

This is an important conclusion, as the unambiguous results from the controlled conditions of this experiment provide support for field experience where drainage is hard to quantify but sometimes assumed.

High frequency SDI in particular has the potential to reduce drainage [13] in sandy soil [125]. Whilst the data for irrigation frequency could not be combined into a single ANOVA for statistical reasons, there was a large consistent response across tape types (M.SDI and C.SDI), soil types and evaporative conditions (Phases) that collectively support the argument that increased frequency reduces drainage under SDI. The reduction in drainage with increased irrigation frequency was of the same order as the response to tape type.

The relationship between drainage and evaporation has been well documented under subsurface drip irrigation in the field. Generally, for a given input of water, if soil evaporation (or ET) is reduced then drainage must increase [65]. With the controlled conditions of the glasshouse experiment, it was possible to quantify the split between drainage and $E_s$ and its responses to drip type, irrigation frequency or environmental condition. Significantly, increasing irrigation frequency increased soil evaporation (Fig. 8). Over all soil types, tape types and Phases, the increase in frequency from once every second day to four times daily approximately double the amount of water evaporated. This water is available for either $E_s$ or transpiration, when plants are present. Therefore, high irrigation frequency maintained relatively high evaporation rates and kept the soil surface wet [176].

### 5.4.6.2    SOIL WATER

Many authors have emphasized the importance of maintaining relatively constant soil water potential in the range favoring plant growth [197, 199, 207, 253]. This is especially important for crop establishment under SDI [205]. Modified SDI generally had the highest soil moisture content and highest water potential (least negative). The only exception was in Phase 1, at low irrigation frequency, when the conventional SDI had higher soil moisture than the modified SDI. This discrepancy can be accounted for by tunneling seen in the conventional SDI pots. However, in the modified SDI the modification plays an important role to avoid tunneling [178].

The volumetric soil water content and water potential generally increased with irrigation frequency, but this trend could not be tested statistically. It suggests at least that high frequency irrigation maintains higher soil water content, whether under SDI or modified SDI, as others have found [74, 131, 209, 229, 240].

There was also strong indication that the modified SDI combined with high irrigation frequency led to highest water content and least negative potential in the top soil. In every set of soil water data, that is three methods of measuring soil water in three experiment Phases, M.SDI combined with high irrigation frequency had the highest water content.

The soil water response is best illustrated by the volumetric soil water data in Tables 23–25. In these data, the highest water content in C.SDI, under high-frequency irrigation, was approximately the same as in the M.SDI with low frequency irrigation. This result parallels the soil evaporation data, in that greatest upward flux of water seems to be associated with highest soil water content. It also supports the field experiments. That is, while soil water content responded to irrigation frequency with both types of tape, the soil was always wetter with the modified tape. In the field experiment, high frequency irrigation coupled with the modified tape type improved soil moisture and plant fresh weight.

This glasshouse experiment was carried out to develop an understanding of soil physics and irrigation; and to understand how much water (irrigation amount) to apply to rewet the soil, including understanding of the components of the water balance [89, 94, 193].

The barrier under the drip line led to higher water content and potential in the soil above the barrier, and the drainage or downward flux of water was decreased [144, 145]. The KISSS™ product should enable crops to be established with less risk than conventional SDI, without the need for excessive irrigation during establishment. But this may require increased irrigation frequency.

These results suggest that irrigation frequency needs to be adjusted as crops develops. Whilst high upward flux is essential for good plant establishment, it is undesirable for established plants with a low leaf area index. This is because the high upward flux will contribute to wasteful soil evaporation.

### 5.4.6.3 CONCLUSIONS

In this glasshouse study, the two subsurface drip tape types were compared in two soil types, under a range of irrigation frequencies and environmental regimes. Generally, the average soil water content above the emitter under approximate steady state was improved with the modified SDI. Frequent irrigation also appeared to increase surface soil water, especially with the modified SDI. The results of the experiments indicate that the drainage was less under the modified SDI than the conventional SDI and with more frequent irrigation. In other words, the soil evaporation was more in the modified SDI due to improved upward movement of water.

The modified SDI (KISSS™) with frequent irrigation appears to be suitable for the coarse textured soils. The product should enable crops to be established with less risk than conventional SDI, without the need for excessive irrigation before or during establishment.

## 5.5 CONCLUSIONS

Current management of irrigation water in the peri-urban vegetable industry is largely based on surface irrigation, most commonly with overhead sprinklers. Water use efficiency is low and environmental impacts are high [58]. Subsurface drip irrigation has great potential in this situation (e.g., Refs. [51, 212]), but problems of 'tunneling,' variable soil surface water and risky crop establishment need to be addressed [149, 152, 180, 212]. It has been mentioned that despite many successes with SDI [150, 216, 226], germination/establishment remains the major challenge with SDI [213], and that possibly crops cannot be established at all this way [102, 103], meaning that conjunctive irrigation from another source is required along with SDI.

To overcome these problems with SDI, researchers have investigated shallow tape installation and increased irrigation frequency [42, 267]; and modification to the SDI design [49, 271]. None of these approaches has been completely successful, although the wetting pattern of SDI has been influenced by placing a continuous impermeable membrane beneath the drip line, with the aim of inhibiting the downward flux of water from the emitters and providing a broad moisture front rather than a point source [178]. Author hypothesized that this membrane would create a small temporary water table from which the upward flux of water would be greater than in conventional SDI and the drainage less. This hypothesis underpinned the research in this chapter that sought to evaluate a newly developed SDI, which included an impermeable membrane and geotextile (KISSS™).

### 5.5.1  INDUSTRY CONTEXT

The literature review considered issues of vegetable industry related to irrigation, with an emphasis on SDI and the production and irrigation management of lettuce. It was noted that peri-urban vegetable production is expanding in Australia [138] and globally [31], and that there is growing competition for water between vegetable growers and the urban community. The review highlighted increased consumer demand for vegetables in local markets [114]. Rapid growth in the Sydney Basin is associated with an influx of new farmers and a lack of knowledge of irrigation scheduling and management, which is linked to low irrigation efficiency and high environmental impact [58].

Adoption of more efficient systems such as drip irrigation has been low, partly because of a lack of knowledge about adapting systems to the varied soils of the region [58] and the wide range of evaporative demands between cooler and warmer months [14]. The need to replace drip tape after every crop and the likely need for an alternative source of irrigation for establishment are major economic disincentives [52]. However, with the adoption of reduced tillage and semipermanent beds for vegetables [230], there is an opportunity to introduce subsurface drip irrigation with its many advantages, including longer life than surface drip.

It is concluded that SDI is a good option for the vegetable industry, provided the risks of poor plant establishment could be overcome. The industry mainly uses seedling 'transplants' for plant establishment [70], so 'establishment' in the research reported here refers to the survival and early growth of the transplanted seedlings. The research used lettuce, which is the major crop of the Sydney Basin [261].

Past research has shown that the major disadvantage of SDI is the risk of poor establishment [67, 278]. Therefore, the evaluation of the modified SDI (KISSS™) focused on its effects on surface soil water and seedling survival and early growth. Consideration was also given to reducing the drainage, which may be associated with overwatering, which is recommended for good establishment with conventional SDI [99].

**Plant fresh weight in relation to soil water** (autumn 2007-2 tape types, 2 irrigation frequencies)

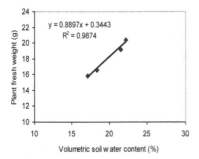

**Plant fresh weight in relation to soil water** (spring 2007-2 tape types, 2 crop factors and 3 irrigation frequencies)

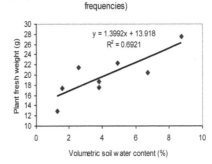

**FIGURE 10** Plant fresh weight response to volumetric soil water content, both seasons.

**Plant weight in relation to soil water** (spring 2007, two tape types and three location within the bed)

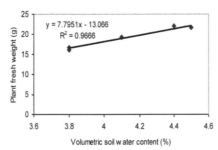

**FIGURE 11** Fresh weight response to variation in volumetric soil water within the bed, spring 2007.

## 5.5.2 MAJOR FINDINGS

### 5.5.2.1 SOIL WATER, CROP ESTABLISHMENT AND EARLY GROWTH

The modification in the subsurface drip tape improved soil water near the soil surface in both field experiments, where pan evaporation ranged between 2 mm/day (autumn) and 6 mm/day (spring), and in two soil types in the glasshouse, where pan evaporation ranged between 2 and 10 mm/day. Soil in the M.SDI was consistently wetter than with the C.SDI at comparable irrigation frequencies and crop factors (irrigation amounts). So the potential advantages of this innovation should be expressed over a range of conditions. In the field, soil water content was also more uniform in the M.SDI treatment. The improved surface soil water regime was associated with improved lettuce crop establishment, and higher leaf appearance rate, leaf size (leaf width and length) and fresh weight. Differences in fresh weight were substantial. The modified SDI system recorded average increases over the conventional SDI of 16% and 25% in the autumn and spring experiments, respectively. If these differences in plant weight continue through to harvest, a reasonable expectation with crops such as lettuce, then the modification to the SDI will result in greater irrigation efficiency (yield per unit of irrigation) to mirror the plant weight response. Management of irrigation with the modified SDI to optimize efficiency is discussed in Section 3.2. Plant weight was also more uniform with the modified SDI than with conventional SDI, offering practical advantages to the farmer at harvest time [228].

The critical importance of surface soil water for establishment is demonstrated by the relationship between soil water content and plant weight at the end of the establishment period in the field experiments (Figs. 10 and 11). In both experiments there was a close relationship between plant weight and soil water, regardless of the source of variation in soil water: tape type, crop factor, or irrigation frequency (Fig. 10) or position in the bed (Fig.11).

In previous work with a modified SDI, the soil was wet more uniformly [178]. The modified drip tape product has undergone extensive development since then, especially in dimensions, and is now sold under a new trade name. This presumably explains why the modified SDI performed better in the present research compared with earlier research [49]. In addition, [49] noted that the Handwood loam soil in their research had good hydraulic properties, which resulted in good crop establishment in C.SDI despite it being drier. Also, they studied seed germination rather than establishment of seedling transplants which may have less critical moisture requirements.

### 5.5.2.2   DRAINAGE

Drainage could not be measured in the field experiment, but the finding that the soil surface was wetter than in conventional SDI, given the same amount of irrigation, suggested that the upward flux was greater and the drainage should be less with the modified SDI. Thus the gain in irrigation efficiency, noted above, arises because more water is transpired (as well as evaporated), and less is drained. Reduced drainage has important environmental implications [122].

The glasshouse experiment provided direct evidence that drainage is reduced in the modified SDI system given the same irrigation and environmental conditions. This was observed regardless of soil type, and in all irrigation frequencies, and under different evaporative demands. Whilst the drainage component of the water balance was significantly reduced, soil evaporation was greater in the modified SDI. Increased soil evaporation is not in itself an indicator of improved irrigation efficiency, but it is important because it means that more of the irrigation water was potentially available for use by establishing seedlings.

Because of poor establishment with conventional SDI, excessive water is often used in an attempt to achieve the near saturation required for germination and establishment [266]. It is not uncommon to find deep drainage losses during the first few irrigations of the season where SDI systems are used to germinate seeds [195]. The present results indicate that the modified SDI evaluated here has the potential to achieve significantly better establishment while also reducing drainage losses.

### 5.5.2.3   IRRIGATION MANAGEMENT: AMOUNT AND FREQUENCY

An objective of this research was to determine if irrigation management with SDI should be varied according to soil type and evaporative demand. From the review of literature, it was concluded that the adequacy of the water supply from SDI to the newly transplanted seedling would be a function of (among other factors) soil hydraulic conductivity (soil type) and evaporative demand [49, 278]. Also, the underlying hypothesis was that the impermeable layer beneath the drip tube would create a temporary water table, and the upward flux of water from a water table will be greater than from unsaturated soil, which is free to drain during and after irrigation [116]. So soil and crop responses to irrigation amount and frequency with modified SDI might be different from earlier studies of these management variables with conventional SDI. Management factors include the amount of irrigation, which is varied according to crop factor (CF), and the irrigation frequency (IF). Both of these were included as treatments in the field experiment, while the glasshouse work focused on irrigation frequency.

In the field experiments, soil water content and plant weight responded to increased crop factor (CF), that is increased irrigation amount, and to increased irrigation frequency (IF). This was so for both the modified SDI and the conventional SDI, the latter being in broad agreement with other studies in light textured soil [13] and particularly on sandy soil [83, 126].

For every combination of CF and IF, the growth of plants with the modified SDI was greater than with conventional SDI. Importantly, at high irrigation frequency (4/day), plant weight with the modified SDI treatment was 10% greater than with conventional SDI, even when given half the amount of irrigation (CF 0.4 *versus* 0.8). When given an equal amount of water (CF 0.8) with frequent irrigation (4/day), the modified SDI resulted in crop fresh weights that were 35% greater than in conventional SDI. This translates directly into improved irrigation efficiency at this stage of the crop. Thus, although the modified SDI with high irrigation frequency had distinct advantages over conventional SDI, including similar seedling growth with half the water, it may still be necessary to have a high *crop factor* as well as frequent irrigation to get the best results.

Results under controlled conditions in the glasshouse confirmed that the upward flux of water to meet the evaporative demand was greater in the modified SDI. They also confirmed that the upward flux was greater with more frequent irrigation. Soil water content and potential were also higher with more frequent irrigation. It is evident with findings of [92]. Over both field and glasshouse experiments, the effect of irrigation frequency was quite consistent, regardless of soil type and evaporative demand, and there was little evidence of an interaction between these variables and tape type. So, despite the predictions from theory that these factors would interact with irrigation management and tape type, in practice the effects were too small to be detected. However, under a wider range of soil types, or under more severe evaporative conditions, differences may emerge. Further research is required to establish the optimum irrigation regime for different soil types and evaporative regimes when using the modified SDI.

The response to frequent irrigation with the modified SDI is in line with the theory that a temporary water table is available for a short period after irrigation, above the impermeable layer. The period is short because the volume of water held is small: If it is assumed that each emitter supplies, a tube 10 cm wide, 50 cm long, to a depth of 3 mm, the volume of water held per irrigation would be 150 cm$^3$. The water requirement in the glasshouse experiment under high evaporation was ~ 800 mL/day, to be supplied by a single emitter.

More frequent irrigation creates this temporary water table more often than less frequent irrigation. It is likely that in higher evaporative demand this temporary reservoir of water is depleted more quickly, necessitating more frequent irrigation. The modified SDI is different from the conventional SDI in that a small amount of water is held against drainage and at zero potential, which allows for a higher rate of upward flux [81].

Frequent water application through SDI has improved water use efficiency as shown by other researchers [74, 240]. These studies suggest that for shallow rooted vegetable crops establishment can be obtained with more frequent irrigations in sandy soil [72].

The results in this research agree with the earlier findings on 'pulse irrigation' [178], but they also show that, with a modified SDI, either fewer pulses, or less irrigation water, may be used to achieve a similar result. Or, in some conditions, even better results may be achieved with a combination of pulse irrigation and a relatively high crop factor.

There were further benefits by increasing irrigation frequency. In the conventional SDI, higher irrigation frequency reduced the severity of tunneling.

The combined field and glasshouse results show that the KISSS™ product should enable crops to be established with less risk than conventional SDI, without the need for excessive irrigation during establishment. However, this may require increased irrigation frequency.

The findings also show that irrigation frequency needs to be adjusted as crops develop. Whilst high upward flux is essential for good establishment, it is undesirable for established plants with a low leaf area index. This is because the high upward flux will contribute to wasteful soil evaporation.

## 5.6  RECOMMENDATIONS FOR FUTURE RESEARCH

### 5.6.1  WATER SAVING USING MODIFIED SDI

Whilst this research showed that the modified SDI improves seedling survival and crop growth during the establishment period, the main focus of the work, it also suggested that irrigation efficiency may be improved. Comparisons with existing sprinkler irrigation and other systems are now required for the duration of the crop to quantify the differences in irrigation efficiency and environmental performance. Irrigation management to optimize crop production, irrigation efficiency and runoff/drainage will be different from irrigation for establishment where the focus is on soil surface water. Research is needed to establish the optimum irrigation frequency for both establishing and established plants. Once crops are established, it should be possible to enjoy the potential benefit of SDI of reduced soil evaporation and increased water use efficiency.

### 5.6.2  DRIP IRRIGATION MANAGEMENT IN DIFFERENT SOIL TYPES

Research is needed to provide guidelines for optimizing irrigation frequency and amount in different soils and climates. Whilst the present research found responses to both irrigation frequency and amount, the interactions with soil type and climate were not clear enough to 'fine tune' recommendations to particular soil types or climates.

An investigation of yield performance of modified drip tape in range of soil type may allow vegetable growers to improve income and saving water. Research on irrigation scheduling for wide range of soil types and vegetable crop would be useful to reduce the environmental risk (runoff and deep drainage). These studies would include an assessment of the importance of more uniform crop maturity with modified SDI, and possible quality improvement in comparison to conventional SDI.

### 5.6.3   NUTRIENT MANAGEMENT WITH SDI

Nutrient application through the modified SDI system is possible to increase nutrient use efficiency and reduce the risk of environmental impacts (deep drainage/nutrient rich runoff) in the Sydney region. A set of best management practices for modified subsurface drip irrigation with fertigation on vegetable production in intensive horticulture is required.

Within the limitations of this research, we can conclude that crop establishment with SDI can be difficult, depending on the tape depth, soil properties and climate. Early designs for subsurface drip irrigation systems were mostly the same as for the surface drip system. But recent design development in the KISSS™ product aimed to improve crop establishment. The results demonstrate that, with this product, the surface soil water is wetter, and more uniform, than with conventional SDI. The improved soil water relations led to improved establishment and plant growth. The results also show that irrigation efficiency can be improved by using the modified SDI, and drainage reduced. More frequent irrigation was confirmed to improve soil surface water and crop growth. The results show that frequent irrigation was just as important with modified SDI as with conventional SDI, although given the same irrigation management the modified product gave superior performance. With modified SDI it may be possible to apply less water and still obtain satisfactory establishment, but it appeared that both frequent irrigation and a high crop factor would be rewarded by substantially higher growth than with conventional SDI. The management required to optimize growth and irrigation efficiency in different soils and environments needs to be developed by further research. The combination of modified SDI and 'pulse irrigation' should provide a useful improvement to horticultural practice, particularly for peri-urban production areas where the competition for water is intense, and the demand for good environmental performance is high.

### 5.7   SUMMARY

Vegetables are grown in the peri-urban zone throughout Australia in diverse soil types and climates. Irrigation allows cropping throughout the year. Competition for water and adverse environmental impacts from irrigation will increasingly influence access to water and the price paid. A review of literature indicated that subsurface drip irrigation (SDI) in Australia has the potential to achieve high water use efficiency and crop yields, as well as reduce drainage and runoff and the associated environmental risks. However, disadvantages of SDI include 'tunneling,' poor soil surface wetting, and risky crop establishment.

The research reported in this chapter, evaluated ways to overcome these problems, including a new product (KISSS™) that has a narrow band of impermeable material below the drip tape, and geotextile above. It was hypothesized that the impermeable layer would create a temporary water table, from which the upward flux of water would be greater than in conventional SDI and the drainage less. This chapter also includes discussions on how irrigation management with the modified SDI should take account of soil type and evaporative demand.

Field experiments at Richmond, NSW were established to compare C.SDI and M.SDI on a sandy soil in autumn (mean pan evaporation 2 mm/day) and spring (mean evaporation 6 mm/day) to investigate lettuce crop establishment. The treatments were two drip tape types (M.SDI, C.SDI) and three irrigation frequencies (1, 2 and 4 times per day). Irrigation application volume was calculated by using a crop factor of 0.4 in autumn; and 0.4 and 0.8 in the spring.

Modified SDI improved crop establishment was compared with conventional SDI. The difference in seedling survival was numerically small but significant ($p<0.05$), indicating a superior environment for establishment in the M.SDI. This was reflected in higher leaf appearance rates in the spring trials. In both experiments, leaves were longer and wider with the M.SDI, and plant fresh weights were greater at the end of the crop establishment period. The differences in fresh weight were substantial, with the M.SDI system recording average increases over the C.SDI of 16% and 25% in the autumn and spring experiments, respectively. Plants were also more uniform with the M.SDI. In both experiments, plant weight was closely related to volumetric soil water content, regardless of the source of variation in water content: tape type, crop factor, irrigation frequency, or location within the plot.

Soil water and plant weight responded to increased irrigation frequency (IF) and crop factor (CF, included in spring only) with both tape types. The effects of CF and IF were additive within tape types. So, while the negative effect of reduced irrigation amount can be offset by increased irrigation frequency, the best growth was obtained where both were high. However, for every combination of CF and IF, plant growth with the modified SDI exceeded the conventional SDI. With the combination of high irrigation frequency (4/day) and a high crop factor (0.8), the modified SDI resulted in a 35% increase in plant fresh weight over conventional SDI. Importantly, at high irrigation frequency (4/day) but with only half the amount of irrigation (CF 0.4 *versus* 0.8), plant weight with modified SDI was similar to conventional SDI (actually 10% greater). Soil water content was also more uniform in the M.SDI treatment.

A glasshouse experiment quantified the components of the water balance under irrigation with conventional and modified subsurface drip irrigation, in sand and sandy loam soils under different evaporation demand. A tension table in the base of each large pot (50x35x5 cm) was used to maintain a suction of −60 cm at the base. Each treatment was subjected to a sequence of different irrigation frequencies, one per two days; and one, two and four per day. Data for drainage and soil water were recorded daily, and averaged over the last three days when daily drainage approached steady-state for any irrigation frequency.

The M.SDI system generally resulted in lower drainage than with the C.SDI, regardless of soil type, irrigation frequency, evaporative demand, and irrigation rate.

As the amount of daily irrigation (I) was known and equal for all treatments, soil evaporation ($E_{soil}$) was estimated from drainage (D) using the simplified soil water balance equation: $E_{soil} = I - D$. Thus soil evaporation was the inverse of drainage. The upward flux of water to meet the evaporative demand was greater in the M.SDI, and it was greater with more frequent irrigation. Soil water content and potential were both higher with the M.SDI. They were also higher with frequent irrigation, as in the field experiment. Overall, the M.SDI had less drainage than conventional SDI, greater upward flux of water (soil evaporation), and wetter surface soils. The findings are consistent with the hypothesis that an impermeable layer beneath the drip tape creates a temporary water table, increasing the upward flux of water.

Both the field and glasshouse experiments showed the benefit of dividing the daily irrigation requirement into smaller, more frequent pulses, for both types of drip tape, regardless of the soil types and climates investigated. Whilst increased irrigation amount and irrigation frequency both increased soil water content and plant growth, the best performance was when both irrigation amount and frequency were high. Frequent irrigation (4/day) was *essential* to obtain the improved crop growth with the M.SDI and a high crop factor in the spring experiment.

These positive responses to tape type and irrigation frequency were obtained at relatively low and high evaporative demand (2, 6 mm/day), and in soils with different texture (coarse sand, sandy loam). So the modified drip tape and more frequent irrigation appear to be reliable, broad recommendations. No specific recommendation can be made on the present data regarding irrigation frequency in relation to evaporative demand, although it might be expected that under very high demand more frequent irrigation will be required unless the modified drip tape can be made to hold a greater volume of water against drainage.

In relation to the first objectives of the study, it is concluded that the modified SDI (KISSS™) improves surface soil water content and uniformity, and has the potential to overcome the plant establishment problems associated with conventional SDI. It does so while saving water and reducing environmental risk (drainage and/or runoff). With respect to research question 2, irrigating with more water, or more frequently, did improve seedling growth, but the modified drip tape (KISSS™) retained an advantage in terms of both establishment and growth at any combination of irrigation amount and frequency. Further research is required to develop guidelines for using the M.SDI in specific soils and climates, especially for heavier-textured soils and more extreme evaporation.

## KEYWORDS

- **Australia**
- **Blaney-Criddle method**
- **bulk density**
- **coefficient of variation, cv**
- **crop coefficient**
- **deep percolation**

- drainage
- emitter flow
- emitter flow variation, $q_{var}$
- EnviroSCAN capacitance probe
- evaporation
- flushing capacity
- glass house
- granular matrix sensor
- Great Britain
- ground water
- gypsum
- gypsum block
- horticulture
- hydraulic conductivity
- hydraulic head
- irrigation
- irrigation depth
- irrigation efficiency
- irrigation frequency
- irrigation scheduling
- Jensen-Haise method
- lettuce
- line-source emitter
- marketable yield
- micro irrigation
- neutron probe
- New South Wales
- nutrient loss
- osmotic potential
- overall agronomic efficiency of water use, $f_{ag}$
- overall irrigation efficiency
- Penman-Monteith equation
- plant spacing
- plastic barrier
- point-source emitter

- pulsing irrigation
- Queensland
- reservoir storage efficiency
- runoff
- salinization
- saturated hydraulic conductivity
- soil conditioner
- soil moisture
- soil permeability
- soil pores
- soil structure
- soil texture
- soil water balance
- soil water potential
- solute movement
- storm water
- surface irrigation
- tensiometer
- total plant water potential
- total water potential
- transpiration efficiency, te
- transplant shock
- tunneling
- uniformity coefficient, uc
- unsaturated hydraulic conductivity
- Vector flow
- vegetables
- Victoria
- volumetric water content
- water balance
- water conveyance efficiency
- water deficit
- water stress
- water use efficiency
- wetting front
- wetting pattern

## REFERENCES

1. ABS (2002) Census of population and housing: classification counts, Australia, 2001, industry of employment by sex, counts of persons, Cat. No: 2022.
2. ABS (2005) Agricultural Commodities – Australia, Cat. No. 7121.
3. Aiken, J.T. (2004) Redeveloping soil mapping key descriptors from generic soil series profile for University of Western Sydney Hawkesbury campus soils. SuperSoil 2004: 3rd Australian New Zealand Soils Conference, 5 – 9 December 2004, University of Sydney, Australia.
4. Alam, M., Trooien, T.P., Lamm, F.R., and Rogers, D.H. (2002) Filtration and maintenance subsurface drip irrigation systems. Publication No: MF-2361. Kansas State University.
5. Alejandro, P., and Eduardo, P. (2001) Managing water with high concentration of sediments for drip irrigation purposes. Proc. 2nd Int'l Symp. Preferential flow, water movement and chemical transport in the environment. Eds. Bosch, D.D., and King, K.W. ASAE publication, 290–292.
6. Allen, R.G., Pereira, L.S., Raes, D., and Smith, M. (1998) Crop evapotranspiration, guidelines for computing crop water requirements. FAO Irrig. And Drain. Paper 56, Roam, Italy.
7. Amali, S., Rolston, D.E., Fulton, A.E., Hanson, B.R., Phene, C.J., and Oster, J.D. (1997) Soil water variability under subsurface drip and furrow irrigation. Irri. Sci. 17(4): 151–155.
8. ASAE, (2001) ASAE Standard S526.2, JAN01, Soil and Water Terminology, ASAE, St. Joseph, Michigan.
9. Assouline, S. (2002) The effects of microdrip and conventional drip irrigation on water distribution and uptake. Soil Sci. Soc. Am. J. 66: 1630–1636.
10. AUSVEG (2004). Domestic Vegetable Industry Snapshot. R&D Meeting, AUSVEG Ltd.
11. Ayars, J.E, Bucks, D.A., Lamm, F.R., and Nakayama, F.S. (2007) Introduction, Microirrigation for crop production design, operation and management, Eds. Lamm, F.R., Ayars, J.E., and Nakayama, F.S. Elsevier, 473–551.
12. Ayars J.E., and Phene, C.J. (2007) Automation, Microirrigation for crop production design, operation and management, Eds. Lamm, F.R., Ayars, J.E., and Nakayama, F.S. Elsevier, 259–284.
13. Ayars, J.E., Phene, C.J., Hutmacher, R.B., Davis, K.R., Schoneman, R.A., Vail, S.S., and Mead, R.M. (1999) Subsurface drip irrigation for row crops: A review of 15 years research at the Water Management Research Laboratory Agric. Water Manag. 42: 1–27.
14. Badgery-Parker, J.J. (1999) A Review of: The Protected Cropping Industry of the Sydney Region, NSW Agriculture.
15. Barber, S.A., Katupitiya, A., and Hickey, M. (2001) Effects of long-term subsurface drip irrigation on soil structure. Proc. of the 10th Australian Agronomy Conference.
16. Barrs, H.D., and Weatherley, P.E. (1962) A reexamination of the relative turgidity technique for estimating water deficits in leaves. Aust. J. Biol. Sci. 15: 413–428.
17. Barth, H.K. (1999) Sustainable and effective irrigation through a new subsoil irrigation system (SIS). Agric. Water Manag. 40: 283–290.
18. Bar-Yosef, B. (1989) Sweet corn response to surface and subsurface trickle P fertigation. Agron. J. 81(3): 443–447.
19. Bar-Yosef, B., and Sheikholslami, M.R. (1976) Distribution of water and Ions in soils irrigated and fertilized from a trickle source. Soil Sci. Soc. Am. J. 40: 575–581.
20. Battam, M.A., Sutton, B.G., and Boughton, D.G. (2003) Soil pits as a simple design aid for subsurface drip irrigation systems. Irri. Sci. 22: 135–141.
21. Ben-Asher, J., and Phene, C.J. (1993) The effect of surface drip irrigation on soil water regime evaporation and transpiration. In Proc. 6th Int'l Conf. Irrig., May 3–4, Tel-Aviv, Israel. 35–42.
22. Bhattari, S.P., Huber, S., and Midmore, D.J. (2004) Aerated subsurface irrigation gives growth and yield benefits to zucchini, vegetable soybean and cotton in heavy clay soils. Ann. Appl. Biol. 144: 285–298.
23. Bierman, P. (2005) Managing irrigation for high-value crops. Fact sheet N: FGV-00649. Cooperative extension services, University of Alaska, USA.
24. Bogle, C.R., and Hartz, T.K. (1986) Comparison of drip and furrow irrigation for muskmelon production. HortScience 21(2): 242–244.

25. Bralts, V.F., and Kesner, C.D. (1983) Drip irrigation uniformity estimation. Transactions of the ASAE 24: 1369–1372.
26. Bralts, V.F., Wu, I.P., and Gitlin, H.M. (1981a) Manufacturing variation and drip irrigation uniformity. Transactions of the ASAE 24(1): 113–119.
27. Bralts, V.F., Wu, I.P., and Gitlin, H.M. (1981b) Drip irrigation uniformity considering emitter plugging. Transactions of the ASAE 24(5): 1234–1240.
28. Brandt, A., Bresler, E., Diner, N., Ben-Asher, I., Heller, J., and Goldberg, D. (1971) Infiltration from a trickle source: I. Mathematical models. Soil Sci. Soc. Am. Proc. 35: 675 – 682.
29. Bresler, E. (1978) Analysis of trickle irrigation with application to design problems. Irri. Sci. 1:3–17.
30. Broner, I., and Alam, M. (1996) Irrigation Subsurface Drip, Fact Sheet No: 4.716 Colorado State University Cooperative Extension.
31. Brook, R.M., and Davila, J.D. (2000) The Peri-Urban Interface: A Tale of Two Cities, School of Agricultural and Forest Sciences, Published by University of Wales and Development Unit, University College, London.
32. Brouwer, C., Prins, K., Kay, M., and Heilbloem, M. (1990) Irrigation water management: irrigation methods, Technical manual No: 5, Available at: www.fao.org.
33. Brown, K.W., Thomas, J.C., Friedman, S., and Meiri, A. (1996) Wetting pattern associated with directed subsurface irrigation. Proc. Of the Int'l Conf. on Evaporation and irrigation scheduling. 3–6 Nov. San Antonio, Texas, USA.
34. Brown, D.A., and Don Scott, H. (1984) Dependence of crop growth and yield on root development and activity. Roots, Nutrients, Water flux, and Plant Growth. Eds. Barber, S.A., and Boulding, D.R. ASA Special Publication No: 49.
35. Bryla, D.R., Banuelos, G.S., and Mitchell, J.P. (2003) water requirements of subsurface drip-irrigated faba bean in California. Irri. Sci. 22: 31–37.
36. Bucks, D.A., and Davis, S. (1986) Historical Development. Trickle irrigation for crop production design, operation and management. Eds. Nakayama, F.S., and Bucks, D.A. Elsevier, 1–26.
37. Bucks, D.A., Erie, L.J., French, O.F., Nakayama, F.S., and Pew, W.D. (1981) Subsurface trickle irrigation management with multiple cropping. Transactions of the ASAE 24(6): 1482–1489.
38. Bureau of Meteorology (2007) Climate data. Available at: www.bom.gov.au
39. Burt, C.M. (1999) Irrigation water balance fundamentals. Proc. Benchmarking Irrig. Sys. Performance using water measurement and water balances. San Luis Obispo, CA, USA. 1–13.
40. Burt, C.M. (2004) Rapid field evaluation of drip and microspray distribution uniformity. Irrigation and drainage system 18: 275–297.
41. Burt, C. M., Clemmens, A. J., Strelkoff, T. S., Solomon, K. H., Bliesner, R. D., Hardy, L. A., Howell, T. A., and Eisenhauer, D.E. (1997) Irrigation performance measures: efficiency and uniformity. J. of Irri. And Drain. Engi. 123(6): 423–442.
42. Burt, C.M., and Styles, S.W. (1994) Drip and micro irrigation for trees, vines, and row crops: Design and management, California Polytechnic State University, San Luis Obispo, California.
43. Camp, C. R. (1998) Subsurface Drip Irrigation: A review. Transactions of the ASAE 41(5): 1353–1367.
44. Camp, C.R., and Lamm, F.R. (2003) Irrigation systems: Subsurface drip. Encyclopedia of Water Science, 560–564.
45. Camp, C.R., Bauer, B.J., and Busscher, W.J. (1997) A comparison of uniformity measures for drip irrigation. Transactions of the ASAE 40(4): 1013–1020.
46. Camp, C.R., Garrett, J.T., Sadler, E.J., and Busscher, W.J. (1993) Microirrigation management for double-cropped vegetables in a humid area. Transactions of the ASAE 36(6): 1639–1644.
47. Camp, C.R., Lamm, F.R., Evans, R.G., and Phene, C.J. (2000) Sub surface drip irrigation – past, present, and future. Proceedings of the 4th Decennial National Irrigation Symposium, Nov. 14–16.
48. Campbell, G.S., and Mulla, D.J. (1990) Measurement of soil water content. Irrigation of agricultural crops. Eds. Stewart, B.A., and Nielsen, D.R. Published by ASA, CSA, and SSSA, USA. 127–142.

49. Charlesworth, B.P., and Muirhead, A.W. (2003) Crop establishment using subsurface drip irrigation: a comparison of point source and area sources. Irri. Sci. 22:171–176.
50. Charlesworth, P. (2005) Soil water monitoring. CSIRO/CRC Irrigation Futures publication. Published by CSIRO-land and Water.
51. Chase, R.G. (1985) Subsurface trickle irrigation in a continuous cropping system. Proc. 3rd Int'l Drip Irrigation Congress, Fresno, CA, USA. 909–914.
52. Christen, E., Ayars, J., Hornbuckle, J., and Hickey, M. (2006) Technology and practice for irrigation in vegetables. DPI NSW.
53. Clark, G.A., and Smajstrla, A.G (1996) Design considerations for vegetable crop drip irrigation systems. HortTechnology 6(3): 155–159.
54. Clark, G.A., Maynard, D.N., and Stanley, C.D. (1996) Drip-irrigation management for watermelon in a humid region, Appl. Eng. Agric. 12(3): 335–340.
55. Clark, G.A., and Phene, C.J. (1992) Automated centralized data acquisition and control of irrigation management system. ASAE Paper No: 92–3021,11 pp.
56. Coelho, E.F., and Or, D. (1996) A parametric model for two-dimensional water uptake by corn roots under drip irrigation. Soil Sci. Soc. Am. J. 60: 1039–1049.
57. Coelho, E.F., and Or, D. (1999) Root distribution and water uptake patterns of corn under surface and subsurface drip irrigation. Plant and Soil 206: 123–136.
58. Cornish, P.S., Yiasoumi, B., and Maheshwari, B. (2005) Urban Region – Peri-urban Horticulture. A scoping study on opportunities for improved application system. Irrigation Matters Series No. 02/05. CRC for Irrigation Future. 24–26.
59. Cornish, P.S., and Hollinger, E. (2002) Managing Water Quality and Environmental Flows in the Hawkesbury-Nepean, Project code: VG98044.
60. Cote, C.M., Bristow, K.L., Charlesworth, P.B., Cook, F.J., and Thorburn, P.J. (2003) Analysis of soil wetting and solute transport in subsurface trickle irrigation. Irri. Sci. 22(3/4): 143–156.
61. Cull, P., Gishford, S., and Johnson, G. (1989) Irrigation management and nutrition on vegetables. Eds. Greenhalgh, W.J., McGlasson, B., and Briggs, T. Proc. of Research and development conference on vegetables, the market and the producer. Acta Horticulturae 247: 223–228.
62. Dalvi, V.B., Tiwari, K.N., Pawade, M.N., and Phirke, P.S. (1999) Response surface analysis of tomato production under micro irrigation, Agric. Water Manag. 41(1): 11–19.
63. Dang, H. (2004) Vietnamese vegetable growers in the Sydney region – farming practices. Rural Industries Research and Development Corporation Publication No. 04/028.
64. Daniells, I., Manning, B., and Pearce, L. (2002) Profile descriptions. NSW Agriculture.
65. Darusman, A.H.K., Loyd, R.S., William, E.S., and Freddie, R.L. (1997) Water flux below the root zone vs. irrigation amount in drip-irrigated corn. Agronomy J. 89(3): 375–379.
66. Davis, S., and Pugh, W.J. (1974) Drip irrigation: Surface and subsurface compared with sprinkler and furrow. In: Proc. 2nd Int'l Drip Irrigation Cong., July, 7–14. San Diego, California, USA. 109–114.
67. Deery, D. (2003) Soil slotting technology for improved crop establishment from sub surface drip irrigation. B.Sc. (Agri) Hon. Thesis. Submitted to Charles Sturt University, Wagga, NSW, Australia.
68. DeTar, W.R. (2004) Using a subsurface drip irrigation system to measure crop water use. Irri. Sci. 23: 111–122.
69. DeTar, W.R., Browne, G.T., Phene, C.J., and Sanden, B.L. (1996) Real-time irrigation scheduling of potatoes with sprinkler and subsurface drip systems. In Proc. Int'l Conf. on Evapotranspiration and Irrigation Scheduling, Nov 3–6, San Antonio, Texas, USA, 812–824.
70. Dimsey, R., and Vujovic, S. (2005) Growing lettuce. Ag notes AG1119, DPI Victoria.
71. Don Scott, H. (2000) Soil physics: Agricultural and environmental applications, Iowa State University Publication.
72. Dukes, M.D., and Scholberg, J.M. (2005) Soil moisture controlled subsurface drip irrigation on sandy soils. Applied Eng. in Agri. 21(1): 89–101.
73. Enciso, J., Jifon, J., and Wiendenfeld, B. (2007) Subsurface drip irrigation of onions: Effects of drip tape emitter spacing on yield and quality. Agric. Water Manag. 92: 126–130.

74. El-Gindy, A.M., and El-Araby, A.M. (1996) Vegetable crop response to surface and subsurface drip under calcareous soil. Evaporation and irrigation scheduling. Proc. of the Inter. Conf. Nov. 3–6 San, Antonio, Texas.

75. FAO (2000) Urban and peri-urban agriculture. Available at: www.fao.org.

76. FAO (1997) Small-scale irrigation for arid zones. Available at: www.fao.org/docrep/W3094E.

77. Fares, A., and Alva, A.K. (2000) Soil water components based on capacitance probes in a sandy soil. Soil Sci. Soc. Am. J. 64: 311–318.

78. Ford, G.W., Smith, I.S., Badawy, N.S., and Roulston, A. (1980) Effects of gypsum of two red duplex wheat soils in Northern Victoria. National Soils Conference, 19 –23 May, Sydney.

79. Freeman, B.M., Blackwell, J., and Garzoli, K.V. (1976) Irrigation frequency and total water application with trickle and furrow systems. Agric. Water Manag. 1: 21–31.

80. Gallardo, M., Jackson, L.E., Schulbach, K., Snyder, R.L., Thompson, R.B., and Wyland, L.J. (1996) Production and water use in lettuce under variable water supply. Irri. Sci. 16: 125–137.

81. Gardner, W.R. (1958) Some steady-state solutions of the unsaturated moisture flow equation with application to evaporation from a water table. Soil Science 85(4): 228–232.

82. Gardner, W.R., and Fireman, M. (1958) Laboratory studies of evaporation from soil columns in the presence of a water table. Soil Science 85(4): 244–249.

83. Ghali, G.S., and Svehlik, Z.J. (1988) Soil-water dynamics and optimum operating regime in trickle irrigated fields. Agric. Water Manag. 13: 127–143.

84. Giddings, J. (2000) Tensiometer tips and Tensiometers need periodic maintenance. Available at: www.dpi.nsw.gov.au

85. Goldhamer, D.A., and Snyder, R.L. (1989) Irrigation scheduling – A guide for efficient on-farm water management. Publication No: 21454, University of California.

86. Goyal, Megh R., and Rivera, L.E. (1985). Trickle irrigation scheduling of vegetables. Drip irrigation in action. Proc. 3rd Int'l Drip/Trickle Irrigation Congress. 18–21 November, Fresno, California, USA. Publications of American Society of Agricultural Engineers, 838–843.

87. Grattan, S.R., Snyder, R.L., and Robinson, F.E. (1988) Yield threshold soil water depletion. Irrigation Scheduling: A guide for efficient on-farm water management. Eds. Goldhamer, D.A., and Snyder, R.L. UC Publication 21454.

88. Greene, R.S.B., and Ford, G.W. (1980) The effect of gypsum on the structural stability of two Victorian soils. National Soils Conference, 19 –23 May, Sydney.

89. Grimes, D.W., Munk, D.S., and Goldhamer, D.A. (1990) Drip irrigation emitter depth placement in a slowly permeable soil. Proc. 3rd Int'l Irrigation Symp. Oct. 28-Nov.1, Phoenix, Arizona, USA. 248–254.

90. Hagin, J., and Lowengart, A. (1996) Fertigation for minimizing environmental pollution by fertilizers. Fertilizer research 43: 5–7.

91. Haines, W.B. (1930) The hysteresis effect in capillary properties and the methods of moisture distribution associated therewith. J. Agric. Sci. 20: 96–105.

92. Haman, D.Z., and Smajstrla, A.G. (2002) Scheduling tips for drip irrigation of vegetables. Publication No: AE259. Florida extension service, University of Florida, USA.

93. Haman, D.Z., and Smajstrla, A.G. (2003) Design tips for drip irrigation of vegetables. Publication No: AE093, Florida Cooperative Extension Service, University of Florida. http://edis.ifas.ufl.edu/AE087 - FOOTNOTE_2#FOOTNOTE_2

94. Hanson, B.R., Hopmans, J.W., and Simunek, J. (2008) Leaching with subsurface drip irrigation under saline, shallow groundwater conditions. Vadose Zone J. 7: 818.

95. Hanson, B.R. (1987) A systems approach to drainage reduction. Calif. Agric. 41(9,10): 19–24.

96. Hanson, B.R. (1996) Fertilizing row crops with drip irrigation. Irrigation Journal. Available at: www.greenmediaonline.com/ij/1996/1296.

97. Hanson, B.R., and May, D.M. (2004) Response of processing and fresh-market onions to drip irrigation. International Symposium on Irrigation of Horticultural crops. Eds. Snyder, R.L. Davis, CA, USA. Acta Horticulturae 664: 399–405.

98. Hanson, B.R., Fipps, G., and Martin, E.C. (2000) Drip irrigation of row crops: what is the state-of-the-art? Proc. 4th decennial Irrigation Symposium. ASAE publication.

99. Hanson, B.R., Schwankl, L.J., Grattan, S.R., and Prichard, T. (1994) Drip irrigation for row crops. University of California, Davis, USA.

100. Hanson, B.R., Schwankl, L.J., Schulbach, K.F., and Pettygrove, G.S. (1997) A comparison of furrow, surface drip, and subsurface drip irrigation on lettuce yield and applied water. Agric. Water Manag. 33: 139–157.

101. Hardy, S. (2004) Growing lemons in Australia- a production manual. Available at: www.dpi. nsw.gov.au

102. Harris, G.A. (2005a) Sub-surface drip irrigation - Crop management, DPI&F Note, Brisbane.

103. Harris, G.A. (2005b) Sub-surface drip irrigation – Advantages and limitations, DPI&F Note, Brisbane.

104. Harris, G.A. (2005c) Sub-surface drip irrigation - System components, DPI&F Note, Brisbane.

105. Harris, G.A. (2005d) Sub-surface drip irrigation - System design, DPI&F Note, Brisbane.

106. Hartz, T.K. (1996) Water management in drip-irrigated vegetable production. HortTechnology 6(3): 165–168.

107. Hartz, T.K. (1997) Effects of drip irrigation scheduling on muskmelon yield and quality. Scientia Horticulturae 69: 117–122.

108. Hartz, T.K. (1999) Water Management in Drip-irrigated vegetable production. Vegetable Research and Information Center. University of California, Davis, CA.

109. Hartz, T.K. (2000) Drip irrigation and fertigation management of celery. Celery Grower Guidelines, Vegetable Research and Information Center, University of California, Davis, USA.

110. Hatfield, J.L., Sauer, T.J., and Rueger, J.H. (2001) Managing soils to achieve greater water use efficiency: A review. Agron. J. 93: 271–280.

111. Haynes, R.J. (1985) Principles of fertilizer use for trickle irrigated crops. Fertilizer Research, 6:235–255.

112. Haynes, R.J. (1990) Movement and transformations of fertigated nitrogen below trickle emitters and their effects on pH in the wetted soil volume. Fertilizer Research 23: 105–112.

113. Hickey, M., and Hoogers, R. (2006) Maximizing returns from water in the Australian vegetable industry-NSW, NSW Agriculture.

114. Hickey, M., Hoogers, R., Singh, R., Christen, E., Henderson, C., Ashcroft, B., Top, M., O'Donnell, D., Sylvia, S., and Hoffmann, H. (2006) Maximizing returns from water in the Australian vegetable crops-National report, NSW Agriculture.

115. Hickley, M. (2005) Projects helps vegetable growers maximize returns from water. Vegiebites – National vegetable industry center newsletter. Eds. Tony, N. DPI, NSW.

116. Hillel, D. (1972) Soil and water- Physical principles and processes, New York and London: Academic press.

117. Hillel, D. (2004) Introduction to environmental soil physics, Elsevier, Amsterdam.

118. Hills, D.J., Tajrishy, M.A.M., and Gu, Y. (1989a) Effects of chemical clogging on drip tape irrigation uniformity. Transactions of the ASAE 32(4): 1202–1206.

119. Hla, A.K., and Scherer, T.F. (2003) Introduction to micro irrigation. Fact sheet No: AE-1243. North Dakota State University.

120. Hoffmann, H. (2007) Efficient irrigation for determinate tomatoes in the Gascoyne River area, Ag Note No: 27/90. Available at: www.agric.wa.gov.au.

121. Hollinger, E. (1998) Links between management of market gardens and storm water losses of sediment, nitrogen and phosphorus. MSc(Hono) thesis, University of Western Sydney, Hawkesbury.

122. Hollinger, E., Cornish, P.S., Baginska, B., Mann, R., and Kuczera, G. (2001) Farm-scale stormwater losses of sediment and nutrients from a market garden near Sydney, Australia, Agric. Water Manag. 47(3): 227–241.

123. Howell, T.A., Bucks, D.A., Goldhamer, D.A., and Lima, J.M. (1986) Irrigation scheduling. Trickle irrigation for crop production design, operation and management. Eds. Nakayama, F.S., and Bucks, D.A. Elsevier, 241–279.

124. Howell, T.A., and Meron, M. (2007) Irrigation scheduling. Microirrigation for crop production design, operation and management. Eds. Lamm, L.R., Ayars, J.E., and Nakayama, F.S. Elsevier, 61–130.

125. Howell, T.A., Schneider, A.D., and Stewart B.A. (1995) Subsurface and surface irrigation of corn –U.S. southern High Plains Proc. Of Fifth Int'l Microirrigation Congress, April 2–6 Orlando, Florida, ASAE, 375–381.

126. Howell, T.A., Schneider, A.D., and Evett, S.R. (1997) Subsurface and surface micro irrigation of corn-Southern High Plains. Transactions of the ASAE 40(3): 635–641.

127. Hulme, T., Grosskopk, T., and Hindle, J. (2002) Agricultural land classification. Agfact AC.25, NSW Agriculture.

128. Hulugalle, N.R., Friend, J.J., and Kelly, R. (2002) Some physical and chemical properties of hard setting Alfisols can be affected by trickle irrigation. Irri. Sci. 21(3): 103–113.

129. Human, J.J., and Grobler, P.J.L. (1990) The influence of different irrigation scheduling methods on the leaf area index, leaf area duration and bulb formation of long-season onions. Eds. Alvino, A. Symposium on scheduling of irrigation for vegetable crops under field condition. Acta Horticulturae 278: 825–832.

130. Imas, P. (1999) Recent techniques in fertigation horticultural crops in Israel. Proc. of Recent trends in nutrition management in horticultural crop. 11–12th Feb. Dapoli, Maharashtra, India.

131. Ismail, S.M., Ozawa, K., and Khondaker, N.A. (2008) Influence of single and multiple water application timings on yield and water use efficiency in tomato (var. First power). Agric. Water Manag. 95: 116 –122.

132. Jackson, L.E., and Bloom, A.J. (1990) Root distribution in relation to soil nitrogen availability in field grown tomatoes. Plant Soil 128: 115–126.

133. Jalota, S.K., Romesh, K., and Ghuman, B.S. (1998) Hydraulic properties of soil: Methods in soil physics, New Delhi: Narosa Publishing House.

134. Janat, M., and Somi, G. (2001) Performance of cotton crop grown under surface irrigation and drip fertigation – Field water efficiency and dry matter distribution. Commun. Soil Sci. Plant Anal. 32(19&20): 3063–3076.

135. Jensen, M.E., Rangeley, W.R., and Dieleman, P.J. (1990) Irrigation trends in world agriculture. Irrigation of agricultural crops. Eds. Stewart, B.A., Nielsen, D.R. Agron. Monogr. 30. ASA, CSSA, and SSSA, Midison, WI, USA. 31–67.

136. Johnson, H., and Stewart, J. (2004) Soils manual. UG practical class, University of Western Sydney.

137. Johnson, H., Dennis, P., and Ronald, V. (1991) Saving water in vegetable gardens. Leaflet No: 2977. Vegetable research information center, University of California, USA.

138. Johnson, N.L., Kelleher, F.M., and Chant, J.J. (1998) The future of agriculture in the peri-urban fringe of Sydney, Ninth Proceedings of the Australian Agronomy Conference.

139. Jordan, W.R. (1983) Whole plant response to water deficits: An overview. Limitations to efficient water use in crop production. Eds. Taylor, H.M., Jordan, W.R., and Sinclair, T.R. Published by ASA, CSA and SSSA.

140. Kamara, L., Zartamn, R., and Ramsey, R.H. (1991) Cotton root distribution as a function of trickle irrigation emitter depth. Irri. Sci. 12: 141–144.

141. Kang, Y., Yuan, B., and Nishiyama, S. (1999) Design of micro irrigation laterals at minimum cost. Irri. Sci. 18: 125–133.

142. Karam, F., Mounser, O., Sarkis, F., and Lahoud, R. (2002) Yield and nitrogen recovery of lettuce under different irrigation regimes. J Appl. Hort 4(2): 70–76.

143. Kinsela, M.N. (1985) Vegetable crops. A Manual of Australian Agriculture. Eds. Reid, R.L. William Hienemann, Melbourne. 138–145.

144. Kirkham, D., and Horton, R. (1990) Managing soil water and chemical transport with subsurface flow barrier II Theoretical. In Agronomy Abstracts, ASA, Madison. 213.

145. Kiuchi, M., Horton, R., and Kaspar, T.C. (1994) Leaching characteristics of repacked soil columns as influenced by subsurface flow barriers. Soil Sci. Soc. Am. J. 58: 745–753.

146. Lal, R., and Shukla, M.K. (2004) Principles of soil physics, Marcel Decker, New York, USA.

147.Lamm, F.R. (2002) Advantages and disadvantages of subsurface drip irrigation. Proc. Int'l Meeting on Advances in Drip/Micro Irrigation, Puerto de La Cruz, Tenerife, Canary Islands, December 2–5.

148.Lamm, F.R., and Aiken, R.M. (2005) Effect of irrigation frequency for limited subsurface drip irrigation of corn. Proc. Irrig. Assoc. Int'l Tech. Conf. Nov. 6–8, 2005, Phoenix, Arizona.

149.Lamm, F.R., and Camp, C.R. (2007) Subsurface drip irrigation. Microirrigation for crop production design, operation and management, Eds. Lamm, F.R., Ayars, J.E., and Nakayama, F.S. Elsevier, 473–551.

150.Lamm, F.R., and Trooien, T.P. (2005) Dripline depth effects on corn production when crop establishment is nonlimiting. Applied Engineering in Agriculture 21(5): 835–840.

151.Lamm, F.R., Trooien, T.P., Manages, H.L., and Sunderman, H.D. (2001) Nitrogen fertilization for subsurface drip irrigated corn. Transactions of the ASAE 44: 533–542.

152.Lamont, J.W., Orzolek, D.M., Harper, K.J., Jarrett, R.A., and Greaser, L.G. (2002) Drip Irrigation for Vegetable Production, Penn State Agricultural Research and Cooperative Extension Fact sheet.

153.Lantzke, N. (2007) Irrigating vegetables on sandy soils. AgNote: 66/95. Available at: www.agric.wa.gov.au.

154.Lazarovitch, N., Simunek, J., and Shani U. (2005) System dependent boundary condition for water flow subsurface source. Soil Sci. Soc. Am. J. 69:46–50.

155.Lazarovitch, N., Warrick, A.W., Furman, A., and Simunek, J. (2007) Subsurface Water Distribution from Drip Irrigation Described by Moment Analyzes. Vadose Zone J. 6: 116–123.

156.Letey, J. (1985) Irrigation uniformity as related to optimum crop production-additional research is needed. Irri. Sci. 6: 253–263.

157.Levin, I., van Rooyen, P.C., and van Rooyen, F.C. (1979) The effect of discharge rate and intermittent water application by point-source irrigation on the soil moisture distribution pattern. Soil Sci. Soc. Am. J. 43: 8–16.

158.Ley, T.W., Stevens, R.G., Topielec, R.R., and Neibling, W.H. (2006) Soil water monitoring and measurements, Washington State University, Publication No: PNW0475.

159.Li, Y., Wallach, R., and Cohen, Y. (2002) The role of soil hydraulic conductivity on the spatial and temporal variation of root water uptake in drip-irrigated corn. Plant and Soil 243: 131–142.

160.Lindsay, C.A., Sutton, B.G., and Collis-George, N. (1989) Irrigation scheduling of subsurface drip irrigated salad tomatoes. Eds. Greenhalgh, W.J., McGlasson, B., and Briggs, T. Proc. of Research and development conference on vegetables, the market and the producer. Acta Horticulturae 247: 229–232.

161.Locascio, J.S. (2005) Management of irrigation for vegetables: past, present, future, Hort Technology 15(3): 482–485.

162.Loch, R.J., Grant, C.G., McKenzie, D.C., and Raine, S.R. (2005). Improving plants' water use efficiency and potential impacts from soil structure change - Research Investment Opportunities. Final report to the National Program for Sustainable Irrigation. CRCIF Report number 3.14/1. Cooperative Research Center for Irrigation Futures, Toowoomba.

163.Lof, H. (1976) Water use efficiency and competition between arid zone annuals especially the grasses *Phalaris minor* and *Hordeum murinum*. Agric. Res. Rep. No: 853, The Netherlands.

164.Loomis, R.S. (1983) Crop manipulations for efficient use of water: An overview. Limitations to efficient water use in crop production. Eds. Taylor, H.M., Jordan, W.R., and Sinclair, T.R. Published by ASA, CSA and SSSA.

165.Lubana, P.P.S., and Narda, N.K. (2001) Modeling soil water dynamics under trickle emitters – A review. J. Agric. Eng. Res. 78(3): 217–232.

166.Machado, R.M.A., and Oliveira, M.R. (2003) Comparison of tomato root distributions by minirhizotron and destructive sampling. Plant Soil 255(1): 375–385.

167.Magen, H. (1995) Fertigation: An overview of some practical aspects. Fertilizer News of India 40(12): 97–100.

168. Maheshwari, B., Plunkett, M., and Singh, P. (2003) Farmers' perceptions about irrigation scheduling in the Hawkesbury-Nepean catchment. Australasia Pacific Extension Network Conference, 26–18 Nov., Hobart, Tasmania.

169. Maheshwari, B., and Simmons, B. (2003) Impacts of urbanization on irrigation water in the Hawkesbury-Nepean catchment, Australia. Water for a sustainable world-Limited supplies and expanding demand. Proc. 2nd Int'l Conf. on Irrigation And Drainage, Eds. Clemmens, A.J., and Anderson, S.S. 12–15 May, Phoenix, Arizona, USA.

170. Marr, C.W. (1993) Fertigation of vegetable crops Publication No: MF-1092. Kansas State University.

171. Martinez, H.J.J., Bar-Yosef, B., and Kafkafi, U. (1991) Effect of subsurface and subsurface fertigation on sweet corn rooting, uptake, dry matter production and yield. Irri. Sci. 12: 153–159.

172. McDougall, S. (2002) Adapting to change: enhancing change skills through collaboratively developing an integrated pest and disease management strategy for lettuce. Project No: VG98048.

173. McMullen, B. (2000) SOILpak for vegetable growers NSW Agriculture.

174. Meek, B.D., Ehlig, C.F., Stolzy, L.H., and Graham, L.E. (1983) Furrow and trickle irrigation: effects on soil oxygen and ethylene and tomato yield. Soil Sci. Soc. A. J. 47: 631–635.

175. Mehta B.K., and Wang Q.J. (2004) Irrigation in a variable landscape: Matching irrigation systems and enterprises to soil hydraulic characteristics, Final Report, Department of Primary Industries, Victoria.

176. Meshkat, M., Warner, R.C., and Workman, S.R. (2000) Evaporation reduction potential in an undisturbed soil irrigated with surface drip and sand tube irrigation. Transactions of the ASAE 43(1): 79–86.

177. Metin-Sezen, S., Yazar, A., Canbolat, M., Eker, S., and Celikel, G. (2005) Effect of drip irrigation management on yield and quality of field grown green beans. Agric. Water Manag. 71(3): 243–255.

178. Miller, M.L., Charlesworth, P.B., Katupitiya, A., and Muirhead, W.A. (2000) A comparison of new and conventional sub surface drip irrigation systems using pulsed and continuous irrigation management. In: Proc. Nat'l Conf. Irrig. Assoc. Australia. May 23–25 Melbourne, Australia.

179. Miyazaki, T. (2006) Water flows in soil. CRC Press Taylor and Francis Group, 27–28.

180. Mizyed, N., and Kruse, E.G. (1989) Emitter discharge evaluation of subsurface trickle irrigation systems. Transactions of the ASAE 32(4): 1223–1228.

181. Mmolawa, K., and Or, D. (2000a) Water and solute dynamics under drip-irrigated crop: experiment and analytical model. Transactions of the ASAE 43(6): 1597–1608.

182. Mmolawa, K., and Or, D. (2000b) Root zone solute dynamics under drip irrigation: A review. Plant and soil 222: 163–190.

183. Montagu, K.D., Conroy, J.P., and Francis, G.S. (1998) Root and shoot response of field grown lettuce and broccoli to compact subsoil. Australian Journal Agricultural Research 49: 89–97.

184. Munoz-Capena, R., Dukes, M.D., Li, Y.C., and Klassen, W. (2005) Field comparison of tensiometer and granular matrix sensor automatic drip irrigation on tomato. HortTechnology 15(3): 584–590.

185. NEH - National Engineering Handbook-Irrigation, (1991) Soil-plant-water relationships, Soil Conservation Service, Department of Agriculture, USA.

186. Neufeld, J., Davison, J., and Stevenson, T. (1993) Sub-surface drip irrigation. Fact sheet No: 97–13. University of Nevada.

187. Nguyen, V. (2000) Asian vegetable industry in New South Wales. Access to Asian vegetable. RIRDC newsletter. Issue 29.

188. Nightingale, H.I., Phene, C.J., and Patton, S.H. (1985) Trickle irrigation effects on soil-chemical properties. 3rd Int'l Drip Irrigation Congress, Fresno, CA, USA. 730–735.

189. Noborio, K., Horton, R., and Tan, C.S. (1999) Time domain reflectometry probe for simulation measurement of soil matric potential and water content. Soil Sci. Soc. Am. J. 63: 1500–1505.

190. NSW Agriculture (2002) Horticulture in the Sydney drinking water catchment. Published by NSW Agriculture.

191.O'Neill, M., Pablo, R., and Begay, T. (2002). Development and evaluation of drip irrigation for North Western New Mexico. 36th Annual Progress Report, New Mexico State University.

192.Or, D. (1996) Drip irrigation in heterogeneous soils: steady state field experiments for stochastic model evaluation. Soil Sci. Soc. Am. J. 60: 1339–1349.

193.Or, D., and Coelho, F.E. (1996) Soil water dynamics under drip irrigation: Transient flow and uptake model. Transactions of the ASAE 39(6): 2017–2025.

194.Owens, G., Wicks, C., Dunker, G., Diczbalis, Y. A., and Bowman, L.G. (2003) Tensiometers-their use and management, Ag Note No: 200/748. Available at: www.primaryindustry.nt.gov.au

195.Patel, N., and Rajput, T.B.S. (2007) Effect of drip tape placement depth and irrigation level on yield of potato. Agric. Water Manag 88: 209–223.

196.Phene, C.J. (1996) Shovel versus computer. Irrigation Business Tech. 4(3): 6.

197.Phene, C.J., Allee, C.P., and Pierro, J.D. (1989) Soil matric potential sensor measurements in real time irrigation scheduling. Agric. Water Manag.16: 173–185.

198.Phene, C.J., and Beale, O.W. (1976) High frequency irrigation for water nutrient management in humid regions. Soil Sci. Soc. Am. J. 40: 430–436.

199.Phene, C.J., and Howell, T.A. (1984) Soil sensor control of high frequency irrigation systems. Transactions of the ASAE 27(2): 392–396.

200.Phene, C.J., and Phene, R.C. (1987) Drip irrigation systems and management. ASPAC Food and Fertilizer Tech. Ctr., Taiwan, Ext. Bull. No: 244.

201.Phene, C.J., Davis, K.R., Hutmacher, R.B., and McCormick, R.L. (1987) Advantages of subsurface irrigation for processing tomato. Eds. Sims, W.L. Davis, California. Acta Horticulturae 200:101–114.

202.Phene, C.J., Davis, K.R., Hutmacher, R.B., Bar-Yosef, B., Meek, D.W., and Misaki, J. (1991) Effect of high frequency surface and subsurface drip irrigation on root distribution of sweet corn. Irri. Sci. 12: 135–140.

203.Pier, J.W., and Doerge, T.A. (1995) Concurrent evaluation of agronomic, economic and environmental aspects of trickle irrigated watermelon production. J. Environ. Qual. 24: 79–86.

204.Pitts, D., Peterson, K., Gilbert, G., and Fastenau, R. (1996) Field assessment of irrigation system performance. Appl. Engr. Agric. 12(3): 307–313.

205.Plaut, Z., Rom, R., and Meiri, A. (1985) Cotton response to subsurface trickle irrigation. Proc. 3rd Int'l Drip/Trickle Irrigation Congress, Fresno, CA, USA. 916–920.

206.Post, S.E.C., Peck, D.E., Brendler, R.A., Sakovich, N.J., and Waddel, L. (1985) Evaluation of nonoverlapping, low flow sprinklers. Proc. 3rd Int'l Drip/Trickle Irrigation Congress, Fresno, CA, USA. 294–305.

207.Pogue, W.R., and Pooley, S.G. (1985) Tensiometric management of soil water. Proc. 3rd Int'l Drip/Trickle Irrigation Congress, Fresno, CA, USA. 761–766.

208.Prathapar, S.A., and Qureshi, A.S. (1999) Mechanically reclaiming abandoned saline soils: A numerical evaluation Research report No: 30, International Water Management Institute, Colombo, Sri Lanka.

209.Provenzano, G. (2007) Using HYDRUS-2D simulation model to evaluate wetted soil volume in subsurface drip Irrigation systems. J. Irrig. and Drain. Eng. 133(4): 342–349.

210.Prunty, L., and Casey, F.X.M. (2002) Soil water retention curve using a flexible smooth function. Vadose Zone J. 1: 179–185.

211.Qassim, A., and Ashcroft, B. (2001) Estimating vegetable crop water use with moisture – accounting method. Agriculture notes No–AG1192. DPI, Victoria.

212.Qassim, A. (2003) Subsurface irrigation: A situation analysis. International program for technology and research in irrigation and drainage. DPI, Victoria.

213.Raine, S.R., and Foley, J.P. (2001) Application systems for cotton irrigation-Are you asking the right questions and getting the answer right? In: Proc. Nat'l Conf. Irrig. Assoc. Australia. May 23 25 Melbourne.

214.Raine, S.R., Foley, J.P., and Henkel, C.R. (2000) Drip irrigation in the Australian cotton industry: A scoping study. NECA publication no: 179757/1. National Center for Engineering in Agriculture, University of Southern Queensland, Toowoomba.

215. Ritchie, J.T. (1983) Efficient water use in crop production: Discussion on the generality of relations between biomass production and evapotranspiration. Limitations to efficient water use in crop production. Eds. Taylor, H.M., Jordan, W.R., and Sinclair, T.R. Published by ASA, CSA and SSSA.

216. Roberts, T.L., White, S.A., Warrick, A.W., and Thompson, T.L. (2008) Tape depth and germination method influence patterns of salt accumulation with subsurface drip irrigation. Agric. Water Manag. 95(5): 669–677.

217. Rogers, M.E. (1988) An Agricultural Study of the Greater Sydney Region, NSW Agriculture.

218. Romano, N., Hopmans, J.W., and Dane, J.H. (2002) The soil solution phase, Methods of soil analysis-Part 4 Physical methods Eds. Dick, W.A. SSA, Madison, USA, 692–699.

219. Rubeiz, I.G., Oebker, N.F., and Stroehlein, J.L. (1989) Subsurface drip irrigation and urea phosphate fertigation for vegetables on calcareous soils. J. Plant Nutrition 12(12): 1457–1465.

220. Ruskin, R. (2005) Subsurface drip irrigation and yields. Available at: www.geoflow.com/agriculture.

221. Sadler, E.J., Camp, C.R., and Busscher, W.J. (1995) Emitter flow rate changes by excavating subsurface drip irrigation tubing. Proc. Fifth Int'l Microirrigation congress, Eds. Lamm, F.R. ASAE, St. Joseph, Michigan, 2–6 April, Orlando, FL, USA. 763–768.

222. Sakellariou-Makrantonaki, M., Kalfountzos, D., and Vyrlas, P. (2002) Water saving and yield increase of sugar beet with subsurface drip irrigation. The Int. J. 4(2–3): 85–91.

223. Sankara, G.H., and Yellamanda, T. (1995) Efficient use of irrigation water. Kalyani publishers, New Delhi.

224. Sammis, T.W. (1980) Comparison of sprinkler, trickle, subsurface and furrow irrigation methods for row crops. Agron. J. 72(5): 701–704.

225. Schwankl, L.J., Edstrom, J.P., and Hopmans, J.W. (1996) Performance of micro irrigation systems in almonds Proc. Seventh Int'l Conf. on Water and Irrigation. Tel Aviv, Israel, 123–132.

226. Schwankl, L.J., Grattan, S.R., and Miyao, E.M. (1990) Drip irrigation burial depth and seed planting depth effects on tomato germination. Proc. Third International Irrigation Symp. Oct. 28–Nov.1 Phoenix, Arizona. ASAE, St. Joseph, Michigan. 682–687.

227. Schwankl, L.J., and Hanson, B.R. (2007) Surface drip irrigation. Microirrigation for crop production, Eds. Lamm, F.R., Ayars, J.E., and Nakayama, F.S. Elsevier, 431–472.

228. Schwankl, L.J., Schulback, K.F., Hanson, B.R., and Pettygrove, S. (1993) Irrigating lettuce with buried drip, surface drip and furrow irrigation system: Overall performance. Management of irrigation and drainage system. Eds. Allen, R.G. 21–23 July, Utah, USA.

229. Segal, E., Ben-Gal, A., and Shani, U. (2006) Root water uptake efficiency under ultra-high irrigation frequency. Plant and Soil 282: 333–341.

230. Senn, A.A., and Cornish, P.S. (2000) An example of adoption of reduced cultivation by Sydney's vegetable growers. In Soil 2000: New Horizons for a New Century. Australian and New Zealand Joint Soils Conference. Volume 2: Oral papers. Eds. Adams, J.A., and Metherell, A.K. 3–8 December, Lincoln University. New Zealand Society of Soil Science. 263–264.

231. Senn, A.A. (2001) Sustainable intensive horticulture in the Sydney Basin. Report No: DD0624.97, NSW Agriculture.

232. Shani, U., Xue, S., Gordin-Katz, and Warrik, A.W. (1996) Soil-limiting flow from subsurface emitters - Pressure measurements. J. Irrig. and Drain. Engin. 122(5): 291–295.

233. Sharmasarkar, F.C., Sharmasarkar, S., Held, L.J., Miller, S.D., Vance, G.F., and Zhang, R. (2001) Agroeconomic analysis of drip irrigation for sugarbeet production. Agron. J. 93: 517–523.

234. Shaviv, A., and Sinai, G. (2004) Application of conditioner solution by subsurface emitters for stabilizing the surrounding soil. J. Irrig. and Drai. Engi. 11/12: 485- 490.

235. Shaviv, A., Ravina, I., and Zaslavsky, D. (1987) Field evaluation of methods of incorporating soil conditioners. Soil and tillage Research 9: 151–160.

236. Shock, C.C., Barnum, J.M., and Seddigh, M. (1998) Calibration of water-mark soil moisture sensors for irrigation management. Proc. Int'l Irrig show, San Diego, California, USA, 139–146.

237. Shock, C.C., Feibert, E.B.G., and Saunders, L.D. (2004) Plant population and nitrogen fertilization for subsurface drip-irrigated onion. HortScience 39(7): 1722–1727.

238. Shock, C.C., Feibert, E.B.G., and Saunders, L.D. (2005) Onion response to drip irrigation intensity and emitter flow rate. HortTechnology 15 (3): 652–659.

239. Shock, C.C., Flock, R., Feibert, E., Shock, C.A., Pereira, A., and Jenson, L. (2005) Irrigation monitoring using soil water tension. Factsheet No: EM 8900, Oregon state University.

240. Silber, A., Xu, G., and Wallach, R. (2003) High irrigation frequency: the effect on plant growth and on uptake of water and nutrients. Fertilization strategies for field vegetable production. Eds. Tremblay, N. Acta Horticulturae 627: 89–96.

241. Singh, D.K., and Rajput, T.B.S. (2007) Response of lateral placement depths of subsurface drip irrigation on okra (*Abelmoschus esculentus*) Int. J. Plant Prod. 1(1): 73–84.

242. Solomon, K. (1993) Subsurface drip irrigation: product selection and performance. Subsurface Drip Irrigation: Theory, Practices and Applications. Eds. Jorsengen, G.S., and Norum, K.N. CATI Publication No. 9211001.

243. Solomon, K.H. (1984b) Yield related interpretation of irrigation uniformity and efficiency measures. Irri. Sci. 5: 161–172.

244. Solomon, K.H. (1985) Global uniformity of trickle irrigation. Transactions of the ASAE 28(4): 1151–1158.

245. Souza, C.F., and Matsura, E.E. (2003) Multi-wire time domain reflectometry (TDR) probe with electrical impedance discontinuities for measuring water content distribution. Agric. Water Manag. 59: 205–216.

246. Steel, G.D.R., and Torrie, J.H. (1960) Principles and procedures of statistics. McGraw-Hill publishing, USA.

247. Stirzaker, R., Etherington, R., Lu, P., Thomson, T., and Wilkie, J. (2005) Improving irrigation with wetting front detectors. RIRDC publication No: 04/176.

248. Storke, P.R., Jerie, P.H., and Callinan, A.P.L. (2003) Subsurface drip irrigation in raised bed tomato production – Soil acidification under current commercial practice. Australian J. of Soil Research 41: 1305–1315.

249. Sutton, B.G., and Merit, N. (1993) Maintenance of lettuce root zone at field capacity gives best yield with drip irrigation, Scientia Horticulturae 56: 1–11.

250. Sydney Water (2007) Available at: www.sydneywater.com.au

251. Tanji, K.K., and Hanson, B.R. (1990) Drainage and return flows in relation to irrigation management, In: Irrigation of Agricultural Crops, Agronomy No. 30, Eds. Stewart, B.A., and Nielsen, D.R. American Society of Agronomy, Madison, WI, USA, 1057–1088.

252. Thompson, L. (2005) Adoption of permanent subsurface drip irrigation for forage production to enhance the economical and environmental sustainability of Victoria's dairy industry. Project No: GF3/036.

253. Thompson, R.B., Gallardo, M., Valdez, L.C., and Fernandez, M.D. (2004) Using plant water status to define threshold values for irrigation management of vegetable crops using soil moisture sensors. Agric. Water Manag. 88(1–3): 147–158.

254. Thompson, T.L. and Doerge, T.A. (1995a) Nitrogen and water rates for subsurface trickle irrigated romaine lettuce. HortScience 30(6): 1233–1237.

255. Thompson, T.L., and Doerge, T.A. (1995b) Nitrogen and water rates for subsurface trickle –irrigated collard, mustard and spinach. HortScience 30(7): 1382–1387.

256. Thompson, T.L., and Doerge, T.A. (1996) Nitrogen and water interactions in subsurface trickle –irrigated leaf lettuce: Plant response. Soil Sci. Soc. Am. J. 60: 163–168.

257. Thompson, T.L., Doerge, T.A., and Godin, R.E. (2002) Subsurface drip irrigation and fertigation of broccoli: Yield, quality and nitrogen uptake. Soil Sci. Soc. Am. J. 66: 186–192.

258. Thorburn, P.J., Cook, F.J., and Bristow, K.L. (2003) Soil dependant wetting from trickle emitters: implications for system design and management. Irri. Sci. 22: 121–127.

259. Thorburn, P., Biggs, J., Bristow, K., Horan, H., and Huth, N. (2003) Benefits of subsurface application of nitrogen and water to trickle irrigated sugarcane. 11th Australian Agronomy Conference paper.

260. Titley, M. (2000) Australian lettuce production and processing-An overview. Proc. of Australian lettuce industry conference, 6–8 June, NSW.

261.Tony, N. (2004) Field lettuce production. Agfact H8, NSW Agriculture.

262.Trimmer, W., and Hanson, H. (1994) Irrigation scheduling. Publication No: 288. Oregon State University, USA.

263.Turner, N.C. (1981) Techniques and experimental approaches for the measurement of plant water status. Plant and Soil 58: 339–366.

264.Upchurch, D.R., Wanjura, D.F., and Mahan, J.R. (1990) Automating trickle-irrigation using continuous irrigation canopy temperature measurements. Symposium on scheduling of irrigation for vegetable crops under field condition. Eds. Alvino, A. Acta Horticulturae 278: 299–324.

265.van Keulen, H. (1975) Simulation of water use and herbage growth in arid regions. Pufoc, Wageningen, The Netherlands.

266.Vazquez, N., Pardo, A., Suso, M.L., and Quemada, M. (2005) A methodology for measuring drainage and nitrate leaching of mineral nitrogen from arable land. Plant Soil 269: 297–308.

267.Vazquez, N., Pardo, A., Suso, M.L., and Quemada, M. (2006) Drainage and nitrate leaching under processing tomato growth with drip irrigation and plastic mulching. Agriculture ecosystem and environment 112: 313–323.

268.Wallace, E. (2000) Vegetable Growers' Handbook. A Basic Guide for the Commercial Production of Vegetable Crops in New Zealand. Agro-Research Enterprises, Havelock North.

269.Wang, F., Kang, Y., and Liu, S. (2006) Effects of drip irrigation frequency on soil wetting pattern and potato growth in North China Plain. Agric. Water Manag. 79 (3): 248–264.

270.Warrick, A.W., and Shani, U. (1996) Soil-limiting flow from subsurface emitters-Effect on uniformity. J. Irrig. And Drain. 122(5): 296–300.

271.Welsh, F., Douglas, K., Urs, P., and Byles, D.J. (1995) Enhancing sub surface drip irrigation through vector flow™. Proc. of 5th International Microirrigation Congress April 2 – 6, Orlando, USA.

272.Wendt, C.W., Onken, A.B., Wilke, O.C., Hargrove, R.S., Bausch, W., and Barnes, L. (1977) Effect of irrigation systems on the water requirement of sweet corn. Soil Sci. Soc. Am. J. 41: 785–788

273.Wenkert, W. (1983) Water transport and balance within the plant: An overview. In: Limitations to efficient water use in crop production. Eds. Taylor, H.M., Jordan, W.R., and Sinclair, T.R. Published by ASA, CSA and SSSA.

274.Whalley, W.R., Leeds-Harrison, P.B., Joy, P., and Hoefsloot, P. (1994) Time domain reflectometry and tensiometry combined in an integrated soil water monitoring system. J. Agric. Engn Res. 59: 141–144.

275.Williams, J., Prebble, R.E., Williams, W.T., and Hignett, C.T. (1983) The influence of texture, structure and clay mineralogy on the soil moisture characteristics. Australian J. Soil Research 21: 15–32.

276.Wu, P., Gitlin, H.M., Solomon, K.H., and Saruwatari, C.A. (1986) Design principles-Trickle irrigation for crop production. Eds. Nakayama, F.S., and Bucks, D.A. Elsevier, 53–92.

277.Wu, I.P., Barragan, J., and Bralts, V.F. (2007) Field performance and evaluation. Microirrigation for crop production design, operation and management, Eds. Lamm, F.R., Ayars, J.E., and Nakayama, F.S. Elsevier, 357–387.

278.Zimmer, A., McFarland, M.J., and Moore, J. (1988) Upward free water movement from buried trickle emitters. Proc. of Annual International Summer Meeting of the ASAE. June 26–29. South Dakota. p.16.

279.Zur, B. (1996). Wetted soil volume as a design objective in trickle irrigation. Irri. Sci. 16(3): 101–105.

# CHAPTER 6

# MECHANICS OF CLOGGING IN MICRO IRRIGATION SYSTEM

VISHAL KESHAVRAO CHAVAN, P. BALAKRISHNAN,
SANTOSH DESHMUKH, and M. B. NAGDEVE

## CONTENTS

## 6.1  INTRODUCTION

Indian agriculture mainly depends upon vagaries of monsoon rains, which are unevenly distributed in space and time and not adequate to meet the moisture requirement of the crops for successful farming. India with only 2.4 percent of the world's total area and 4 percent of the total available fresh water supports about 17 percent of the world's population. The agricultural sector consumes over 80 percent of the available water in India for irrigation of crops and will continue to be the major water-consuming sector due to the intensification of agriculture. The 2006 estimates by the Ministry of Water Resources, Government of India indicate that by the year 2050, India needs to increase by five times more water supplies to industries, while its drinking water demand will double and irrigation demand will rise by 50 percent. In India, out of 172.6 Mha of cropped area, only 76.82 Mha area is under irrigation, which means only 44.51 percent of the cropped area is irrigated. Further, considering the fact that the population of the country is estimated to reach 1.4 billion by 2020 with the food requirement of 280 Mtons, the agricultural sector must grow by 4 percent and augment about 3–4 Mtons per year. Though the ultimate irrigation potential of the country has been assessed at 140 Mha planned to be achieved by 2050, even after achieving the same, approximately half the cultivated land would still remain rain-fed, and therefore, water would continue to be the most critical resource limiting agricultural growth.

The water resources of the country are varied and limited, but still most of the area is irrigated using the conventional methods of irrigation with the efficiency of 35–40 percent. Considering the daunting task of achieving the food production targets, it is imperative that efficient irrigation methods like drip/trickle/micro/mini-sprinkler and sprinkler irrigation systems are adopted in large scale for judicious use and management of water to cope up with increasing demand for water in agriculture in order to enhance and accelerate the agricultural production in the country.

Drip irrigation, also called trickle or micro or daily irrigation, is a localized irrigation method that slowly and frequently provides water directly to the plant root zone [4] and is the most efficient irrigation method with an application efficiency of >90 percent. However, not until the innovation of polyethylene plastics in the 1960s did drip irrigation begin to gain momentum. Traditionally, irrigation had relied upon a broad coverage of water to an area that may or may not contain plants. Promoted for water conservation, drip irrigation does just the opposite. It applies small amounts of water (usually every two or three days) to the immediate root zone of plants. In drip irrigation, water is delivered to individual plants at a low pressure and delivery rate to specific areas or zones in the landscape or garden. The slow application promotes a thorough penetration of the water to individual plant root zones and reduces potential runoff and deep percolation. The depth of water penetration depends on the length of time the system is allowed to operate and the texture of the soil.

The suitability of any irrigation system mainly depends upon its design, layout and performance. Due to its merits and positive effects, drip irrigation has become rapidly popular in India and also the state governments are promoting drip irrigation on a large scale by providing subsidy. The advantage of using a drip-irrigation system is that it can significantly reduce soil evaporation and increase water use efficiency by creating a low, wet area in the root zone. World over, the studies indicate that drip irrigation results in 30–70 percent water saving and yield increase by about 40–100 percent or even more compared to surface irrigation methods. Due to water shortages in many parts of the world today, drip irrigation is becoming quite popular. In 2000, more than 73 percent of all agricultural fields in Israel were irrigated using drip irrigation systems and 3.8 Mha worldwide were irrigated using drip irrigation systems. By 2008, total world agriculture area was 1,628 Mha and 277 Mha was under irrigation and 6 Mha were drip irrigated. In India, there has been a tremendous growth in the area under micro irrigation during the last 15 years. In India, the area under drip increased from a mere 1,500 ha in 1985 to 70,859 ha in 1991–1992 and at present, around 0.35 Mha area is under drip irrigation with the efforts of the Governments of India and the States. The National Committee on Plasticulture Applications in Horticulture (NCPAH), Ministry of Agriculture, Government of India (GoI) has estimated a total of 27 Mha area in the country with potential of drip irrigation application. Therefore, there is a vast scope for increasing the area under drip irrigation.

This chapter discusses the research results on the clogging mechanism and initial adsorption mechanism in micro irrigation system.

## 6.1.1  CLOGGING

Due to limited water resources and environmental consequences of common irrigation systems, drip irrigation technology is getting more attention and playing an important role in agricultural production, particularly with high value cash crops such as greenhouse plants, ornamentals and fruits. Therefore, use of drip irrigation systems is rapidly increasing around the world. Despite its advantages, in drip irrigation system, emitter clogging is one of the major problems, which can cause large economic losses to the farmers. Emitter clogging is directly related to the quality of the irrigation water, which includes factors such as suspended solid particles, chemical composition and microbes, and also insects and root activities within and around the tubing can also cause problems. The major operational difficulties in drip irrigation method arise from the clogging of dripper, which reduces the efficiency and crop yield.

Emitter clogging continues to be a major problem in micro irrigation systems. For high-valued annual crops and perennial crops, where the longevity of the system is especially important, emitter clogging can cause large economic losses. Even though information is available on the factors causing clogging, control measures are not always successful. These Problems can be minimized by appropriate design, installation, and operational practices. Reclamation procedures to correct clogging increase maintenance costs, and unfortunately, may not be permanent. Clogging problems often discourage the operators, and consequently cause abandonment of the system and return to a less efficient irrigation application method.

Emitter clogging is directly related to the quality of the irrigation water, which includes factors such as suspended particle load, chemical composition, and microbial type and population. Insect and root activities within and around the tubing can cause similar problems. Consequently, these factors dictate the type of water treatment or cultural practices necessary for clogging prevention. Clogging problems are often site-specific and solutions are not always available or economically feasible. No single foolproof quantitative method is available for estimating the clogging potential. However, by analyzing the water for some specific constituents, possible problems can be anticipated and control measures formulated.

Most tests can be made in the laboratory. However, some analyzes must be made at the sampling sites because rapid chemical and biological changes can occur after the source of water is introduced into the drip irrigation system. Water quality can also change throughout the year so that samples should be taken at various times over the irrigation period. These are further rated in terms of an arbitrary clogging hazard ranging from minor to severe. Clogging problems are diminished with lower concentrations of solids, salts and bacteria in the water. Additionally, clogging is aggravated by water temperature changes.

The causes of clogging differed based on emitter dimension [1, 2] and positions in lateral. De Kreij et al. [3] found that the tube emitter system with laminar flow suffered more severe clogging than the labyrinth system with turbulent flow, because laminar flow is predisposed to clogging. Emitter clogging was recognized as inconvenient and one of the most important concerns for drip irrigation systems, resulting in lowered system performance and water stress to the nonirrigated plants [8]. Partial and total plugging of emitters is closely related to the quality of the irrigation water, and occurs as a result of multiple factors, including physical, biological and chemical agents [5]. Favorable environmental conditions in drip irrigation systems can cause rapid growth of several species of algae and bacteria resulting in slime and filament build up, which often becomes large enough to cause biological clogging [5].

On the other hand, some of the bacterial species may cause emitter clogging due to the precipitation of iron, manganese and sulfur minerals dissolved in irrigation water. Filtration, chemical treatment of water and flushing of laterals are means generally applied to control emitter clogging. Physical clogging can be eliminated with the use of fine filters and screens. Emitter clogging is directly related to irrigation water quality, which is a function of the amount of suspended solids, chemical constituents of water and microorganism activities in water. Therefore, the above-mentioned factors have a strong influence on the precautions that will be taken for preventing the plugging of the emitters. During irrigation some clogging due to microorganism activities take place in cases when wastewater is used

In micro irrigation systems that are characterized by a number of emitters with narrow nozzles, irrigation uniformity can be spoilt by the clogging of the nozzles with particles of chemical character. Chemical problems are due to dissolved solids interacting with each other to form precipitates, such as the precipitation of calcium carbonate in waters rich in calcium and bicarbonates [9]. In locations where the amount of the ingredients as dissolved calcium, bicarbonate, iron, manganese and magnesium are excessive in irrigation water, the emitters are clogged by the precipitation of these solutes [7]. Chemical precipitation can be controlled with acid injection. However, biological clogging is quite difficult to control. Chlorination is the most common practice used in the prevention and treatment of emitter clogging caused by algae and bacteria [6, 10]. Calcium hypochlorite, sodium hypochlorite and particularly chlorine are the most common and inexpensive treatments for bacterial slimes and for inhibition of bacterial growth in drip irrigation systems. However, continuous chlorination would increase total dissolved solids in the irrigation water and would contribute to increased soil salinity [7].

### 1.1.2 NATURE AND SCOPE OF MECHANICS OF CLOGGING

The various components of drip irrigation are made up of plastic and polymer materials and due to their flexibility and other advantages over metals. There are more than 200 various components and materials used in drip irrigation installed at farm level. But, the major component coming in contact of water includes: Poly tube (material: linear low density poly ethylene), dripper (material: poly propylene), pipes (material: high density poly ethylene and polyvinyl chloride) and silicon diaphragm in emitters (material: silicon). It is found that the initiation of clogging starts at molecular level. Considering this, it is very necessary to study the clogging mechanism and initial adsorption mechanism in drip irrigation system.

The phenomenon of adsorption is the collection or accumulation of one substance on the surface of another substance. In adsorption mainly the surface of solids is involved and accumulated substances remain on the surface. Adsorption therefore, is said to be a surface phenomenon as it occurs because of attractive forces exerted by atoms or molecules present at the surface of the adsorbent. These attractive forces may be of two types: i) physical forces (cohesive forces or Vander Waals forces) ii) chemical forces (chemical bond forces). Thus, an attempt on study of adsorption parameters of these materials of silicon diaphragm, poly vinyl chloride (PVC), poly propylene (PP), high density poly ethylene (HDPE) and linear low density poly ethylene (LLDPE) used in drip irrigation would give insight into which material is more susceptible for adsorption and possible solutions to reduce clogging mechanism in the initial stage itself.

Considering all the above aspects, the present study was conducted in micro irrigated field. Partial and total plugging of emitters is closely related to the quality of the irrigation water, and occurs as a result of multiple factors, including physical, biological and chemical agents [5].

Chemical problems are due to dissolved solids interacting with each other to form precipitates, such as the precipitation of calcium carbonate in waters rich in calcium and bicarbonates [9]. In locations where the amount of the ingredients as dissolved calcium, bicarbonate, iron, manganese and magnesium are excessive in irrigation water, the emitters are clogged by the precipitation of these solutes [7]. Chlorination is the most common practice used in the prevention and treatment of emitter clogging caused by algae and bacteria [6, 10].

### 6.2 MATERIAL AND METHODS

### 6.2.1 THEORETICAL PREDICTION OF INITIAL ADSORPTION MECHANISM OF CATIONS AND ANIONS

Freundlich and Langmuir equations are the equations generally used to study the monolayer and multilayer adsorption of cations and anions. In this study, the Freundlich and Langmuir equations were used to predict the initial adsorption mechanism of cations and anions on the surface of different polymers. The adsorption data was summarized using these equations. The **Freuindlich Equation** is given below:

$$x = (a\ c)^{1/n} \tag{1}$$

where: $x$ = is the amount of adsorbed cations (mg.kg$^{-1}$ material) and c = residual solution metal concentration (mg.L$^{-1}$) in equilibrium solution. Two parameters 'a' and '1/n' are calculated using the linear transformation:

$$\text{Log } x = \text{Log } a + (1/n) \text{ Log } c \tag{2}$$

**Langmuir Equation** is defined below:

$$x = kb\ c/ (1+ kc) \tag{3}$$

The parameters b and k are calculated using the linear transformations

$$c/x = c/b + 1\ (k/b) \tag{4}$$

where: b = the maximum monolayer adsorption capacity (mg of metal. kg$^{-1}$ of material) and k = being an affinity coefficient related to the bonding energy (L.mg$^{-1}$).

## 6.2.2 EXPERIMENTAL SETUP

For the purpose of this research, a stock solution of about 2,000 ppm (EC 3996 us at 30.2°C and pH 8.05 at 30.2°C) of MgSO$_4$, K$_2$SO$_4$, NaHCO$_3$ and NaCl was prepared. The 1.6 g of each MgSO$_4$ K$_2$SO$_4$, NaHCO$_3$, NaCl were mixed in 1 L of distiled water (TDS = 0) and stock solution of 2000 ppm were made. After preparing 2000 ppm of stock solution, again solutions of 2000, 1600, 1200, 1000, 800, 400, 200,100 and 50 ppm were prepared. For preparing 2000 ppm stock solution, 1.6 g each of MgSO$_4$, K$_2$SO$_4$, NaHCO$_3$, and NaCl were added to 1 L of distiled water.

After preparing the solution, 25 g each of different granules of plastic materials of PVC, HDPE, LLDPE, PP and silicon diaphragm were added in poly bottles. All these poly bottles were then kept in mechanical shaker of 160 rpm for 2 h and then suspension was filtered to get a clear aliquot and salts content (cations and anions present in solution of total dissolved salts, i.e., calcium and magnesium ions) and this was determined with the help of flame photometer. The difference between the amounts of total salts in the solution after shaking and amount initially present was taken as the amount of salts adsorbed by material on the walls of drip component. Then the readings for TDS (total dissolved solid), for different samples were taken by conductivity meter for 24, 48, 72, 7 days and 15 days. Adsorption of NaCl on different materials was determined separately. Conductivity meter was used for identifying the TDS of NaCl and flame photometer was used for predicting the ions of calcium, magnesium, carbonates, and sulfate. Experiment setup is shown in Figs. 1 and 2.

**FIGURE 1**   Poly bottles on shaker.

**FIGURE 2**   Measuring TDS and EC.

## 6.3   RESULTS AND DISCUSSION

Adsorption characteristic curves for component of drip irrigation with respect to PP, HDPE, LLDPE, and Silicon Diaphragm were studied and they are plotted in Fig. 3 using Eqs. (1)–(4): the Langmuir and Freundlich equations. Various adsorption characteristics for different materials for different days were different. Most of the materials indicated similar adsorption characteristics with respect to Langmuir and Freundlich adsorption curves. Adsorption equation for Langmuir with respect to $c/x$ and adsorption and graded concentration of plotted graph was observed and there was no positive corelation between adsorbed material and equilibrium concentration for silicon diaphragm PVC, PP, and LLDPE.

The results on adsorption of TDS by silicon diaphragm, PVC, PP, HDPE and LLDPE materials in drip irrigation system showed different adsorption characteristics for different periods of 1, 2, 3, 7 and 15 days (Table 1). The Freundlich adsorption 'a' values for silicon diaphragm (SD) were 30.1, 27.7, 25.9, 16.9 and 17.9 ppm; for PVC were 30.2, 48.1, 173.4, 192.3 and 299.9 ppm; for PP were 2.2, 5.9, 13.4, 50.1 and 80.4 ppm; for HDPE were 8.9, 7.2, 29.3, 85.1 and 132.4 ppm; and for LLDPE were 51.6, 69.7, 125.9, 205.6 and 263.6 ppm. Overall, it can be stated that for TDS, there was no positive corelation in case of Langmuir adsorption characteristics as indicated by wide scattering of data in the plots of the graphs with lower values of $R^2$ (0 to 0.45).

However, in case of Freundlich parameters, most interesting inference can be drawn as there was positive corelation regarding adsorption of dissolved salts on the surfaces of plastic materials and silicon diaphragm ($R^2 = 0.23$ to $0.86$). Thus, only the Freundlich adsorption characteristics were found existing for total dissolved salts (TDS).

The Freundlich adsorption parameters presented in Fig. 2 revealed that the adsorption maxima 'a' of TDS over the period of time showed decreasing trend for silicon diaphragm ($r^2 = 0.876$), while for other materials indicated increasing trend. The trend adsorption maxima 'a' was LLDPE> PVC> HDPE> PP> SD for overall 15 days. With strong corelation for LLDPE ($r^2 = 0.965$), PVC ($r^2 = 0.943$), HDPE ($r^2 = 0.884$) and PP ($r^2 = 0.873$), it can be interpreted that multilayer adsorption took place on the plastic surfaces. Further, LLDPE was found to be the most susceptible material among all the materials tried. The adsorption characteristics for Freundlich equations for different materials are presented in Table 1.

**TABLE 1**   Langmuir and Freundlich adsorption characteristics.

| Materials | Days | Langmuir parameters $c/x = 1/kb + (1/b)*c$ | | | | | Freundulich parameters $Log(x) = loga + (1/n)*logc$ | | | |
|---|---|---|---|---|---|---|---|---|---|---|
| | | Slope 1/b | Admax 1/slope mg/kg | Intercept 1/kb | AC slope/intercept L/mg | $R^2$ | Intercept loga | a | Slope 1/n | $R^2$ |
| Silicon Diaphragm | 1 | 0 | 0 | 1.405 | 0 | 0.064 | 1.48 | 30.13 | 0.41 | 0.463 |
| | 2 | 0 | 0 | 1.672 | 0 | 0.129 | 1.44 | 27.67 | 0.44 | 0.508 |
| | 3 | 0 | 0 | 2.599 | 0 | 0.455 | 1.41 | 25.88 | 0.50 | 0.843 |
| | 7 | 0 | 0 | 2.217 | 0 | 0.185 | 1.23 | 16.90 | 0.60 | 0.815 |
| | 15 | 0 | 0 | 2.689 | 0 | 0.239 | 1.25 | 17.86 | 0.62 | 0.856 |
| PVC | 1 | 0 | 0 | 0.903 | 0 | 0 | 1.48 | 30.20 | 0.35 | 0.228 |
| | 2 | −0.001 | −1000.00 | 1.898 | −0.001 | 0.089 | 1.68 | 48.08 | 0.37 | 0.476 |
| | 3 | −0.001 | −1000.00 | 2.874 | 0 | 0.091 | 2.24 | 173.38 | 0.20 | 0.235 |
| | 7 | −0.002 | −500.00 | 4.098 | 0 | 0.136 | 2.28 | 192.31 | 0.22 | 0.350 |
| | 15 | −0.003 | −333.33 | 7.021 | 0 | 0.199 | 2.48 | 299.92 | 0.19 | 0.431 |
| PP | 1 | 0.004 | 250.00 | 0.063 | 0.063 | 0.199 | 0.33 | 2.15 | 0.63 | 0.657 |
| | 2 | 0 | 0 | 0.825 | 0 | 0.104 | 0.77 | 5.93 | 0.61 | 0.700 |
| | 3 | −0.001 | −1000.00 | 1.300 | −0.001 | 0.339 | 1.13 | 13.43 | 0.52 | 0.827 |
| | 7 | −0.002 | −500.00 | 3.048 | −0.001 | 0.451 | 1.70 | 50.12 | 0.41 | 0.707 |
| | 15 | −0.001 | −1000.00 | 3.845 | 0 | 0.139 | 1.93 | 84.53 | 0.40 | 0.501 |
| HDPE | 1 | 0 | 0 | 0.602 | 0 | 0.243 | 0.95 | 8.85 | 0.56 | 0.320 |
| | 2 | 0.001 | 0 | 0.848 | 0.002 | 0.232 | 0.86 | 7.19 | 0.67 | 0.540 |
| | 3 | 0 | 0 | 1.795 | 0 | 0.030 | 1.47 | 29.31 | 0.53 | 0.485 |
| | 7 | 0 | 0 | 3.490 | 0 | 0.076 | 1.93 | 85.11 | 0.41 | 0.516 |
| | 15 | 0.001 | 1000.00 | 5.221 | 0 | 0.155 | 2.12 | 132.43 | 0.38 | 0.530 |

**TABLE 1** *(Continued)*

| Materials | Days | Langmuir parameters $c/x = 1/kb + (1/b)*c$ | | | | | Freundulich parameters $Log(x)=loga+(1/n)*logc$ | | | |
|---|---|---|---|---|---|---|---|---|---|---|
| | | Slope 1/b | A d m a x 1/slope mg/kg | Intercept 1/kb | AC slope/ intercept L/mg | $R^2$ | Intercept loga | a | Slope 1/n | $R^2$ |
| | 1 | 0 | 0 | 5.687 | 0 | 0.175 | 1.71 | 51.64 | 0.57 | 0.769 |
| | 2 | 0 | 0 | 7.082 | 0 | 0.279 | 1.84 | 69.66 | 0.54 | 0.796 |
| LLDPE | 3 | −0.001 | −1000.00 | 9.133 | 0 | 0.285 | 2.10 | 125.89 | 0.47 | 0.739 |
| | 7 | −0.001 | −1000.00 | 11.670 | 0 | 0.303 | 2.31 | 205.59 | 0.41 | 0.675 |
| | 15 | −0.001 | −1000.00 | 14.600 | 0 | 0.331 | 2.42 | 263.63 | 0.40 | 0.671 |

The results in Table 2 indicate that, among the various plastic and silicon diaphragm materials, the clogging due to adsorption mechanism over a period of 15 days was maximum on LLDPE (11.10%) followed by PP (5.82%), HDPE (4.91%), silicon diaphragm (3.27%) and PVC (3.03%) in that order. The clogging rates in case of LLDPE were 5.28 percent (i.e., 1.91 times more), 6.19 percent (2.26 times more), 7.83 percent (3.39 times more) and 8.07 percent (3.36 times more) over PP, HDPE, silicon diaphragm and PVC materials, respectively. Thus, the LLDPE material was found more susceptible to clogging.

The overall percentage adsorption of NaCl on LLDPE ranged from 0.08 to 0.44 from day 1 to day 15. The percentage adsorption was maximum (0.44%). Similarly for PVC the maximum percentage adsorption was (0.38%). In the same way maximum percentage adsorption for PP, HDPE, and LLDPE were (0.37%), (0.38%), (0.44%) respectively (Table 3). It was observed that the maximum percentage adsorption (0.44%) took place on the surface of LLDPE compared to other materials. This revealed that LLDPE was more susceptible for clogging.

Table 4 indicated the adsorption rates of calcium, magnesium, calcium carbonate, sulfate on the surface of silicon diaphragm, PVC, PP, HDPE and LLDPE. The adsorption rate of calcium over a period of 15 days was maximum on HDPE (26.32 ppm and 0.35%) followed by silicon diaphragm (12.94 ppm and 0.17%), LLDPE (10.12 ppm and 0.13%) and PVC (9.11 ppm and 0.12%), respectively. There was no adsorption effect of calcium on PP at all. The adsorption rate of calcium was maximum at the concentration of 804 ppm (1.51%) on silicon diaphragm, 220 ppm (4.14%) on PVC, 113 ppm (3.58%) on HDPE and 220 ppm (0.65%) on LLDPE, respectively.

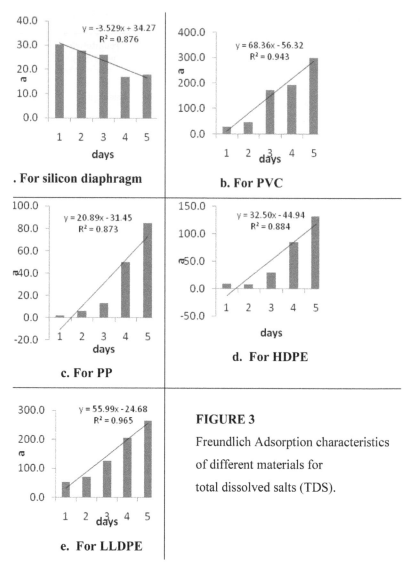

**FIGURE 3**  Freundlich Adsorption characteristics of different materials for total dissolved salts (TDS). (a) For silicon diaphragm; (b) For PVC; (c) For PP; (d) For HDPE; (e) For LLDPE.

The adsorption rate of magnesium over a period of 15 days was maximum on silicon diaphragm (38.04 ppm and 0.50%) followed by HDPE (28.19 ppm and 0.37%), PP (23.93 ppm and 0.31%), PVC (18.4 ppm and 0.24%) and LLDPE (15.33 ppm and 0.20%), respectively. The adsorption rate of magnesium was maximum at the concentration of 240 ppm (8.69%) on silicon diaphragm, 116 ppm (1.06%) on PVC, 70 ppm (1.76%) on PP, 220 ppm (1.11%) on HDPE and 220 ppm (8.69%) on LLDPE, respectively.

**TABLE 2**  Adsorption of TDS for different materials.

| No. of days | Overall percentage of adsorption on different materials | | | | | | | | | |
| | Silicon dia-phragm | | PVC | | PP | | HDPE | | LLDPE | |
| | Total, ppm | % | Total, ppm | % | Total, ppm | % | Total, ppm | % | Total, ppm | % |
| 1 | 116 | 1.51 | 85 | 1.12 | 93 | 1.22 | 120 | 1.59 | 526 | 6.94 |
| 2 | 126 | 1.64 | 130 | 1.71 | 236 | 3.09 | 173 | 2.28 | 582 | 7.68 |
| 3 | 153 | 1.99 | 144 | 1.89 | 257 | 3.37 | 253 | 3.42 | 636 | 8.40 |
| 7 | 219 | 2.85 | 184 | 2.41 | 328 | 4.30 | 308 | 4.06 | 703 | 9.28 |
| 15 | 251 | 3.27 | 231 | 3.03 | 444 | 5.82 | 372 | 4.91 | 841 | 11.10 |

**TABLE 3**  Percent adsorption of NaCl.

| No. of days | Overall percentage of adsorption on different materials | | | | |
| | Silicon diaphragm | PVC | PP | HDPE | LLDPE |
| 1 | 0.14 | 0.08 | 0.09 | 0.09 | 0.08 |
| 2 | 0.21 | 0.14 | 0.16 | 0.17 | 0.13 |
| 3 | 0.28 | 0.21 | 0.21 | 0.23 | 0.19 |
| 7 | 0.33 | 0.31 | 0.31 | 0.30 | 0.28 |
| 15 | 0.39 | 0.38 | 0.37 | 0.38 | 0.44 |

The adsorption rate of calcium carbonate over a period of 15 days was maximum on PVC (13.2 ppm and 0.17%) followed by HDPE (3 ppm and 0.04%). There was no adsorption effect of calcium carbonate on silicon diaphragm, PP and LLDPE at all. The adsorption rate of calcium carbonate was maximum at the concentration of 2140 ppm (0.62%) on PVC and 805 ppm (0.37%) on HDPE, respectively.

The adsorption rate of sulfate over a period of 15 days was maximum on PVC (99.91 ppm and 1.31%) followed by PP (82.55 ppm and 1.08%), HDPE (79.9 ppm and 1.05%), silicon diaphragm (55.54 ppm and 0.72%) and LLDPE (41.36 ppm and 0.55%) respectively. The adsorption rate of sulfate was maximum at the concentration of 1620 ppm (1.03%) on silicon diaphragm, 810 ppm (5.93%) on PVC, 805 ppm (1.94%) on HDPE and 2130 ppm (0.67%) on LLDPE respectively. Overall it concluded that magnesium, sulfate and calcium which affect more on the above materials whereas there was least effect of carbonate on the surface of the materials.

**TABLE 4**   Effect of different ions on different materials.

| S. no. | Materials | Calcium | | Magnesium | | Carbonate | | Sulphate | |
|---|---|---|---|---|---|---|---|---|---|
| | | $A_d$ | % | $A_d$ | % | $A_d$ | % | $A_d$ | % |
| 1 | Silicon diaphragm | 12.94 | 0.17 | 38.04 | 0.50 | 0 | 0 | 55.54 | 0.72 |
| 2 | PVC | 9.11 | 0.12 | 18.4 | 0.24 | 13.2 | 0.17 | 99.1 | 1.31 |
| 3 | PP | 0 | 0 | 23.93 | 0.31 | 0 | 0 | 82.55 | 1.08 |
| 4 | HDPE | 26.32 | 0.35 | 28.19 | 0.37 | 3 | 0.04 | 79.9 | 1.05 |
| 5 | LLDPE | 10.12 | 0.13 | 15.33 | 0.20 | 0 | 0 | 41.36 | 0.55 |

$A_d$ = Adsorption.

## 6.4   SUMMARY

The research study in this chapter clearly indicated that LLDPE material (linear low density polyethylene), which is used for tubing in drip irrigation technology, was more susceptible for clogging compared to silicon diaphragm, PVC, PP and HDPE and hence the same could be avoided in the material construction of any component of drip irrigation system. Therefore, drip components like emitters and tubings should be made up of PVC, silicon diaphragm, HDPE and PP materials in that order of preference to reduce the clogging problems in drip irrigation system.

## KEYWORDS

- adsorption rate
- Clogging
- drip irrigation
- emitter
- India
- LLDPE
- polluted water
- silicon diaphragm

## REFERENCES

1. Ahmed, B.A.O., Yamamoto, T., Fujiyama, H., and Miyamoto, K., 2007, Assessment of emitter discharge in micro irrigation system as affected by polluted water. *Irrigation Drainage System* **21**: 97–107.
2. Chigerwe, J., Manjengwa, N., and Van der Zaag, P., 2004, Low head drip irrigation kits and treadle pumps for smallholder farmers in Zimbabwe: a technical evaluation based on laboratory tests. Phys. Chem. Earth 29, pp. 1049–1059.
3. De Kreij, C., Van der Burg, A.M.M., and Runia, W.T., 2003, Drip irrigation emitter clogging in Dutch greenhouses as affected by methane and organic acids. *Agriculture Water Management* **60**: 73–85.

4. Evans, R.G., 2000, Micro Irrigation. Washington State University, Irrigated Agriculture Research and Extension Center, 24106 North Bunn Road Prosser, WA 99350, USA.
5. Gilbert, R.G., Nakayama, F.S., Bucks, D.A., French, O.F., and Adamson, K.C., 1981, Trickle irrigation: emitter clogging and flow problems. *Journal of Agricultural Water Management* **3**: 159–178.
6. Hills, D.J., Brenes, M.J., 2001, Microirrigation of wastewater effluent using drip tape. *Journal of Applied Agriculture Engineering* **17**(3): 303- 308.
7. Hills, D.J., Navar, F.M., and Waller, P.M., 1989, Effects of chemical clogging on drip-tape irrigation uniformity. *Transactions of American Society of Agriculture Engineers (ASABE),* **32**(4): 1202–1206.
8. Povoa, A.F., and Hills, D.J., 1994, Sensitivity of micro irrigation system pressure to emitter plugging and lateral line perforations. *Transaction of American Society of Agriculture Engineers* **37**(3): 793–799.
9. Wu, F., Fan, Y., Li, H., Guo, Z., Li, J., and Li, W., 2004, Clogging of emitter in subsurface drip irrigation system. *Transaction of China State Agriculture Engineering,* **20**(1): 80–83.
10. Yuan, Z., Waller, P.M., and Choi, C.Y., 1998, Effect of organic acids on salt precipitation in drip emitters and soil. *Transactions of American Society of Agriculture Engineers,* **41** (6): 1689–1696.

## CHAPTER 7

# WATER MOVEMENT IN DRIP IRRIGATED SANDY SOILS

ERIC SIMONNE, APARNA GAZULA, ROBERT HOCHMUTH,
and JIM DEVALERIO

## CONTENTS

*The authors acknowledge the financial support from the University of Florida Agricultural Experiment Station, the University of Florida Cooperative Extension Service, and the USDA T-STAR and On-Farm SARE grant programs. The authors also thank all the cooperating growers, industry partners, and University of Florida's Institute of Food and Agricultural Science (UF/IFAS) research and Extension personnel (Mace Bauer, Michael Dukes, John Duval, George Hochmuth, Elizabeth Lamb, Gene McAvoy, Rafael Muñoz-Carpena, Teresa Olczyk, Ed Skvarch, and David Studstill) for their valuable contributions to this project.

## 7.1  INTRODUCTION

Plasticulture is a technology consisting of the combined use of raised beds, drip irrigation and polyethylene mulch. In Florida, USA [7], this technology is widely used in vegetable crop production, such as: tomato (Solanum lycopersicum; 13,000 ha), bell pepper (Capsicum annuum; 7,500 ha), eggplant (Solanum melongena; no official statistic), watermelon (Citrullus lanatus; 10,500 ha), muskmelon (Cucumis melo; no official statistic), summer and zucchini squash (Cucurbita pepo; 5,000 ha), and strawberry (Fragaria x ananassa; 4,000 ha). These crops are grown mainly on the loamy soils of the Florida Panhandle, the deep sandy soils of North and Central Florida, the spodic soils of North-east and South West Florida and the calcareous soils of South Miami-Dade County (Fig. 1).

Because of the low water holding capacity (<10%) and the low organic matter content (<2%) of Florida sandy soils, keeping nutrients in the root zone (and out of the waterways) requires the specific knowledge of what portion of the soil is wetted during water application through drip irrigation, and how much water can be held in the root zone.

Therefore, this chapter includes how to: (1) Determine the size and shape of the wetted zone, (2) Incorporate our results into a rule for splitting irrigation events, and (3) Develop and evaluate an educational program to help North Florida watermelon and strawberry growers who use drip irrigation adopt irrigation and nutrient best management practices.

### 7.1.1  ADVANTAGES OF PLASTICULTURE

The main advantages of plasticulture include [38]: flexible application of water, nutrients and chemicals through the drip tape; crop earliness; weed control; effective soil fumigation; uniformity of water application; and double (or triple) cropping. In addition, drip irrigation is a low-volume method of irrigation compared to overhead irrigation or subsurface irrigation [5]. Typically, vegetable crop water requirements (cm of water depth) are 89–102 for seepage irrigation, 46–51 for overhead irrigation, and 30–38 for drip irrigation. Plasticulture is also seen today as a means of reducing the risk of foodborne illnesses when harvested plant parts are isolated from soil contamination by the polyethylene mulch and when irrigation water does not directly contact the harvested plant parts [39]. Therefore, plasticulture offers advantages during field production, postharvest period, and helps conserve natural resources. However, plasticulture is more expensive than bare-ground production. Plasticulture requires continuous maintenance of the drip-irrigation system, and demands higher management skills for the correct injection and operation of the drip-irrigation system. Plasticulture also creates the need for polyethylene mulch and drip tubing disposal at the end of the cropping season [24, 31].

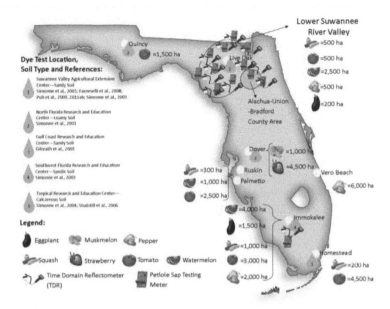

**FIGURE 1** Major vegetable producing regions of Florida and location of dye tests and on-farm demonstrations conducted between 2000 and 2013 [22]; and unofficial regional acreage based on industry knowledge and state statistics, "Ewing, J. 2013. 2012 Florida agriculture by the numbers, US Dept. Agric. – Natl. Agric. Stat. Serv. – Fla. Field Office, http://florida-agriculture.com/brochures/P-01304.pdf."

## 7.1.2   BEST MANAGEMENT PRACTICES

Plasticulture is also a best management practice (BMP) in Florida because drip irrigation allows for better control of water and nutrient application [9, 10, 18]. The U.S. Clean Water Act of 1977 [41] mandated that all states in USA: (a) Develop a classification of water bodies based on their intended use (such as drinking, shellfish production, fishing, swimming, irrigation, or navigation); (b) Develop narrative or numerical criteria for water quality parameters; and (c) Develop a corrective process to restore the water quality of impaired water bodies. As the list of impaired water bodies in Florida was developed by the Florida Department of Environmental Protection, the Florida Watershed Restoration Act of 1999 [12] was passed to coordinate restoration efforts. As stated in the Clean Water Act, the maximum amount of pollutants that a water body can accept and still maintain its water quality parameters consistent with its intended use is termed Total Maximum Daily Load (TMDL; [11]). The measurement and allocation of loads to specific point sources or specific categories of nonpoint sources of pollution in a water basin are established in the Basin Management Action Plan (BMAP). Cultural practices implemented at the field level to reduce nutrient loss below the root zone are called BMPs.

### 7.1.3  BEST MANAGEMENT PRACTICE (BMP) FOR VEGETABLE CROPS IN FLORIDA

Because vegetables are grown with intensive fertilization and irrigation, the Florida Department of Agriculture and Consumer Services, the Florida Department of Environmental Protection, the five water management districts in Florida, and the University of Florida's Institute of Food and Agricultural Science (UF/IFAS) have cooperated to draft, test, and adopt BMPs for vegetable crops grown in Florida ([9]; http:// bmp.ifas.ufl.edu/). By definition, BMPs are specific cultural practices that maintain crop profitability, while reducing the environmental impact of production. Initially, the BMP program was designed to be a voluntary, incentive-based program. Landowners officially enroll in the BMP program by signing a notice of intent to implement and by following a BMP implementation plan that was specifically designed for their land use. By law, farmers with land enrolled in the BMP program receive a presumption of compliance with water quality standards and become eligible for cost-share programs available to offset the cost of BMP implementation. Modifications of the Florida Watershed Restoration Act in 2005 required that landowners whose land is not enrolled in the BMP program must develop their own monitoring program and show that the discharge coming out of their land is in compliance with water quality standards. Because of the costs and burden associated with this provision, landowners are choosing the BMP program over the self-monitoring option. The BMP program has become "quasi voluntary" but it is better the think about BMPs as a way to farm more efficiently.

### 7.1.4  CURRENT FERTILIZATION AND IRRIGATION RECOMMENDATIONS FOR VEGETABLE CROPS GROWN WITH PLASTICULTURE IN FLORIDA

The BMPs for vegetable crops are based on the University of Florida's Institute of Food and Agricultural Science's recommendations for vegetable production in Florida [25], including fertilization [22] and irrigation [6]. Based on soil test results, typical fertilization recommendations for vegetable crops grown with plasticulture consist of a base rate and a supplemental application allowance. In the base rate: (a) 0% to 50% of the total N and K, and 100% of the P and micronutrients are preplant incorporated in the raised bed; and (b) the remaining N and K are injected through the drip tape following established schedules. Supplemental fertilizer allowance consists of the application of additional fertilizer: (a) after a leaching rain event; (b) when the results of petiole sap testing or tissue analysis are "low"; and/or (c) when the harvest season is extended. Leaching rain events are defined as 7.7 cm of rain occurring in 3 days or 10.2 cm occurring in 7 days [6]. Sampling protocols and thresholds for the interpretation of sap testing and tissue analysis are available for most vegetable crops [19, 25]. Typical irrigation recommendations for vegetable crops grown in Florida consist of:

   (a)  Calculating a daily target water application rate based on crop age and actual crop evapotranspiration (ETc, [1, 6]);
   (b)  Fine tuning the target water amount based on real-time soil moisture measurements (using tensiometers, Time Domain Reflectometry probes or other soil moisture sensors);
   (c)  Determining the contribution of rainfall;

(d)  Knowing the total amount of water that can be held in the root zone and developing a rule for splitting irrigation amounts; and

(e)  Keeping records of irrigation system use, performance and maintenance.

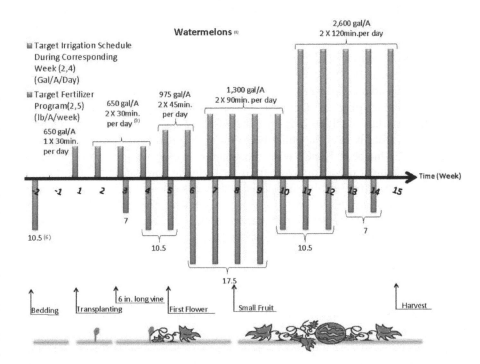

**FIGURE 2**   University of Florida's irrigation and fertilization recommendations for watermelon grown in Florida. ([1] Olson, S.M., P.J. Dittmar, P.D. Roberts, S.E. Webb and S.A. Smith.2012. Cucurbit production in Florida, HS725, Electronic Database Information System, Univ. of Fla., Gainesville, FL. http://edis.ifas.ufl.edu/pdffiles/cv/cv12300.pdf for complete recommendations for watermelon production in Florida. [2] Standard bed spacing for watermelon is 8-ft. (244 cm); 1 A=5,445 linear bed feet (1 ha = 4,098 m of bed). [3] Target volumes were converted to estimated time assuming 100% irrigation efficiency and 24 gal/100 lbf/hr (309 L/100 m/hr) drip-tape flow rate. [4] Adjustments to the target injection schedule are made based on daily soil moisture measurements. [5] Adjustments to the target fertilization schedule are made based on weekly sap testing or tissue sampling.).

Fertilization recommendations may be achieved with any combination of appropriate fertilizer sources. Irrigation recommendations may be achieved using a wide range of flow rates and emitter spacings. Based on the flow rates, drip tapes are classified as low flow (<248L/100 m/hr), medium flow (248–373 L/100 m/hr) or high flow (>373L/100 m/hr). Typical drip-tape emitter spacings used in vegetable production are 10.2, 20.4, 30.6, and 45.7 cm. Hence, fertilization and irrigation may be achieved with a wide range of practices. For example, recommended fertilization and irrigation schedules for watermelon are shown on Fig. 2.

Because drip irrigation applies nutrients and water near the drip tape, implementing fertilization and irrigation recommendations correctly requires the use of appropriate units to express water and nutrient application rates, the knowledge of the conventions used in the recommendations, and correct calculations. For example, a fertilizer rate of 100 kg/ha applied to a bare ground crop will be done using the broadcast method. With this method, each square meter in the field receives the same amount (100 kg/10,000 $m^2$, Note: 10,000 $m^2$ = one ha) of fertilizer. This approach cannot be used in plasticulture as it would lead to fertilizer application everywhere in the field including between the beds. Plants cannot access nutrients between the beds and risk of leaching is high.

Another example is the expression of irrigation amount. Typical units of irrigation are vertical amounts of water. This unit well represents the application of water when overhead, subsurface or flood irrigations are used because these irrigation methods apply water uniformly over the entire field. When drip irrigation is used, no water is applied between the beds. Hence, when plasticulture is used, the mindset of water and nutrient management needs to switch from fertilizing and irrigating field surfaces to fertilizing and irrigating lengths of mulched beds. Consequently, recommendations for vegetable crops grown in Florida are based on a standard bed spacing for each crop. Bed spacing is the distance between the centers of two adjacent raised beds. In original units (and SI units), standard bed spacings are [22]:

(a)  4 ft (122 cm) for strawberry,
(b)  5 ft (153 cm) for muskmelons,
(c)  6 ft (183 cm) for tomato, pepper, eggplant, and summer squash, and
(d)  8 ft (244 cm) for watermelon.

The use of standard bed spacing also determines the length of bed in one acre (lbf = linear bed feet). This is calculated by dividing the number of square feet in one acre (1 acre = 43,560 sq ft; 1 ha = 10,000 $m^2$) by the bed spacing. In the original units of the recommendation, bed spacings of 4, 5, 6, and 8 ft (122, 153, 183 and 244 cm, respectively) correspond to 10,890, 8,712, 7,260, and 5,445 lbf (8,197, 6,534, 5,465, and 4,098 linear meters of bed, lbm), respectively. In this chapter, the symbol "A" is used to represent the length of bed found in one acre planted at standard bed spacing. Hence, a recommendation of "**X lbs/A**" represents an amount of X lbs of fertilizer applied uniformly to a length of bed for a crop planted at the standard bed spacing. Because in commercial agriculture alley ways are left unplanted every third or sixth bed and because crops may be planted at spacings other than the standard bed spacing, the actual number of linear bed feet at standard bed spacing needs to be calculated for each field.

Based on their experience and understanding of the conventions used in the recommendations, vegetable growers in Florida may have different levels of expertise and proficiency at managing nutrient and water applications (Table 1). A state-wide implementation of water and nutrient BMPs requires that the educational programs and efforts conducted by UF/IFAS Extension and state agency personnel be at the field level. Ideally, all fertilization and irrigation practices should be consistent with Level 5 (Table 1). When excess irrigation is used, nutrients tend to leach, and crops tend to have "low" levels of nutrients.

**TABLE 1** Levels of nutrient and water management for vegetable crops grown with plasticulture.

| Management level | | Nutrient management | Irrigation |
|---|---|---|---|
| Level | Rating | method | scheduling method |
| 0 | None | Guessing | Guessing |
| 1 | Very low | Soil testing and still guessing | Using the "feel and see" method |
| 2 | Low | Soil testing and implementing "a" recommendation. Unsure how to use the information | Using systematic irrigation (example: 2 h every day from transplanting to harvest) |
| 3 | Intermediate | Soil testing and understanding how to use the information to implement the recommendation | Using a soil moisture measuring tool to start irrigation |
| 4 | Advanced | Soil testing, understanding how to use the information to implement the recommendation, and monitoring crop nutritional status to adjust fertigation schedules | Using a soil moisture measuring tool to schedule irrigation and apply amounts based on a budgeting procedure. |
| 5 | Recommended | Soil testing, understanding how to use the information to implement the recommendation, monitoring crop nutritional status to adjust fertigation schedules, and practicing year-round nutrient management. Together, these represent following nutrient BMPs. | Using together a water use estimate based on crop plant stage of growth, a measurement of soil moisture, determining rainfall contribution to soil moisture, having a guideline for splitting irrigation and keeping irrigation records. Together, these represent following irrigation BMPs. |

Because, historically, fertilizer cost represents only 10% to 15% of production costs, growers may select fertilizer rates above the recommended one as a convenient, preventive means to avoid yield reductions. While crop yields are usually maintained, nutrient leaching below the root zone tends to occur too. Hence, a reduction in nutrient losses below the root zone will only be achieved through education on irrigation management combined with monitoring of crop nutritional status. Education is more effective when it involves hands-on demonstrations and one-on-one learning experiences.

### 7.1.5 METHODS FOR QUANTIFICATION OF WATER MOVEMENT IN THE SOIL

Water movement in the soil is a rather abstract concept because it is a process that takes time, occurs underground, and depends on soil type. The most common methods used to quantify the movement of water in the soil use radioactive isotopes (deuterated water), chemical tracers (typically halogens) and colored substances (dyes). Radioisotope labeled water is often used for the study of ground water movement when soil sampling is not practical [23]. Bromide (Br⁻) is used as a tracer to study water and solute transport because its background concentration is low and it is not adsorbed by negatively charged soil minerals. Hence, Br⁻ moves with water [13]. Chloride ions (Cl⁻) behave similarly in the soil, but Cl⁻ is ubiquitous in the environment and is often used in fertilizers. This technique involves application of the chemical tracer, soil sampling, and laboratory analyzes. When they do not get adsorbed by soil colloids, dyes such as Brilliant Blue FCF (C.I. Food Blue 2) are a valuable tool for visualizing water flow patterns. Because it is nontoxic to plants and animals, has a water solubility of 200 kg.m⁻³, and is relatively inexpensive ($13/L), Brilliant Blue FCF is well suited for field use [15]. Depending on pH, the dye is either neutral or dissociates to a mono or bivalent anion but it keeps its color [14]. Dye-tracer studies in the field using Brilliant Blue FCF showed that in the top 0–25 cm layer, the degree of dye coverage tended to be larger for the lower irrigation intensities indicating that water flow in the top soil took place through a relatively great proportion of the pores in the soil matrix [16]. The relative speed of movement between chemical tracers and dyes has been determined. Batch adsorption experiments conducted in laboratories suggested that the dye forms ion pairs with $Ca^{2+}$. Field experiment showed that the movement of Brilliant Blue FCF dye is slightly retarded compared with those of the conservative tracer Br⁻ and iodine ions (I⁻; [14]). Nevertheless, because of natural field heterogeneity and the fact that soil sampling and chemical analyzes are not needed, dyes have been used in field experiments on tillage [27], fumigation [17], herbicide applications [2], fertigation management [29, 30], and nutrient leaching [8, 42].

The use of dyes as a teaching tool to improve water and nutrient management relies on the assumption that the movement of the dye during irrigation closely represents that of nutrients, in particular nitrate-nitrogen ($NO_3$-N). When soluble $KNO_3$ was injected at a rate of 45 kg.ha⁻¹ of N (which is twice the highest recommended weekly injection rates of N for most vegetable crops in Florida) together with the dye, $NO_3$-N concentration was significantly greater in the wetted area and in the dyed area (18 mg.kg⁻¹ soil for both) than below it (3 mg.kg⁻¹) for irrigation volumes ranging between 152 and 610 L/100 m [37]. These results suggest that (1) $NO_3$-N did not move faster than the water (it did not "get ahead of the water front") and (2) $NO_3$-N did not get "left behind where it was injected" since differences in $NO_3$-N concentration in the dyed area and above it were not significantly significant. Hence, the movement of the Brilliant Blue FCF may be used as a visual tool to represent the movement of $NO_3$-N in field conditions.

Few reports are available documenting water movement in mulched beds of soil in Florida. In soils with an impermeable layer in Quincy, FL and in Hendry County (Fig. 1), water moved by gravity until it reached the impermeable layer, then moved horizontally until complete emitter-to-emitter coverage was achieved [34]. In a Krome Very Gravely Loam soil of south Miami-Dade County (Fig. 1; with bed rock at the 18–25 cm depth), increasing irrigation volume from 261 to 1,764 L/100 m (using drip tapes with flow rates of 261 to 435 L/100 m/hr (medium to high flow-rate range) and emitter spacings ranging between 10.2 and 30.5 cm), did not have a practical effect on the depth and length of the wetted zone. Because the gravely texture of this soil type does not allow the formation of compact beds, the dimensions of the wetted zone ranged randomly between 11.5 and 20 cm [35, 40]. The determination of how much water can be held in the wetted zone requires the quantification of the vertical and horizontal water movement in mulched beds.

## 7.2 MATERIALS AND METHODS

This report covers a research and Extension project conducted in North Florida between 2000 and 2013, which consisted of two types of activities. First, a series of large, replicated dye tests were conducted at the UF/IFAS Suwannee Valley Agricultural Extension Center in Live Oak, FL on a Blanton-Foxworth-Alpin complex (formerly classified as Lakeland fine sand). Then, on-farm educational demonstrations were conducted on the joint management of irrigation and nutrients (Fig. 1).

### 7.2.1 REPLICATED TRIALS

Replicated dye tests were conducted on newly formed raised beds without plants. They involved a single irrigation event and a disruptive amount of digging. In each test, treatments consisted of duration of irrigation and drip-tape type (flow rate and emitter spacing; Table 2). Each combination of drip tape and duration of irrigation was replicated four times. Irrigation systems consisted of a well, pump, back-flow prevention device, fertilizer injector (model DI16–11, Dosatron, Clearwater, FL), 150-mesh screen filter, 138 kPa pressure regulator, and drip tape (Fig. 3). Dye tests consisted of pressurizing the system, injecting the dye (Terramark SPI High Concentrate, ProSource One, Memphis, TN) at a 1:49 (v:v) dye:water dilution ratio for the first 30 min (which corresponded approximately to 1.24 L/100 m), irrigating for the selected length of time (including the time needed for the dye injection phase), digging longitudinal and transverse sections of the raised bed, taking measurements and making notes on the shape of the colored zone. Selected irrigation durations ranged between 1 and 8 hours.

(a) Photo credit: Eric Simonne, University of Florida.    b. Photo credit: Bob Hochmuth, University of Florida)

**FIGURE 3** Typical irrigation system consisting a well, pump, back-flow prevention device, fertilizer injector (model DI16–11, Dosatron, Clearwater, FL), 150-mesh screen filter, 138 kPa pressure regulator; and drip tape used in replicated dye tests (a) (a. Photo credit: Eric Simonne, University of Florida.) and on-farm demonstrations (b) (b. Photo credit: Bob Hochmuth, University of Florida)).

At the end of each irrigation duration, the drip tape was cut off approximately 7 m from the end of the operating drip tape. Hence, the length of drip tape used was reduced after each irrigation duration. If not compensated, this would increase the operating pressure, which would also change drip-tape flow rate, thereby affecting the results. In order to keep a constant operating pressure throughout the duration of a test, drip tapes identical (in flow rate and length) to the ones used in the test were installed in a portion of the field outside the test area (Fig. 4). The total length of drip tape actively irrigating was maintained constant during the whole test by opening a length of drip tape identical to the one that was closed. Digging was done immediately after the last irrigation duration and measurements of the colored zone depth (D), width (W) and length (L; emitter-to-emitter coverage) were taken on 8 consecutive emitters for D and L and 2 consecutive emitters for W in each replication (Fig. 5). For each drip tape type, the total amount of water applied was calculated by multiplying the nominal drip-tape flow rate by the duration of operation. Responses of D, W and L to irrigation volume applied were determined by regression analysis [32].

**TABLE 2**   Flow rate and emitter spacing of drip tapes used in dye tests conducted in 2002 and 2003 at the UF/IFAS Suwannee Valley Agricultural Extension Center, Live Oak, FL on a 7-m deep Blanton-Foxworth-Alpin complex (formerly Lakeland fine sand).

| Date | Drip tape[1] | | Irrigation duration |
|---|---|---|---|
| | Flow rate | Emitter spacing | |
| | (L/hr/100 m) | (cm) | (hr) |
| 3 Mar. 2002 | 309(c) | 30.5 | 1, 2, 4, 8 |
| 9 May 2002 | 309(c) | 30.5 | 1, 2, 4, 8 |
| 3 Dec. 2002 | 309(c) | 30.5 | 1, 1.5, 2, 2.5 |
| 4 Dec. 2002 | 206(e), 425(e) | 10.2 | 1, 2, 3, 4 |
| | 516(c) | 20.3 | |
| | 219(b), 335(b), 387(a), 516(c) | 30.5 | |
| 25 Mar. 2003 | 309(c) | 30.5 | 5, 6, 7 |
| 3 Dec. 2003 | 206(e), 425(e), 516(c) | 10.2 | 1, 2, 4, 8 |
| | 258(c), 516(c) | 20.3 | |
| | 219(b), 309(b), 309(e), 335(c), 387(a), 464(e), 516(b) | 30.5 | |

[1] Manufacturer: (a) Chapin; (b) Eurodrip; (c) John Deer (formerly Roberts);

(d) Netafim; (e) Queen Gil.

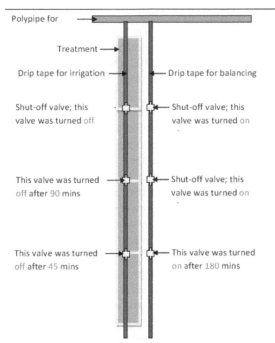

**FIGURE 4** Schematic representation of treatment plots (in green) showing drip tape with shut-off valves (blue line) to create irrigation treatment durations of 45, 90, 180 and 240 min. An additional drip tape with shut-off valves (red line) was used to balance the irrigation pressure by turning on a section of equal length as each section of the treatment tape was turned off (Credit: Bee Ling Poh, University of Florida).

**FIGURE 5** Overview of a dye test conducted on a Blanton-Foxworth-Alpin complex (formerly Lakeland fine sand) at the UF/IFAS Suwanee Valley Agricultural Extension Center, Live Oak, FL, showing longitudinal (left) and transverse (right) sections of the bed (Photo credit: Eric Simonne, University of Florida).

## 7.2.2   ON-FARM DEMONSTRATIONS

Based on the results of the replicated trials, demonstrations involving the joint management of irrigation and nutrients were conducted with commercial strawberry growers in the Alachua-Bradford-Union Counties and with the tomato, bell pepper, muskmelon and watermelon growers in the Lower Suwannee River Valley area (Tables 3 and 4; [21], [36]).

The 15-ha strawberry industry of Alachua-Union-Bradford Counties dates back to the beginning of the century (Fig. 1). In this area of North Florida, winter temperatures have historically prevented growers from targeting the early season market window of December and January. The unpredictable, yet frequent freezing events in the fall require overhead irrigation for frost protection, hoping that lowest night temperatures do not go below −3 to −4°C, which would be damaging to the strawberries and/or the leaves of the strawberry plants. Hence, growers in these counties establish strawberry fields in the late summer and manage water and nutrients throughout the fall and early winter for a targeted full bloom period in March and April. This marketing window coincides with the end of the strawberry season in Central Florida and with the coveted spring direct marketing opportunities for growers. If the spring is dry and insect and disease pressures are low, harvest season often lasts until mid-June.

During the 2007–2013 growing seasons, seven strawberry farms (ranging from 0.25 to 2.5 ha in size) in the region were targeted for irrigation and fertilization management education, and ultimately BMP adoption through weekly farm visits by the local County and Regional Extension Agents. On these farms, plasticulture with drip irrigation was used with the exception of one grower who used granular fertilizer applied under plastic and overhead irrigation. Based on experience, this grower applies approximately 2/3 of the fertilizer preplant, and adds the remaining through the top of the bed by punching holes in plastic near the middle of the growing season. Soils in these farms have an impermeable spodic layer at a depth that varies between 30 and 75 cm. For economic reasons, none of these farms have continued to use soil fumigants. In the late 1990's, irrigation and fertilization practices were largely empirical and rated as "very low" as described in Table 1.

With the combined use of fumigation, plasticulture, drip irrigation and transplants, and the recent shift to seedless (triploid) watermelon production, the 2,300 ha, $30 million watermelon industry in North Florida targets the May-June shipping market window. Growers typically follow production recommendations [26]. In the late 1990's, irrigation and fertilizer practices were rated "low" as described in Table 1. As an attempt to reduce labor required, growers applied preplant incorporated fertilizer at approximately 50% or more of the N and K and 100% of the P and micronutrients, and started injections "when the plants ran out of fertilizer."

**TABLE 3**  Guidelines for conducting a dye test in a commercial field.

| Step No. | What To Do | Notes |
|---|---|---|
| | **Preparation before the Dye Test** | |
| 1 | Prepare the field and form the raised beds as will be done for vegetable production. | Read drip tape information provided by the manufacturer; record nominal flow rates, operating pressure, and maximum bed length. If two drip tapes are used on each bed, calculate the "apparent flow rate" by multiplying by two the drip tape nominal flow rate. |
| | In many instances, this will include the use fumigants, herbicide or preplant fertilizer. Follow labels for reentry intervals | |
| | | Flush irrigation system as recommended |
| 2 | Isolate a section of the field where the dye test will be performed. | Select a representative area of the field close to a water source. Area selected should be between 300 to 600 m of bed length |
| | Soil type should be representative of that of the field. | |
| 3 | In some instances, a temporary water source may be needed. Connect the temporary water source to the section of the field where the dye test will be performed. | Because the field section for the dye test may be much smaller than the entire field, a 3/4 to 1 inch poly pipe may be temporarily used to supply water to this section. |
| 4 | Prepare an injection point for the dye. | The injection point should be close to the field section where the dye will be injected (a few meters). Components of the injection manifold are described in Table 5. |
| 5 | Select a water-soluble dye. | While several dyes are available on the market, satisfactory results have been achieved with the Brilliant Blue FCF dye. Always read and follow the label. Dye costs approximately $13/L. Dyes are highly concentrated and should be used with care. |
| 6 | Mark bed sections 10 to 20 m in length. Sections may be marked with flags or paint. Select a few representative operating times. | Each section will correspond to a different operating time. |
| | | Operating times should represent the different irrigation durations used throughout the season. For example, possible operating times are 1, 2, 3, 4, 6, and 8 hrs. Operating times of 1 to 4 h represent irrigation times, while 6 and 8 h represent excessive irrigations or irrigations needed to apply fumigants. |
| 7 | Flush the system; check for leaks | Essential if system was left idle for few days after installation. |
| | **Execution of the Dye Test** | |
| 8 | Bring stop watch, data collection form and pen, a knife, a tape measure and shovels. | Stop watch will be used to keep track of irrigation times; data collection form will be used to record time and irrigation volume applied (based on water meter readings); the tape measure will be used to measure the wetting zone. |

**TABLE 3**   *(Continued)*

| Step No. | What To Do | Notes |
|---|---|---|
| 9 | Bring a 20-L bucket | May be used to predilute concentrated dye. |
| 10 | Turn on the water and pressurize system. | Read and record pressures at gauges. |
|  |  | Check and repair leaks in the field |
| 11 | Start injecting dye once system has been charged | Read and record time and initial reading on water meter |
| 12 | At preselected times (see step 6), cut and tie the tapes at premarket spots (see step 6) | Sections receiving the shortest irrigation times should be placed at the farthest end of the test. |
|  |  | Monitor pressure changes at gauges 1, 2, and 3 after each section is tied. |
| 13 | For each section, dig a transverse (perpendicular to the bed axis) and a 4-foot longitudinal (parallel to the bed axis) section. | Holes should be deep enough to see the bottom of the dye. The dye may appear faded immediately after digging, but soil drying will improve contrast. When digging, always select the sides best exposed to direct sun light so that soil will dry faster. |
| 14 | Measure and record the position of the water front for each visible emitter. | Observe the shape of the dye pattern; notice uniformity. Record depth and width on transverse bed sections, and depth and length on the longitudinal bed sections. Note that different numbers of emitters (and therefore, measurements) may be done on the transverse and longitudinal sections. |
| 15 | Disconnect the injection point to allow drip irrigation system to operate normally | The execution of the dye test should not have long-term effects on drip system operation and design. |

Adapted from: Simonne, E., D. Studstill, M. Dukes, J. Duval, R. Hochmuth, E. Lamb, G. McAvoy, T. Olczyk, and S. Olson. 2004. How to conduct an on-farm dye test and how use its results to improve drip irrigation management in vegetable production, EDIS, HS 980, http://edis.ifas.ufl.edu/HS222.

**TABLE 4**   Description of typical injection manifold components used in on-farm dye tests[1].

| Component[2] | Role | Relative importance |
|---|---|---|
| On-off valve | Controls water supply | Essential |
| Pressure gauge #1 | Monitors changes in pressure of water source. In-coming pressure should be 40 to 50 psi. | Practical |
| Back-flow prevention device | Prevents water and dye to be siphoned back into the water source | Essential (mandated by Florida Statutes) |
| In-line faucet | Provides an outside supply of water | Practical when no other water source is available. |
| Water meter | Measures actual water volume applied | Essential |
| In-line screen filter | Reduces risk of clogging. Improves uniformity | Useful |
| Pressure reducer | Maintains operating pressure close to drip-tape manufacturer operating pressure. Ensures proper flow rate and improves uniformity | Essential |

**TABLE 4** *(Continued)*

| | | |
|---|---|---|
| Pressure gauge #2 | Monitors changes in water pressure before the injection point | Practical |
| Injection point (Mazzi- or Dosatron-type) | Injects the dye in the system. Use a 1:50 to 1:100 dilution rate. Insert weight and a filter at the end of the suction line. | Essential |
| Pressure gauge #3 | Monitors changes in pressure at the furthest point of the system | Practical |

[1] Adapted from: Simonne, E., D. Studstill, M. Dukes, J. Duval, R. Hochmuth, E. Lamb, G. McAvoy, T. Olczyk, and S. Olson. 2004. How to conduct an on-farm dye test and how use its results to improve drip irrigation management in vegetable production, Electronic Database Information System, HS 980, http://edis.ifas.ufl.edu/HS222.
[2] From water source to field.

Neither the watermelon nor the strawberry growers monitored soil moisture levels or plant nutritional status at the beginning of this project. The management practices used at that time were not consistent with the BMPs. During the period of 2003–2005, the concept of being able to "see" where water and fertilizer moves in the bed was beginning to be taught by conducting on-farm blue dye demonstrations (approximately 50 in the region), presenting multimedia presentations at grower meetings, using educational posters that showed color pictures of examples of blue dye staining events over a range of irrigation run times, and showing videos of the blue dye process and results (http://vfd.ifas.ufl.edu/gainesville/blue_dye/index.shtml). Farmers were witnessing the raised-bed soil wetting patterns in their own fields following their own practices. In addition, crop $NO_3$-N and K nutritional status was determined in real-time on hundreds of plant petiole samples thanks to sap testing [19]. At the same time, the use of soil moisture devices (such as gypsum blocks, granular matrix sensors, tensiometers, and Time Domain Reflectometers) and their correct placement was demonstrated. Results of dye tests, sap tests, and soil moisture measurements were used by growers to make daily adjustments to the irrigation schedule and fertility programs.

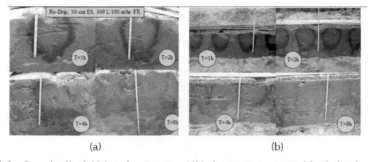

(a) (b)

**FIGURE 6** Longitudinal (6a) and transverse (6b) dye patterns created by irrigations ranging from 1 to 8 h with a 30-cm emitter spacing, 309 L/100 m/hr drip tape on a Blanton-Foxworth-Alpin complex (formerly Lakeland fine sand) at the UF/IFAS Suwanee Valley Agricultural Extension Center, in Live Oak, FL. (the change in color in the soils profile marks the bottom of the plow zone and root zone at the 30-cm depth), (Photo credit: Eric Simonne, University of Florida).

## 7.3   RESULTS AND DISCUSSION

### 7.3.1   SHAPE OF THE WETTED ZONE

The dye patterns in the soil appeared as a 2-to-3-cm thick blue ring surrounding an uncolored section of soil because the dye was injected first followed by clear water (Fig. 6). If the dye was continually injected with all the irrigation water, the entire wetted section would appear blue. The dye was easily distinguishable in the soil at the rate used, but the contrast between the blue dye and the soil color was improved by allowing the wetted zone to dry for 1 to 2 h after digging. Improved contrast was necessary for improving quality of wetting patterns especially when dye injections were followed by long irrigation durations and the large water volumes diluted the dye in the soil. In all tests, the uniformity of the dye patterns was high.

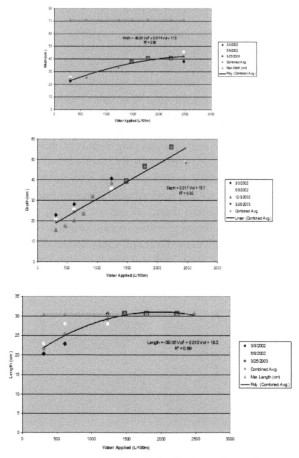

**FIGURE 7**   Responses of the Depth (a; D), Width (b; W) and Length (c; L) of the wetted zone to irrigation volume applied (V) on a Blanton-Foxworth-Alpin complex (formerly Lakeland Fine Sand) during four dye tests conducted at the UF/IFAS Suwannee Valley Agricultural Extension Center in Live Oak, FL(new beds) using a single drip tape with a nominal flow rate of 24 gal/100ft/hr.

## 7.3.2    RESPONSES OF DEPTH, WIDTH AND LENGTH OF THE WETTED ZONE TO IRRIGATION VOLUME APPLIED

When a drip tape with 30-cm emitter spacing and a 309 L/100 m/hr flow rate was used, responses of the depth (D, cm), width (W, cm) and length (L, cm) of the wetted zone to irrigation volume (L/100 m of bed) are shown in the following (Fig. 7):

$$D = 0.017 \, V + 13.7, \, R^2=0.92 \text{ (Fig. 7a)} \tag{1}$$

$$W = -[3 \times 10^{-6} \, V^2 + 0.0174 \, V + 17.6, \, R^2=0.96 \text{ (Fig. 7b)} \tag{2}$$

$$L = -[3 \times 10^{-6} \, V^2 + 0.0130 \, V + 18.2, \, R^2=0.99 \text{ (Fig. 7c)} \tag{3}$$

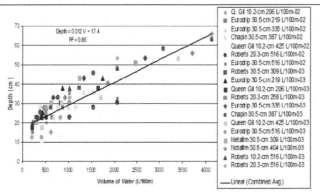

**FIGURE 8**   Responses of the depth (a; D) of the wetted zone to irrigation volume applied (V) on a Blanton-Foxworth-Alpin complex (formerly Lakeland Fine Sand) with multiple drip tape types during dye tests conducted at the UF/IFAS Suwannee Valley Agricultural Extension Center in Live Oak, FL on newly formed raised beds.

These results first showed that the depth of the wetted zone was below the root zone for single irrigation events between 2 to 4 h (Fig. 5). Hence, single irrigation events longer than 2 h are likely to start moving fertilizer below the root zone in a deep sandy soil. These results also showed that while the depth of the wetted zone kept increasing with increasing irrigation volumes, the wetted width tended to remain constant once it reached 40 cm. This means that the practical area of soil wetted under these conditions is 20 cm on each side of the drip tape. Attempts to increase the width of the wetted zone by pulsing irrigation (for example by applying 4 irrigation cycles of 1 hr instead of one irrigation cycle of 4 h) were not successful [28]. These results finally show that complete emitter-to emitter coverage was achieved between 2 and 4 h when emitters were spaced 30 cm apart. These results may also be applied to the adequate positioning of soil moisture devices. For greater sensitivity, the devices should be placed in the zone of the raised bed where soil moisture changes the most. If the probe is placed in an area that tends to be dry (too close to the shoulder), soil moisture measurements will indicate a need for irrigation too often. If the probe is placed too close to the drip tape, soil moisture measurements will indicate that no irrigation is needed when it may be. These results show that soil moisture measurements will be the most representative of root zone moisture when the probe is placed within, but near the maximum wetted width (20 cm on each side of the tape).

When drip tapes with emitter spacings ranging from 10 to 30 cm and flow rates ranging from 206 to 516 L/100 m/hr were used, the response of the Depth (D, cm) of the wetted zone to irrigation volume applied (V, L/100 m) is defined by equation below (Fig. 8):

$$D = 0.012 \, V + 17.04, \, R^2 = 0.86 \quad \text{(Fig. 8)} \tag{4}$$

In both cases (Figs. 7a and 8), the slope of the linear equations were of the same magnitude (0.017 and 0.012 cm/L/100 m, respectively). In practical terms, this means that in the absence of a plant, the depth of the water front increased by 0.012 to 0.017 cm for every one L/100 m of irrigation water applied by drip tapes of any flow rate and any emitter spacing. These results also imply that complete raised-bed wetting (shoulder to shoulder) is not achievable when beds are more than 40-cm wide. This may affect the efficacy of the soil fumigants that remain entirely in the liquid phase. The typical maximum root depth for vegetables grown on this soil type is 30 cm, which is also the tillage depth. The depth of the waterfront reached 30 cm for an irrigation volume of 1,250 L/100 m of bed. This corresponds to an irrigation time of approximately 2 h for a medium-flow drip tape. Hence, when target irrigation volumes are greater than this amount, the calculated daily irrigation volume should be applied in multiple events with each not delivering more than 1,250 L/100 m of bed. When emitters are spaced 30-cm apart, length response to irrigation volume applied showed that complete emitter-to-emitter coverage was achieved after an irrigation volume of nearly 1,500 L/100 m (which corresponded to an irrigation of nearly 4 h with a medium flow-rate drip tape).

**TABLE 5**   Sample irrigation scheduling guidelines based on target watermelon crop water use and amount of water that can be stored in the root zone for different weather conditions.

| Crop stage of growth (Crop factor)[1] | Weeks after transplanting | Cool days | Warm days | Hot days | Very hot days |
|---|---|---|---|---|---|
| | | Daily Class A pan evaporation (inch; mm) | | | |
| | | <0.10 inch or <2.54 mm | 0.10 to 0.20 inch or 2.54–5.08 mm | 0.20 to 0.30 inch or 5.08–7.62 mm | >0.40 inch or>10.16 mm |
| | | Estimated Irrigation Volume[2] | | | |
| | | [Number of 1-hr Daily Irrigation Events], Gallons/A/day (L/ha/day) | | | |
| 1 | 1–2 | 295   (2,758) [1] | 582 (5,442) [1] | 872 (8,153) [1] | 1,163 (10,874) [1] |
| 2 | 3–4 | 580   (5,423) [1] | 1,160   (10,846) [1] | 1,740 (16,269) [2] | 2,324   (21,729) [2] |
| 3 | 5–11 | 1,015 (9,490) [1] | 2,033   (19,009) [2] | 3,048 (28,499) [3] | 4,064   (37,998) [3] |

**TABLE 5**   *(Continued)*

| Crop stage of growth (Crop factor)[1] | Weeks after transplanting | Cool days | Warm days | Hot days | Very hot days |
|---|---|---|---|---|---|
| | | Daily Class A pan evaporation (inch; mm) | | | |
| | | <0.10 inch or <2.54 mm | 0.10 to 0.20 inch or 2.54–5.08 mm | 0.20 to 0.30 inch or 5.08–7.62 mm | >0.40 inch or>10.16 mm |
| | | Estimated Irrigation Volume[2] [Number of 1-hr Daily Irrigation Events], Gallons/A/day (L/ha/day) | | | |
| 4 | 12 | 1,308 (12,230) [1] | 2,614 (24,441) [2] | 3,916 (36,615) [3] | 5,227 (48,872) [4] |
| 5 | 13 | 1,015 (9,490) [1] | 2,033 (19,009) [2] | 3,048 (28,499) [3] | 4,064 (37,998) [3] |

[1] Crop evapotranspiration was estimated by multiplying daily Class A Pan evaporation (Ep) by a crop factor developed for watermelon by Di Gioia, F., E. Simonne, D. Jarry, M. Dukes, R. Hochmuth, and D. Studstill. 2009. Real-time drip-irrigation scheduling of watermelon grown with plasticulture. Proc. Fla. State Hort. Soc. 122:212–217.
Crop factor values were 0.24, 0.48, 0.84, 1.08 and 0.84 for crop stage of growth 1 to 5, respectively. See Olson and Santos (2012) for description of growth stages.

[2] One hour of irrigation applies 1300 gal/A (12,155 L/ha) of water on 8-ft (2 m) centers when a drip tape with nominal flow rate of 24 gal/100 ft/hr (298 L/100 m/hr) is used. Volumes assume 90% irrigation efficiency. Conversion factor for gallons/A/day to L/ha/day is 9.35. Conversion factor for gallons/100 ft/hr to L/100 m/hr is 12.42.

Similar results were observed with a 10-cm emitter spacing, low-flow drip tape (Fig. 9). The waterfront did not move as fast after one hour of operation compared to the medium-flow drip tape. This was expected, since the low-flow drip tape applied 206 L/100 m/hr whereas the medium flow one applied 309 L/100 m/hr. However, some dye was still present between two consecutive emitters with the medium flow-rate tape (Fig. 6b, T=8 hrs) whereas none was present with the 10-cm emitter spacing tape even with a lower flow rate (Fig. 9, T=8 hrs). While this observation was beyond the scope of this project, it suggests that flow rate and emitter spacing also affect the movement of the waterfront.

## 7.3.3   APPLICATION TO IRRIGATION SCHEDULING AND EARLY CHANGES IN INDUSTRY PRACTICES

The results of these replicated dye tests were used to update the recommendations and resulted in immediate practice changes in the vegetable industry. Recommendations now include a real-time estimate of crop water use and a maximum volume of water that can be applied before splitting irrigation [6]. For example with watermelon, using the crop factors and class A pan evaporation measurements developed in North Florida [4], estimated irrigation volumes and corresponding number of 1-hr irrigation events are available for different weather conditions and crop stage of growth (Table 5). No recommendations exist regarding flow rates or emitter spacings. However, based on the time needed to reach complete emitter-to-emitter coverage, the use of 45 cm as an emitter spacing was discontinued by the vegetable industry. Based on the width of the wetted zone (40 cm for sandy soils; 13 to 20 cm in gravely soils), two drip tapes are being used when complete bed width wetting is needed. Based on the rate of vertical water movement in sandy soils, irrigation recommendations now include a maximum volume of water to be applied in a single event (Table 5).

### 7.3.4    ON-FARM DEMONSTRATIONS AND FURTHER CHANGES IN INDUSTRY PRACTICES

Educational programs were based on practical, visual, small group and one-on-one educational events. Typically, a dye test was conducted on a commercial farm, and regular visits by the Regional and/or County Extension Agent focused on visualizing soil moisture, the daily measurement of soil moisture and the weekly assessment of plant nutritional status through petiole sap analysis (Fig. 10). While initially the agent was doing all the work themselves, the growers were learning how to do it on their own.

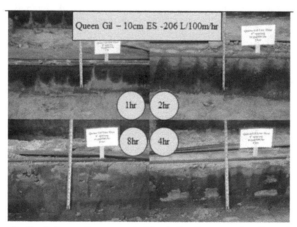

**FIGURE 9**    Longitudinal dye patterns created by irrigations ranging from 1 to 8 h with a 10-cm emitter spacing, 206 L/100 m/hr drip tape on a Blanton-Foxworth-Alpin complex (formerly Lakeland fine sand) at the UF/IFAS Suwanee Valley Agricultural Extension Center, near Live Oak, FL. (the change in color in the soils profile marks the bottom of the plow zone and root zone at the 30-cm depth) (photo credit: Eric Simonne, University of Florida)

Fig. 10a.

Fig. 10b.

Fig. 10c.

Fig. 10d.

**FIGURE 10**    The four components of on-farm BMP education used between 2000 and 2013 in North Florida: (a) Longitudinal section showing the dye movement on a commercial summer squash (Cucurbita pepo L.). (The difference in depth movement in those three emitters indicates partial clogging possibly caused by ineffective maintenance program); (b) TDR probe that allows instantaneous and roaming measurements of soil moisture for the adjustment of the irrigation schedule, (c) cooperating grower measuring NO3-N and K petiole concentrations using sap-test meters used for the adjustment of the fertigation schedule, and (d) educational poster. (Photo credit: Bob Hochmuth (a),(b),(c); Aparna Gazula (d); University of Florida).

BMP adoption by the strawberry growers in the Alachua-Bradford-Union county region followed the same pattern in all of the 21 farmer/crop year events during the seven years. With help from the Extension Agents on how to properly sample petioles, extract the sap, calibrate the meter, and interpret the results, growers understood the meaning of the results and whether or not how they needed to modify their fertility and irrigation programs. Typically, sap test results were "low" for $NO_3$-N during early crop growth stages. These measurements were used to add fertilizer in amounts required to get the N within the recommended UF/IFAS range. The exception to this scenario is the farmer who only used granular fertilizer and overhead irrigation. Sap testing in that case mainly documented deficiencies in N after the fertilizer was leached from high rainfall or from frost protection events.

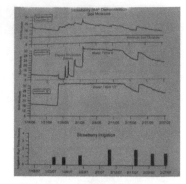

**FIGURE 11** Extension programs helping North Florida vegetable growers to improve nutrient and water management: Left (a): Changes in sap $NO_3$-N concentration in a commercial watermelon field (source: Bob Hochmuth, unpublished data) and; Right (b): Changes in soil moisture at the 10, 20 and 30 cm depths below a commercial strawberry field and corresponding days of irrigation (Source: Mace Bauer, unpublished data).

Initially, the interpretative petiole sap thresholds used for the growers in North Florida were those that had been developed for strawberries grown in Central Florida [33]. The data collected and the knowledge gained during the first years of this project were used to refine the existing thresholds for the specific growing season used in the Alachua-Union-Bradford county strawberry producing region [20]. After the initial teaching and the adoption of recalibrated interpretative thresholds, the levels of $NO_3$-N were diagnosed as "at risk of being low" only 10% (15 of 155 events) of the time. Also, a step-by-step, practical guide was developed for the correct calculation and injection of fertilizer through the drip tape [3]. In some advanced demonstrations, changes in soil moisture were measured with probes buried at the 4, 8 and 12 inches. The Extension Agent and the growers made irrigation decisions together (Fig. 11.b).

Having successfully communicated water and fertilizer placement principles through blue dye demonstrations, additional efforts were needed to accomplish behavioral adoption of BMPs with respect to farmers using sap testing as a decision making tool for fertilizer application scheduling (Table 6). One hundred and 50 nine sap testing events were performed by the Extension Agents on the seven strawberry farms from 2007 to 2013. Weekly sap-testing measurements were used to diagnose crop nutritional status (Fig. 11a). The effort was necessary because the equipment is expensive for small farmers (about $800 for the two sap meters ($NO_3$-N and K) and $750 for one portable TDR probe) and requires a fair amount of skill to be used effectively. During the seven years, only one farmer purchased his own sap-testing equipment. Consequently, farmers often relied on Extension assistance in monitoring sap nutrient levels.

**TABLE 6** Summary of growers changes in irrigation and fertilization practices based on educational activities offered by the University of Florida Extension Agents and conducted at several commercial vegetable farms in the Lower Suwannee Valley. (Soils were deep sandy soils on all sites).

| Vegetable Operation (Crop, Annual plantings) | Typical Empirical Irrigation and Fertilization Practices (Before Educational Programs) | Educational Events Conducted: # Farms with petiole sap testing (PST), soil moisture monitoring (SMM) and dye tests (DT) | | | New practices (After Educational Programs) | Estimated Reduction in Fertilizer Applied and in Irrigation Water Used |
|---|---|---|---|---|---|---|
| | | PST | SMM | DT | | |
| 4 tomato growers (25 ha) | High preplant fertilizer and high volume irrigation early in season. Exact amounts not known. | 4 | 4 | 2 | Reduced preplant fertilizer, fertigation based on sap testing, reduced early season irrigation, shorter but more frequent daily irrigation events. | 55 kg/ha of N and K$_2$O. 50% reduction in irrigation the first 4 weeks. |
| 6 pepper growers (60 ha) | High preplant fertilizer and high volume irrigation early in season. | 6 | 6 | 4 | Reduced preplant fertilizer, fertigation based on sap testing, reduced early season irrigation, shorter but more frequent daily irrigation events. | 55 kg/ha of N and K$_2$O. 50% reduction in first 4 weeks. |
| 25 watermelon growers (2,300 ha) | High preplant fertilizer and high volume irrigation early in season often included long runs on cold nights. | 25 | 25 | 12 | Reduced preplant fertilizer, fertigation based on sap testing, reduced early season irrigation, shorter but more frequent daily irrigation events, eliminated long events at night. | 82 kg/ha of N and K$_2$O per acre. 50% reduction in first 4 weeks. |
| 15 Mixed vegetable growers (150ha) | High preplant fertilizer and high volume irrigation early in season. | 15 | 15 | 7 | Reduced preplant fertilizer, fertigation based on sap testing, reduced early season irrigation, shorter but more frequent daily irrigation events. | 65 kg/ha of N and K$_2$O per acre. 50% reduction in first 4 weeks. |

## 7.4 CONCLUSIONS

Overall, this project has resulted in a significant adoption in irrigation and nutrient BMPs by vegetable growers in North Florida (Table 6). This success was based on the science-based recommendations of the University of Florida, the continuous presence of the Extension Agents and the willingness of the growers to comply with BMP regulation and reduce their input costs. The availability of cost-share programs to offset the cost of purchase of the measuring instruments is poised to help make these cultural practice changes permanent. In the meantime, on-going research efforts aim at determining the level of pollution reduction achieved by the implementation of BMPs.

## 7.5 SUMMARY

High-value vegetable crops such as tomato, bell peppers, watermelon, muskmelon and summer squash are widely grown with high irrigation and fertilization inputs on Florida's sandy soils using plasticulture. The reconciliation of vegetable production with environmental regulation is requiring a better understanding of how water and nutrients move in mulched beds. By law, Florida growers have the choice to (a) join the Best Management Practice (BMP) program and be awarded a presumption of compliance with water quality standards, or (b) develop a monitoring program documenting that their operation does not contribute to the degradation of environmental water quality. Historically, vegetable growers have used irrigation and fertilizer amounts in excess of recommendations because they believed they gained an economic benefit from doing so and they did not understand that excessive irrigation leaches mobile nutrients below the root zone of vegetable crops. Using Brilliant Blue FCF dye as a tracer, a series of replicated trials was established to visualize and quantify the movement of irrigation water in mulched beds. The responses of the depth (D), width (W) and length (L) of the dye front to drip-irrigation volume applied (V) were:

$$D = 0.017 V + 13.7, (R^2=0.92),$$
$$W = -3 \times 10{-6} V^2 + 0.0174 V + 17.6, (R^2=0.96), \text{ and}$$
$$L = -3 x 10{-6} V^2 + 0.013 V + 18.2, (R^2=0.99)$$

when a 30.5-emitter spacing, 309 L/100 m/hr drip-tape flow rate was used between 1 and 8 hr on a Blanton-Foxworth-Alpin complex. These results indicate that (a) the waterfront was moving at a rate of 0.012 and 0.017 cm/L/100 m and (b) the maximum wetted width was approximately 20 cm on each side of the drip tape. These results were also used to fine-tune the irrigation and fertilization recommendations of the University of Florida. Dye tests were also conducted on commercial fields where growers could see how their management practices were affecting water movement in mulched beds and under what circumstances nutrients were moved below the root zone.

The visual experience provided the "eureka moment" farmers needed to make them realize they could be farming more efficiently and that they were probably wasting water and fertilizer. Typically, this realization resulted in the farmer posing the question, "OK, how do I ensure my reductions of fertilizer and watering events do not go below crop demand and I damage the crop"? Seizing the moment, University of Florida Extension Agents helped vegetable growers in North Florida better manage irrigation and fertilization by using sap testing to estimate crop $NO_3$-N and K nutritional status and soil moisture measuring equipment to monitor irrigation events. With specific recommendations, simple tools to measure water and nutrients in real time, and technical help, growers participating in the program have reduced their irrigation and N and K fertilization early in the season by 50% and 25%, respectively. The availability of cost-share programs to offset the cost of purchase of the measuring instruments is poised to help make these cultural practice changes permanent. In the meantime, on-going research efforts aim at determining the level of pollution reduction achieved by the implementation of BMPs in Florida.

## KEYWORDS

- behavior change
- best management practice
- drip irrigation
- dye tracer
- extension
- fertigation
- flow rate
- grower
- leaching
- plasticulture
- polyethylene mulch
- root zone
- sandy soils
- sap testing
- soil moisture sensing
- strawberry
- time domain reflectometry
- University of Florida
- vegetable crops
- watermelon

## REFERENCES

1. Allen, R.G., L.S. Pereira, D. Raes, and M. Smith. 1998. *Crop evapotranspiration - Guidelines for computing crop water requirements*. FAO Irrigation and drainage paper 56. Food and Agriculture Organization of the United Nations, Rome, Italy. http://www.fao.org/docrep/X0490E/x0490e00.htm.
2. Chase, C., W.M. Stall, E.H. Simonne, R.C. Hochmuth, M.D. Dukes and A.W. Weiss. 2006. Nutsedge control with drip-applied 1,3-dichloropropene plus chloropicrin in a sandy soil. HortTechnology 16(4):641–648.
3. DeValerio, J., D. Nistler, R. Hocmuth, and E. Simonne. 2012. Fertigation for vegetables: A practical guide for small fields, HS1206, Electronic Database Information System, Univ. of Fla., Gainesville, FL. http://edis.ifas.ufl.edu/pdffiles/HS/HS120600.pdf

4. Di Gioia, F., E. Simonne, D. Jarry, M. Dukes, R. Hochmuth, and D. Studstill. 2009. Real-time drip-irrigation scheduling of watermelon grown with plasticulture. Proc. Fla. State Hort. Soc. 122:212–217.

5. Dukes, M.D., L. Zotarelli, and K.T. Morgan.2010. Use of irrigation technologies for vegetable crops. Florida. HortTechnology 20(1):133–142.

6. Dukes, M.D., L. Zotarelli, G.D. Liu and E.H. Simonne. 2012. Principles and practices of irrigation management for vegetables, AE260, Electronic Database Information System, Univ. of Fla., Gainesville, FL. <http://edis.ifas.ufl.edu/pdffiles/cv/cv10700.pdf>

7. Ewing, J. 2013. 2012 Florida agriculture by the numbers, US Dept. Agric./Natl. Agric. Stat. Serv. – Fla. Field Office, <http://florida-agriculture.com/brochures/P-01304.pdf> (accessed June13, 2013).

8. Farneselli, M., D.W. Studstill, E.H. Simonne, R.C. Hochmuth, G.J. Hochmuth, and F. Tei. 2008. Depth and width of the wetted zone after leaching irrigation on a sandy soil and implication for nitrate load calculation. Commun. Soil Sci. Plant Anal. 39:1183–1192.

9. Florida Department of Agriculture and Consumer Services. 2005. Water Quality/quantity Best Management Practices for Florida Vegetable and Agronomic Crops. <http://www.floridaagwaterpolicy.com/PDF/Bmps/Bmp_VeggieAgroCrops2005.pdf>

10. Florida Department of Agriculture and Consumer Services. 2006. Best management practices (BMPS) for Florida vegetable and agronomic crops. Florida Department of Agriculture and Consumer Services Rule 5 M-8. 13 Mar. 2006. <http://fac.dos.state. fl.us/faconline jcbapter05. pdf>

11. Florida Department of Environmental Protection. 2005. Total maximum daily loads: Background information. 21 Oct. 2005. <http://www.dep.state . fl.us/water/tmdlj background.htm>

12. Florida Senate. 1999. Florida Watershed Restoration Act (TMDL bill) SB 2282, Florida Statutes Title XXIX, Ch. 403.067. 13 Mar. 2006. <http://www.flsenate.gov/ Statutes/index. cfm?App_mode=Display_ Statute&Search_String=&URL=Ch0403/ SEC067.HTM&Title ->2005->Ch0403>Section%20067#0403.067>

13. Flury, M., and A. Papritz. 1993. Bromide in the natural environment-Occurrence and toxicity. J. Environ. Qual. 22(4):747–758.

14. Flury, M. and H. Flühler. 1995. Tracer Characteristics of Brilliant Blue FCF. Soil Sci. Soc. Am. J. 59: 22–27.

15. German-Heins, J. and M. Flury. 2000. Sorption of brilliant blue FCF in soils as affected by pH ionic strength. Geoderma 97:87–101.

16. Gjettermann, B., K.L. Nielsen, C.T. Petersen, H.E. Jensen, and S. Hansen. 1997. Preferential flow in sandy loam soils as affected by irrigation intensity. Soil Technol. 11(2):139–152.

17. Gilreath J.P., B.M. Santos, and T.N. Motis. 2003. Length of irrigation and soil humidity as basis for delivering fumigants through drip lines in Florida spodosols. Proc. Fla. State Hort. Soc. 116:85–87.

18. Hartz, T.K. 2006. Vegetable production best management practices to minimize nutrient loss. HortTechnology 16(3):398–403.

19. Hochmuth, G.J. 2012. Plant petiole sap-testing for vegetable crops, Circ. 1144, Electronic Database Information System, Univ. of Fla., Gainesville, FL. <http://edis.ifas.ufl.edu/cv004>.

20. Hochmuth, R., D. Dinkin, M. Sweat, and E. Simonne. 2003. Extension programs in North-eastern Florida help growers produce quality strawberries by improving water and nutrient management, HS 956, Electronic Database Information System, Univ. of Fla., Gainesville, FL. <http:// edis.ifas.ufl.edu/HS190>.

21. Hochmuth, R.C., E.H. Simonne, D. Studstill, J. Jones, C. Starling, J. Chandler, and A. Tyree. 2005. On-farm demonstration of soil water movement in vegetable grown with plasticulture. Proc. Fla. State Hort. Soc. 118:184–186.

22. Liu, G.D., E.H. Simonne and G.J. Hochmuth. 2012. Soil and fertilizer management for vegetable production in Florida, HS711, Electronic Database Information System, Univ. of Fla., Gainesville, FL. <http://edis.ifas.ufl.edu/pdffiles/cv/cv10100.pdf>.

23. Mali, N., J. Urbanc, and A. Leis. 2007. Tracing of water movement through the unsaturated zone of a coarse gravel aquifer by means of dye and deuterated water. Environ. Geol. 51:1401–1412.
24. Obreza, T.A. 2011. Maintenance guide for Florida micro irrigation systems, Circ. 1449, Electronic Database Information System, Univ. of Fla., Gainesville, FL. http://edis.ifas.ufl.edu/ SS436
25. Olson, S.M. and B. Santos. 2012. *2012–2013 Vegetable Production Handbook for Florida* (360pp). Vance Publishing, Lenexa, KS.
26. Olson, S.M., P.J. Dittmar, P.D. Roberts, S.E. Webb and S.A. Smith.2012. Cucurbit production in Florida, HS725, Electronic Database Information System, Univ. of Fla., Gainesville, FL. <http://edis.ifas.ufl.edu/pdffiles/cv/cv12300.pdf>.
27. Petersen, C.T., S. Hansen, and H. E. Jensen. 1997. Depth distribution of preferential flow patterns in a sandy loam soil as affected by tillage. Hydrology and Earth System Sciences Discussions 1(4):69–776.
28. Poh, B.L., E. Simonne, R.C. Hochmuth, and D.W. Studstill. 2009. Effect of splitting drip irrigation on the depth and width of the wetted zone in a sandy soil. Proc. Fla. State Hort. Soc. 122:221–224.
29. Poh, B.L., A. Gazula, E.H. Simonne, F. DiGioia, R.C. Hochmuth, and M.R. Alligood. 2011a. Use of reduced irrigation operating pressure in irrigation scheduling. I. Effect of operating pressure, irrigation rate, and nitrogen rate on drip-irrigated fresh-market tomato nutritional status and yield: Implications on irrigation and fertilization management. HortTechnology 21(1):14–21.
30. Poh, B.L., A. Gazula, E.H. Simonne, R.C. Hochmuth, and M.R. Alligood. 2011b. Use of reduced irrigation operating pressure in irrigation scheduling. II. Effect of reduced irrigation system operating pressure on drip-tape flow rate, water application uniformity and soil wetting pattern on a sandy soil. HortTechnology 21(1):22–29.
31. Runyan, C., T. Obreza, T. Tyson, B. Goodman, P. Tacker, R. Yager, J. Thomas, A. Johnson, G. Grabow, B. Smith, and S. Dennis. 2007. Maintenance guide for micro irrigation systems in the Southern Region. SR Regional Water Program, <http://fawn.ifas.ufl.edu/tools/irrigation/citrus/ maintenance/>.
32. SAS. 2008. *SAS/STAT 9.2 User's Guide*. SAS Institute Inc., Cary, NC, USA.
33. Santos, B.M., N.A. Peres, J.F. Price, V.M. Whitaker, P.J. Dittmar, S.M. Olson, and S.A. Smith. 2012. Strawberry production in Florida. CV104, Electronic Database Information System, Univ. of Fla., Gainesville, FL. <http://edis.ifas.ufl.edu/cv134>.
34. Simonne, E.H., D.W. Studstill, R.C. Hochmuth, G. McAvoy, M.D. Dukes and S.M. Olson. 2003. Visualization of water movement in mulched beds with injections of dye with drip irrigation. Proc. Fla. State Hort. Soc. 116:88–91.
35. Simonne, E.H., D.W. Studstill, T.W. Olczyk, and R. Munoz-Carpena. 2004. Water movement in mulched beds in a rocky soil of Miami-Dade County. Proc. Fla. State Hort. Soc. 117:68–70.
36. Simonne, E., D. Studstill, M. Dukes, J. Duval, R. Hochmuth, E. Lamb, G. McAvoy, T. Olczyk, and S. Olson. 2004. How to conduct an on-farm dye test and how use its results to improve drip irrigation management in vegetable production. HS 980, Electronic Database Information System, Univ. of Fla., Gainesville, FL. <http://edis.ifas.ufl.edu/HS222>.
37. Simonne, E., D. Studstill, and R.C. Hochmuth. 2005. Understanding water movement in mulched beds on sandy soils: An approach to ecologically sound fertigation in vegetable production. Acta Hort. 700:173–178.
38. Simonne, E., R. Hochmuth, J. Breman, W. Lamont, D. Treadwell, and A. Gazula. 2012. Drip-irrigation systems for small conventional and organic vegetable farms, HS1144, Electronic Database Information System, Univ. of Fla., Gainesville, FL. <http://edis.ifas.ufl.edu/HS388>
39. Steele, M. and J. Odumeru. 2004. Irrigation water as source of foodborne pathogens on fruit and vegetables. J. Food Protection 67(12):2839–2849.
40. Studstill, D., E. Simonne, T. Olczyk, and R. Muñoz-Carpena. 2006. Water movement in mulched beds in rocky soils of Miami-Dade county, HS1059, Electronic Database Information System, Univ. of Fla., Gainesville, FL. <http://edis.ifas.ufl.edu/HS313>.

41. U.S. Congress. 1977. Clean Water Act. PL 95–217, 27Dec. 1977. U.S. Statutes At Large 91:1566–1611. U.S. Govt. Printing Office, Washington, DC.
42. Zotarelli, L., J.M. Scholberg, M.D. Dukes and R. Muñoz-Carpena. 2007. Monitoring of nitrate leaching in sandy soils. J. Environ. Qual. 36(4):953–962.

# CHAPTER 8

# CROP COEFFICIENTS: SUSTAINABLE TRICKLE IRRIGATED COMMON BEANS[1]

VICTOR H. RAMIREZ BUILES, ERIC W. HARMSEN, and
TIMOTHY G. PORCH

## CONTENTS

[1] We acknowledge the partial financial support from the following sources: USDA-TSTAR Program (TSTAR-100), USDA Hatch Project (Hatch-402), NOAA-CREST (grant NA06OAR4810162), and NASA-EPSCOR (grant NCC5-595).

## 8.1  INTRODUCTION

Common bean is one of the most important sources of protein for more than 300 million of persons in the world, and is the center-piece of daily diet [8]. During 2001–2003 in the U.S.A., the sale from dry bean farming averaged $446 million, ninth among U.S. vegetables, averaging 6.8 pounds per person [49]. In Cuba/Haiti/Dominican Republic, the area planted for beans is 157,000 ha, with a production equal to 141,000 MT [13].

In Puerto Rico, the green-shelled bean production averaged 1,169 tons/year, with an increase related to the release of the genotype Morales, which is currently the most popular white-seed bean variety in Puerto Rico [10], during the period from 2000 to 2003. The variety Morales has been widely accepted in Puerto Rico for its good yield characteristics, and resistance to bean common mosaic virus (BCMV) and to bean rust races prevalent in Puerto Rico [9]. In 2002, the total irrigated area in Puerto Rico was 15,782 ha [38].

Because irrigated agriculture consumes large quantities of water, it is necessary to improve the estimates of water application rates. The development of methods to estimate crop water requirements is especially critical on small islands where utilization of water supplies by urban and industrial sectors continues to increase and where fresh water supplies are limited.

In Puerto Rico, irrigation research has focused on irrigation systems and water use. Harmsen [26] reviewed evapotranspiration studies in Puerto Rico during the previous 50 years. The review revealed that crop coefficients for beans have never been determined in Puerto Rico, and the studies related with direct water consumption carried out on the island were in sugarcane, grass spp., plantain and rice using the water balance method. Since Hamsun's review, studies have focused on irrigation rates as a function of pan evaporation [22].

Goyal and Gonzalez [22] estimated water requirements for green bean and other crops using the Blaney-Criddle reference ET method. Recently, Harmsen et al. [27] recalculated and made corrections to pan evaporation coefficients that are used to estimate reference evapotranspiration ($ET_o$). Harmsen and Gonzalez [27] also developed a computer program for estimating crop evapotranspiration in Puerto Rico (PRET).

One of the most critical steps in irrigation scheduling is the quantification of the crop water requirement. One way of estimating the water requirement of the crop is by multiplying the reference evapotranspiration by the crop coefficient. The resulting crop evapotranspiration estimate is equivalent to the crop water requirement. The FAO has provided the methodology for estimating the crop water requirement, described in the "Drainage and Irrigation Papers 24 and 56 [3, 18]." In these reports, they introduced and described in detail the crop coefficient ($K_c$), basal crop coefficients ($K_{cb}$ and $K_e$) and stress coefficient ($K_s$).

The FAO approach for estimating crop water requirements has been used throughout the world. Sheng Li et al. [45] estimated crop water requirements and identified timing and magnitude of water deficits for corn (*Zea mays* L.), soybean (*Glycine max* L.) and sorghum (*Sorghum bicolor* L.). Villalobos et al. [50] used direct application of the Penman-Monteith equation to calculate crop $ET_c$ in two commercial crops of garlic (*Allium sativum* L.) grown in Córdoba-Spain. Lin-Li et al. [33] measured $K_c$ and evapotranspiration using a gravimetric Lysimeter and the Penman-Monteith methods in wheat and maize under the semiarid conditions of Northern China.

The $ET_c$ and crop coefficients for bean differ owing to genotype, developmental stage, plant density, stress intensity, and agronomic practices [7, 15, 34, 37]. Consequently, it is necessary to evaluate the $ET_c$ and crop coefficients for local conditions and local varieties, dominant crop management practices, and for the influence of drought stress. Due to the variation in crop development rates between locations and years, thermal-based indices have been used to relate crop coefficient curves more directly to phenological development [29, 34].

In this chapter, authors discuss the research studies to: (1) Estimate the evapotranspiration rates for two common bean genotypes, with and without drought stress; (2) Derive crop coefficients for common bean in Puerto Rico; (3) Derive the crop stress coefficient; and (4) Relate the crop coefficients with easily measurable indices.

## 8.2  MATERIALS AND METHODS

### 8.2.1  LOCATION

This research was conducted at the Agricultural Experiment Station of the University of Puerto Rico at Juana Diaz, PR, located in south central PR (18°01'N latitude and 66°22'W longitude, elevation 21 m above mean sea level), which has been classified as a semiarid climatic zone [24]. Average annual rainfall is 33 inches (838 mm) and the average rainfall during the months of January, February and March are only 0.78, 0.72 and 0.86 inches (or 19.8 mm, 18.3 mm and 21.8 mm), respectively. The annual average, minimum and maximum air temperatures are: 26.22°C, 21.33°C, 31.05°C, respectively. The daily average minimum and maximum reference evapotranspiration are 4.3, 3.4 and 5.5 mm/day [26]. The dominant soil is San Anton Clay Loam. Tables 1 and 2 summarize the principal agronomic practices and soil physical characteristics.

### 8.2.2  EXPERIMENTAL PROCEDURE

This research was carried out during the early months of the year during 2006 and 2007, the dry season (January to April) in southern coast of Puerto Rico. The soil moisture in the root zone during this period was controlled by irrigation.

The lysimeters were installed in June of 2005, and the field was planted with beans during July–October, 2005. During 2006 and 2007, two common bean genotypes that exhibited differing responses to drought stress were planted: Morales [9], the most widely grown small white bean in Puerto Rico, bred for yield and disease resistance, and drought susceptible; and SER 16, a small red bean bred by CIAT (Colombia) by Dr. Steve Beebe for drought tolerance. Both genotypes have a type II growth habit. The seed density of planting was 14.0 plant m$^{-2}$ for Morales and 6.5 plant m$^{-2}$ for SER 16 (the differences in plant density between genotypes was due to insufficient seed supplies of SER 16). Fertilizer (16-4-4, NPK) was applied at a rate of 560 pounds per hectare and weeds were controlled through cultivation and herbicide application. The site was selected for soil uniformity. The experiment was arranged in a randomized complete block design.

**TABLE 1** Agronomic and management practices during the 2006 and 2007 field experiments, at Juan Diaz- PR. Note: DOY = Date of the year.

| Parameter | Unit | 2006 | 2007 |
|---|---|---|---|
| Sowing | | | |
| Emergence | DOY | 33 | 17 |
| Plant density | DOY | 38 | 23 |
| Fertilization | plants.m$^{-2}$ | 13.6 (Morales) | 13.2 (Morales) |
| Irrigation system | lb.ha$^{-1}$ | 6.4 (SER16) | 6.0 (SER16) |
| Harvest | — | 560, NPK (16-4-4) (DOY 62) | 560, NPK (16-4-4) (DOY 52) |
| Days to Maturity | DOY | Drip irrigation | Drip irrigation |
| | Days | 110–111 | 101–102 |
| | | 75 | 78 |

**TABLE 2** Soil physical characteristics at field site at Juana Diaz, PR field site [48].

| Soil physical characteristics | Units | Soil depth, cm | | | |
|---|---|---|---|---|---|
| | | 0–20 | 20–40 | 40–60 | 60–80 |
| Bulk density | (g.cm$^{-3}$) | 1.35 | 1.56 | 1.61 | 1.61 |
| Sand | % | 30.25 | 30.08 | 20.94 | 24.23 |
| Silt | % | 44.28 | 43.79 | 26.74 | 19.27 |
| Clay | % | 25.47 | 26.13 | 52.32 | 56.50 |
| FC | (m$^3$.m$^{-3}$) | 0.30 | 0.31 | 0.35 | 0.35 |
| WP | (m$^3$.m$^{-3}$) | 0.18 | 0.20 | 0.21 | 0.21 |
| TAW | (mm) | 24 | 22 | 28 | 28 |

FC = Field Capacity (Moisture content at 0.33 bar); WP= Wilting Point (Moisture content at >15 bar); TAW = Total available water = $1000*[FC – WP]*Z_i$; and Zr = Rooting depth (m).

### 8.2.3 MICRO IRRIGATION PRACTICES

Prior to imposing the drought stress, irrigation was applied to maintain the moisture content at field capacity, with irrigation applied two times per week using a drip irrigation system. The soil moisture content was monitored before and after each irrigation with volumetric moisture content readings ($\theta_v$) using a Profile probe type PR2 sensor (Delta-T Devices, Ltd.). Two soil probe access tubes per treatment were placed at 0–20 cm and 20–40 cm depths.

The volumetric moisture content ($\theta_v$) at field capacity (FC) was measured with a profile probe type PR2 sensor (Delta-T Devices, Ltd.). An area of 1 m$^2$ and depth of 50 cm was selected to measure the soil field capacity. Two access tubes were installed (0–20 cm and 20–40 cm), the area was saturated, and the area was then covered with black polyethylene plastic. After three days of free drainage, the $\theta_v$ was measured and the reading was assumed to be the moisture content at field capacity. Additionally, undisturbed core samples were taken to calibrate the sensor readings.

Drought stress was applied at the beginning of the reproductive phenological stage known as R1 (one blossom open at any node). The drought stress plot received a water application equivalent to 25% of total available water (TAW = FC – WP), corresponding to the drought stress level (DSL, Table 3). The irrigation application rates and the rainfall during the experiment are listed in Table 4.

**TABLE 3**   Field capacity measured directly in the field with the profile probe type PR2 sensor (Delta-T Devices Ltd).

| Soil depth, cm | FC[††] | WP | DSL[†] |
|---|---|---|---|
| | | m³.m⁻³ | |
| 0–20 | 0.38 | 0.18 | 0.23 |
| 20–40 | 0.31 | 0.20 | 0.23 |

† DSL: Drought Stress level, that corresponds with the 25% of the TAW.
†† FC: Volumetric moisture content at field capacity measured with the Delta-T Profile probe.

## 8.2.4   CROP EVAPOTRANSPIRATION (ET$_c$)

### 8.2.4.1   WATER BALANCE METHOD

For this method, we used drainage type lysimeters. The drainage lysimeter has been used successfully in evapotranspiration studies [12, 31, 42], and can provide satisfactory estimates of water use over 3- and 4-day intervals [16]; where the evapotranspiration is given by:

$$ET_c = P + I - RO - DP + (\Delta S) \tag{1}$$

where ET$_c$ is the crop evapotranspiration, P is precipitation, I is irrigation, RO is surface runoff, DP is deep percolation below the root zone, and $\Delta S$ is the change in root zone moisture storage, (all units are in mm). The change $\Delta S$ was converted to equivalent depth of water in mm by multiplying the lysimeter moisture contents by a conversion factor of 0.22 m².mm⁻¹; and measured values of RO and DP were converted to equivalent depth of water in mm by dividing by a lysimeter conversion factor of 0.22 L.mm⁻¹. Twelve drainage lysimeters were installed in the experimental field in 2005. We planted two lysimeters with SER 16 and four lysimeters with Morales in 2006; and three each with one per water level in 2007.

The soil in the lysimeter was encased in round polyethylene containers with an exposed soil surface of 0.22-m² and 0.8-m depth. The containers were sufficiently deep to accommodate the plant roots. The lysimeters were located within the plots measuring 7-m wide x 61-m long, with the long dimension oriented in the direction of the prevailing wind.

In order to achieve similar conditions inside and outside the lysimeters, the following procedure was followed for each lysimeter: i) Soil was removed from the location of the lysimeter in 0.25-m (12 inch) depth intervals. The soil sample from each depth interval was stockpiled separately; ii) The polyethylene containers were placed in the hole; iii) A 20-cm layer of gravel was placed in the bottom of the polyethylene tank and a 1.25 inch (30 cm) PVC tube was placed in the bottom to remove the percolated water during operation; iv) the stock piled soil was placed in the container in the reverse order that the soil was excavated. Each layer was carefully compacted until the original 0.25-m layer thicknesses were achieved. After the container was full, the surface runoff collector and the access tube to measure the volumetric moisture content were installed. The runoff collector consists of a small tank (0.20 m deep) connected to the lysimeter with a plastic gutter.

**TABLE 4**   Irrigation dates and volumes of the various treatments during 2006 and 2007, Juana Diaz-PR [41].

| Date | Growing State | Without drought stress | With drought stress | Rainfall |
|------|------|------|------|------|
| | | Irrigation (mm) | | (mm) |
| | | 2006 | | |
| 14 February | $V_1$ | 21.0 | 19.4 | 3.1 |
| 17 February | $V_2$ | 18.8 | 19.9 | 7.1 |
| 22 February | $V_3$ | 30.9 | 31.6 | 2.7 |
| 25 February | $V_4$ | 3.4 | 3.5 | 0.0 |
| 27 February | $V_5$ | 12.4 | 12.1 | 0.0 |
| 3 March | $V_8$ | 19.5 | 20.0 | 0.0 |
| 11 March | $R_1$* | 15.3 | 0.0 | 56.1 |
| 14 March | $R_2$ | 24.1 | 6.4 | 0.0 |
| 16 March | $R_2$ | 0.0 | 5.1 | 34.0 |
| 25 March | $R_4$ | 22.3 | 0.0 | 37.3 |
| 29 March | $R_5$ | 32.8 | 16.6 | 2.6 |
| 8 April | $R_8$ | 8.4 | 0.0 | 106.2 |
| 11 April | $R_9$ | 14.5 | 3.6 | 0.0 |
| **Total** | | **223.4** | **138.2** | **249.1** |
| **Water deficit level** | | | **18.0 %** | |
| | | 2007 | | |
| 24 January | $V_1$ | 9.71 | 8.21 | 0.0 |
| 31 January | $V_2$ | 21.9 | 15.3 | 0.0 |
| 1 February | $V_2$ | 0.0 | 22.8 | 0.0 |
| 5 February | $V_3$ | 25.0 | 25.7 | 0.0 |
| 7 February | $V_3$ | 26.0 | 22.5 | 0.0 |
| 13 February | $V_4$ | 40.3 | 14.2 | 0.0 |
| 15 February | $V_5$ | 27.3 | 29.1 | 0.0 |
| 21 February | $V_6$ | 24.7 | 21.2 | 1.5 |
| 24 February | $R_1$* | 10.8 | 0.0 | 0.0 |
| 26 February | $R_2$ | 12.7 | 0.0 | 0.0 |
| 1 March | $R_3$ | 29.9 | 10.1 | 0.0 |
| 5 March | $R_4$ | 34.2 | 22.5 | 0.0 |
| 6 March | $R_4$ | 0.0 | 9.3 | 0.0 |
| 9 March | $R_5$ | 60.2 | 19.6 | 0.0 |
| 12 March | $R_6$ | 27.3 | 13.2 | 0.0 |
| 15 March | $R_6$ | 31.9 | 0.00 | 0.7 |
| 20 March | $R_7$ | 15.4 | 15.4 | 0.4 |
| 23 March | $R_8$ | 14.5 | 8.6 | 19.7 |
| 28 March | $R_8$ | 0.0 | 0.0 | 13.9 |
| 30 March | $R_9$ | 0.0 | 0.0 | 17.5 |
| **Total** | | **379.7** | **248.3** | **53.7** |
| **Water deficit level** | | | **30.3%** | |

*. Drought stress beginning.

where

$V_1$: Completely unfolded leaves at the primary leaf node;

$V_2$: First node above primary leaf node;

$V_3$: Three nodes on the main stem including the primary leaf node. Secondary branching begins to show from branch of $V_1$;

$V_n$: n-nodes on the main stem including the primary leaf node;

$R_1$: One blossom open at any node;

$R_2$: Pods at ½-long at the first blossom position.

$R_3$: Pods at 1 inch long at first blossom position;

$R_4$: Pods 2 inches long at first blossom position;

$R_5$: Pods 3 plus inches long, seeds discernible by feel;

$R_6$: Pods 4.5-inch long spurs (maximum length). Seeds at least ¼ inch long axis;

$R_7$: Oldest pods have fully developed green seeds. Other parts of plant will have full-length Pods with seeds near same size;

$R_8$: Leaves yellowing over half of plant, very few small new pod/blossom developing, small pods may be drying. Points of maximum production has been reached;

$R_9$: Mature, at least 80% of the pods showing yellow and mostly ripe.

Daily rainfall was measured for each lysimeter with a manual rain gauge and compared with an automated tipping bucket rain gauge (WatchDog™-Spectrum Technology, Inc.) located within the "reference condition" area. The irrigation was measured using a cumulative electronic digital flow meter (GPI, Inc.), and was recorded manually at the beginning and end of each irrigation event every three or four days. Two flow meters were placed in the irrigation supply lines, one on the well-watered treatment supply line and the second on the drought stress treatment water supply line.

Runoff and depth percolation from each lysimeter were collected periodically (every three or four days). Water from RO and DP was removed from the collection containers periodically by means of a small vacuum pump (Shurflu-4UN26, 12 V, 4.5GPM). The depth of water in the soil profile was related to the soil moisture content as follows:

$$S_i = \Sigma \, (\theta_{v,i,0-10} \, Z_{0-10} + \theta_{v,i,10-20} \, Z_{10-20} + \dots \dots \theta_{v,i,50-60} \, Z_{50-60}) \qquad (2)$$

where, $S_i$ is the depth of soil water on day i [mm]; $\theta_{v,i}$ is the volumetric soil moisture content on day i; and Z is the thickness of the soil layer (10 cm). The depth intervals are specified for each of the six layer considered, for example, 0–10 indicates the soil interval from 0 to 10 mm.

## 8.2.4.2  PENMAN-MONTEITH MODEL

The crop evapotranspiration was also estimated using meteorological and crop data, with *Penman-Monteith* model [3]; using direct measurement of canopy and aerodynamic resistances during the whole growing season. For this purpose, four automatic weather stations were located within the experimental plots as follows: genotype Morales without drought stress genotype SER 16 with drought stress, genotype SER 16 without drought stress, genotype Morales with drought stress. Each weather station was equipped with: "Kipp and Zonen B.V. net radiometer (spectral range 0.2–100 μm)," wind direction and wind speed with wind sensor-Met one 034B-L at 2.2 m; air temperature and relative humidity with HMP45C temperature and relative humidity probe at 2.0 m; soil temperature with TCAV averaging soil thermocouple probe at 0.08 m and 0.02 m depth, soil heat flux using soil heat flux plates at 0.06 m depth; and a volumetric soil moisture content with a CS616 water content reflectometers at 0.15 m depth. Six data points per minutes were collected by each sensor and stored in a CR10X data logger (Campbell scientific, Inc.).

The *Penman-Monteith* model described by Monteith and Unsworth [36], Allen et al. [3], and Kjelgaard and Stockle [32] was used to calculate the latent heat flux ($\lambda E$), which was then divided by the latent heat of vaporization ($\lambda$) to obtain $ET_c$.

$$\lambda E = \frac{\Delta(R_n - G) + \rho_a C_p \dfrac{VPD}{r_a}}{\Delta + \gamma\left(1 + \dfrac{r_s}{r_a}\right)} \tag{3}$$

where, $\lambda E$ is Latent heat flux (Wm$^{-2}$); $R_n$ is net radiation (Wm$^{-2}$), $G$ is soil heat flux (Wm$^{-2}$); $VPD$ is vapor pressure deficit (kPa); $\Delta$ is slope of saturation vapor pressure curve (kPa∘C$^{-1}$) at air temperature; $\rho_a$ is density of air (Kgm$^{-3}$); $Cp$ is specific heat of air (J Kg$^{-1}$∘C$^{-1}$); $\gamma$ is psychometric constant (kPa∘C$^{-1}$); $r_a$ is the aerodynamic resistance (s m$^{-1}$); and $r_s$ canopy resistance to vapor transport (s m$^{-1}$). The aerodynamic resistance ($r_a$) is the resistance to the transport of heat and water vapor from the evaporating surface into air above the canopy and was estimated with the Perrier Eqs. (3) and (5).

$$r_a = \frac{Ln\left[\dfrac{(Z_m - d)}{Z_{om}}\right] Ln\left[\dfrac{(Z_h - d)}{Z_{oh}}\right]}{K^2 u_z} \tag{4}$$

where, $Z_m$ is the height of wind measurements [m]; $z_h$ is the height of the humidity measurements [m]; d is the zero displacement height [m]; $Z_{om} = 0.123h =$ is the roughness length governing momentum transfer of heat and vapor [m]; $Z_{oh} = 0.1\, Z_{om} =$ is roughness length governing transfer of heat and vapor [m]; $K$ is the von Karman`s constant [=0.41]; $u_z$ is the horizontal wind speed (m s$^{-1}$) at height $z$; and $h$ is the canopy height (m). The canopy height was measured for each genotype once per week, and polynomial models were developed to estimate daily values of $h$ as a function of the day of the year (DOY). The $r_a$ was calculated at one-minute time intervals.

The canopy resistance ($r_s$) describes the resistance of vapor flow through a transpiring crop and evaporation from the soil surface, which depends on climatic factors and available soil water. Bulk surface resistance was calculated using the equation proposed by Szeicz and Long [47] and recommended by Allen et al. [3]:

$$r_s = \frac{r_L}{LAI_{active}} \tag{5}$$

where, $LAI_{active}$ is the active leaf area index (m$^2$-leaf area/m$^{-2}$-soil surface) and is equal to 0.5 times the leaf are index; and $r_L$ is the stomatal resistance (m s$^{-1}$) which is the total resistance from cell surfaces to the exterior leaf surfaces [51] and is one of the most sensitive elements in the evapotranspiration under drought stress conditions. The $r_L$ was measured several times during the day from 7:00 to 17:00 in order to obtain a reasonable average value for each phenological growing phase, for each genotype and water level. Two leaf porometers were used: an AP4-UM-3 (Delta-T Devices Ltd) during 2005 and a model SC-1 (Decagon Devices, Inc.) during 2006. The observations were recorded once per week.

The *LAI* was estimated using a nondestructive method, which estimates the leaf area (LA) in cm$^2$ using the maximum single leaf width (W) in cm. The models for LA were as follows:

LA = 9.35(W) – 20.32 for SER16, and

LA = 7.80(W) – 14.59 for Morales

Then according to the plant density, the LAI was estimated on a weekly basis.

### 8.2.4.3   REFERENCE EVAPOTRANSPIRATION

The reference evapotranspiration (ET$_o$) corresponds to the evapotranspiration from a reference crop (e.g., alfalfa or grass) under reference conditions. It is common to use a hypothetical grass reference, with a constant canopy height, canopy resistance and albedo, under well-watered conditions [3]. For this study, one automatic weather station (WatchDog-900ET, Spectrum Technologies, Inc.) was placed within a field planted with a reference crop (grass-*Panicum maximum* and *Clitoria termatea* L.) with enough fetch and sufficient water supply during the period of the experiment, and adjacent to the experimental area. The canopy height was maintained close to 0.15 cm throughout the growing seasons. The automatic weather station measured basic weather information including: solar radiation, temperature, humidity, wind speed and direction every 10 min. ET$_o$ was calculated using the Penman-Monteith equation [3] recommending by FAO-56 equations and standardized by the American Society of Civil Engineer-ASCE:

$$ET_o = \frac{0.408\Delta(R_n - G) + \gamma \frac{C_n}{T+273} U_2(e_s - e_a)}{\Delta + \gamma(1 + C_d U_2)}[\ ] \qquad (6)$$

where, $ET_o$ is the standardized reference crop evapotranspiration for short (ET$_{os}$) or tall (ET$_{rs}$) surfaces (mm.d$^{-1}$ for daily time steps or mm.h$^{-1}$ for hourly time steps); $R_n$ is net radiation at the crop surface (MJ.m$^{-2}$.d$^{-1}$ or MJ.m$^{-2}$.h$^{-1}$); $G$ is soil heat flux density (MJ.m$^{-2}$d$^{-1}$ or MJ.m$^{-2}$h$^{-1}$); $T$ is mean daily temperature at 1.5 to 2.5 m height (°C); $U_2$ is wind speed at 2 m height (m s$^{-1}$); $e_s$ is saturation vapor pressure (kPa); $e_a$ is actual vapor pressure at 1.5 to 2.5 m height (kPa); $\Delta$ is the slope of the saturation vapor pressure-temperature curve (kPa °C$^{-1}$); $\gamma$ is the psychrometric constant (kPa °C$^{-1}$); $C_n$ is the numerator constant that changes with the reference type and calculation time step; and $C_d$ is the denominator constant that changes with the reference type and calculation time step. The $C_n$ incorporates the effect of the aerodynamic roughness of the surface (i.e., reference type), while the $C_d$ incorporates the effects of bulk surface resistance and aerodynamic roughness of the surface.

When the ET$_o$ is derived from weather data, it is necessary to verify the quality and integrity of the data. Allen [2] proposed the difference between daily minimum air temperatures (T$_{min}$) with daily average dew point temperature (T$_{dew}$) as a reference parameter. In this study, if the difference = [T$_{min}$ – T$_{dew}$] was greater than 3°C, then the conditions were considered to be "nonreference" [30], and $ET_o$ was not calculated. Dew point temperature was estimated as a function of the actual vapor pressure, using the Tetens equation.

## 8.2.5 SINGLE CROP COEFFICIENT

The crop coefficient ($K_c$) accounts for the effects of characteristics that distinguish the field crop from the reference crop [3], is a commonly used approach for estimation of consumptive use of water by irrigation, represents the ET under a high level of management and with little or no water or other stresses [4], and is equal to the ratio of the crop evapotranspiration to the reference evapotranspiration.

$$K_c = \frac{ET_c}{ET_o} \tag{7}$$

The crop effects can be combined into one single coefficient (Eq. (7)) or it can be split into two factors describing evaporation from the soil and transpiration from the leaves. As the soil evaporation fluctuates daily as a result of rainfall or irrigation, the single crop coefficient expresses only the time-averaged (multiday) effects of crop evapotranspiration. In determining crop coefficients for a crop season, four stages of crop growth are normally considered [3], which depend on phenological stages, and can be described by a $K_c$-curve that includes the variation of the coefficient during the whole growing season. The $K_c$ curve is comprised of four straight line segments that represent: the initial period ($K_{c,ini}$), the development period ($K_{c\,dev}$), the midseason period ($K_{c,mid}$) and the late season period ($K_{c,end}$). $K_{c,ini}$ represents the period until approximately 10% of the ground is covered by vegetation ($f_c$). $K_{c,mid}$ defines the value for $K_c$ during the peak period for the crop, which is normally when the crop is at "effective full cover," considered to be at the initiation of flowering (R1) in this research. The $K_{c\,end}$ has a sloping line that connects the end of the midseason period with the harvest date [3, 4].

The cover fraction ($f_c$) is a function of vegetation type, ground cover, plant density, canopy architecture, and environmental stresses like drought. In this study, weekly $f_c$ measurements were collected for each genotype and water condition.

## 8.2.6 DUAL CROP COEFFICIENTS

The dual crop coefficients are the basal crop coefficient ($K_{bc}$) and soil evaporation coefficient ($K_e$). The coefficients $K_{bc}$ and $K_e$ relate the potential plant transpiration and soil evaporation, respectively, to the crop evapotranspiration. Measurement of $K_e$ and $K_{cb}$ were made using the FAO-56 approach [3], as given below:

$$ET_c = (K_{cb} + K_e)\,ET_o \tag{8}$$

$$K_{cb} = (ET_c / ET_o) - K_e \tag{9}$$

The soil evaporation coefficient ($K_e$) was estimated as a function of field surface wetted by irrigation ($f_{ew}$). The $f_{ew}$ were estimated as a minimum value between the fraction of the soil that is exposed to sunlight and air ventilation and serves as a source of soil evaporation, and the fraction of soil surface wetted by irrigation or precipitation ($f_w$, Eqs. (11) and (12)), and were measured twice per week.

$$K_e = f_{ew} K_c \tag{10}$$

$$f_{ew} = min \ (1-f_c; f_w) \tag{11}$$

and for drip irrigation:

$$f_{ew} = min \ [(1-f_c); \ (1 - 0.67*f_c)*(f_w)] \tag{12}$$

If the water source was drip irrigation, the $f_w$ was estimated as a cover crop fraction, and for days with rain it was equal to 1.0.

$$f_w = 1 - \frac{2}{3} f_c \tag{13}$$

## 8.2.7 CROP STRESS FACTOR

The crop stress factor ($K_s$) is an important coefficient because it helps to distinguish which crops are sensitive to water deficit conditions [43]. The crop stress factor is a function of the average soil moisture content or matrix potential in a soil layer. It can be estimated by empirical formulas based in soil water content or relative soil water. The $K_s$ was determined throughout the crop season for each study plot. The crop stress factor, as described by Allen et al. [3], has a value between 0 and 1. A value of 1 indicates stress-free conditions (e.g., water is readily available for plant use), whereas a value of zero indicates no available water for plant use. As normally applied, the crop stress factor is equal to 1 until the depletion of water reaches some critical depletion. For example, for dry bean the critical value could be between 45–50% of total available water [3]. Total available water is defined as the field capacity moisture content minus the wilting point moisture content (TAW, Eq. (15)). After depletion exceeds the critical value, the crop stress factor drops linearly until reaching zero at the wilting point moisture content. The $K_s$ was estimated according to FAO-56 methodology as given below:

$$Ks = \frac{TAW - Dr}{TAW - RAW} = \frac{TAW - Dr}{(1 - p)TAW} \tag{14}$$

where, $TAW$ is total available water referring to the capacity of a soil to retain water for plant use (mm); $D_r$ is the root zone depletion (mm); $RAW = FC - PWP =$ is the readily available soil water in the root zone (mm); p is the fraction of TAW that the crop can extract from the root zone without suffering water stress, and

$$TAW = 1000(\theta_{FC} - \theta_{WP})Z_t \tag{15}$$

where, $\theta_{FC}$ is the water content at field capacity (m$^3$ m$^{-3}$); $\theta_{WP}$ is the water content at wilting point (m$^3$ m$^{-3}$); and $Z_t$ is the rooting depth (m). $RAW$ is estimated as follows:

$$RAW = pTAW \tag{16}$$

where, $p$ is the average fraction of total available soil water ($TAW$) that can be depleted from the root zone before moisture stress (reduction in $ET$) occurs. In this study, $p$ was estimated using Eq. (17).

$$p = [0.45 + 0.004(5 - ET_c)] \tag{17}$$

## 8.3   RESULTS AND DISCUSSION

### 8.3.1   EVAPOTRANSPIRATION

Weather conditions prevailing during the two years are shown in Table 5, and are compared with the long-term record presented by Goyal and Gonzalez [29] and Harmsen et al. [27]. Figure 1 shows the seasonal variation in daily climatic elements for the Fortuna-Experiment Station, Juan Diaz, PR. The growing period from February to April in 2006 was cooler and wetter compared to the same period in 2007, but the solar radiation was higher.

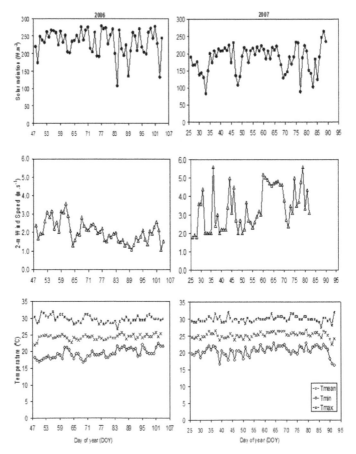

**FIGURE 1**   Daily climatic parameters for the 2006 and 2007 seasons at the Fortuna Experiment Station-Juan Diaz, PR.

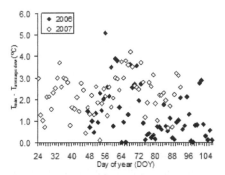

**FIGURE 2** Difference in minimum temperature and daily average dew temperature, for reference evaluation in the $ET_o$ estimation during 2006 and 2007.

The 2007 was warmer than the long-term average, with higher values of $T_{min}$, $T_{max}$ and $T_{mean}$. In 2006, 18 [18] rainfall events were recorded, wetter than the long-term average. Rainfall totals greater than 5.0 mm were registered on 8 days (DOY: 48 = 7.1 mm; 63 = 23.2 mm; 64=32.4 mm; 75=34.0 mm, 78=35.0 mm, 89=61.2 mm; 93=37.5 mm and 96 = 7.2 mm). During 2007, 13 [13] rainfall data were recorded during the experiment, but rainfall on just 3rd day was greater than 5.0 mm (DOY: 80 = 17.4 mm; 86 = 8.8 mm, and 88 = 8.8 mm). The drought stress treatment in 2006 was started on DOY 70 (March 11) and 2007 on DOY 55 (February 24).

Six days in 2006 and 10 days in 2007 were determined to be "nonreference" conditions for $ET_0$ estimation, where the $[T_{min} - T_{average,dew}]$ was >3.0°C (Fig. 2), which indicates a lack of well-watered conditions. The values for these days were corrected using methodology presented by Allen et al. [3].

When the daily mean value of surface sensible heat flux (H) is negative, Berengena and Gavilán [11] reported that the advection intensity could be quantified using the evapotranspiration fraction = $ET/R_n$, for H <0 and $ET/R_n$ > 1.0.

In this study, all the crop and references evapotranspiration ($ET_c$ and $ET_o$) estimations with the P-M model registered $[ET/R_n]$ <1.0 in both years. When the reference correction was made, small changes in the $ET_0$ were observed (0.1 mm.day⁻¹). Berengena and Gavilán [11] reported that the P-M reference evapotranspiration can give appropriate estimates of $ET_o$ even under strong adjective conditions.

The $ET_o$ rate for 2006 varied between 2.4 to 6.5 mm.day⁻¹ with a mean of 4.3 mm.day⁻¹. In 2007, the $ET_o$ rate varied between 2.2 to 6.3 mm.day⁻¹ with a mean of 4.0 mm.day⁻¹. The lower value for 2007 may be attributable to the lower solar radiation. Harmsen et al. [27] estimated the long-term $ET_o$ rates for January, February, March and April as 3.4, 3.8, 4.5 and 4.9 mm.day⁻¹ respectively, with a mean of 4.1 mm.

The $ET_c$ measured by the drainage lysimeters for Morales without drought stress totaled 211 mm in 2006 and 215 mm in 2007, compared with 172.2 mm and 190.0 mm, respectively using the P-M model. The $ET_c$ measured by the drainage lysimeters for SER 16 totaled 142.0 mm in 2006, and 152.5 mm in 2007 and with PM-Model 147.2 and 166.3 mm, respectively (Table 6). The lower $ET_c$ values for SER 16 were associated with the lower plant density, compared with Morales. The water requirements for dry bean for a 90 to 100-day season ranges from 350 to 500 mm depending upon the soil, climate and cultivar [37]. For a 122-day season, Calvache et al. [15] reported a crop water requirement of 447 mm for dry bean.

The low seasonal crop evapotranspiration values in this study are associated with: short crop season (75 and 78 days in 2006–2007, respectively), low plant density, climatic factors (low evaporative demand), and the irrigation system (drip), that dismiss the soil evaporation. Muñoz-Perea et al. [37] reported genotypic differences in $ET_c$ of 318 mm for NW63 and 457 mm for Othello under well water conditions, and 270 mm for Othello to 338 mm for Common Pinto beans under drought stress in Kimberly Idaho conditions.

During the 2006 growing season, SER 16 without drought stress reached maturity earlier than the stressed treatment, which induced the high $ET_c$ rates at the end of the season. Adams et al. [1] reported that dry bean required 25 to 30 mm of water per week (3.6 to 4.3 mm.day$^{-1}$); and the dry bean water use rates increased from 1.3 to 6.3 mm.day$^{-1}$, during pod development [40]. In this study, the $ET_c$ increased from 0.7 mm.day$^{-1}$ in vegetative growing phase to 5.1 mm.day$^{-1}$ during pod filling for Morales in 2006 without drought stress, and 0.6 mm.day$^{-1}$ to 4.6 mm.day$^{-1}$ in 2007. For SER 16, $ET_c$ increased from 0.4 mm.day$^{-1}$ in the vegetative growing phase to 5.1 mm.day$^{-1}$ in pod filling phase in 2006, and from 0.3 mm.day$^{-1}$ to 6.7 in 2007 (Fig. 3).

**TABLE 5** Mean daily weather conditions during the experiment at the Fortuna Experimental Station (Juana Diaz, PR), measured under reference conditions and compared with long-run means (1960–1987).

| | Jan | Feb | Mar | Apr |
|---|---|---|---|---|
| **2006** | | | | |
| Min. Air temperature (°C) | nd‡ | 18.2 | 19.5 | 20.5 |
| Max. Air temperature (°C) | nd | 30.7 | 29.4 | 30.1 |
| Mean. Air temperature (°C) | nd | 24.7 | 24.2 | 24.8 |
| Solar Radiation (W.m$^{-2}$) | nd | 200.6 | 224.2 | 243.2 |
| Relative humidity (%) | nd | 65.0 | 72.6 | 74.7 |
| Wind speed (m.s$^{-1}$) | nd | 2.2 | 3.1 | 2.9 |
| Rainfall (mm) | nd | 13.5 | 191.2 | 57.7 |
| **2007** | | | | |
| Min. Air temperature (°C) | 19.9 | 20.3 | 21.3 | 17.8 |
| Max. Air temperature (°C) | 29.5 | 29.9 | 30.2 | 30.6 |
| Mean. Air temperature (°C) | 24.7 | 25.0 | 25.3 | 23.9 |
| Solar Radiation (W.m$^{-2}$) | 181.6 | 181.2 | 184.8 | 254.7 |
| Relative humidity (%) | 66.7 | 66.9 | 66.5 | 57.5 |
| Wind speed (m.s$^{-1}$) | 2.9 | 2. 8 | 3.1 | 2.8 |
| Rainfall (mm) | 1.8 | 1.4 | 46.9 | 0 |
| **(1960-1987)†** | | | | |
| Min. Air temperature (°C) | 22.6 | 18.6 | 18.9 | 19. 8 |
| Max. Air temperature (°C) | 29.7 | 29.7 | 30.2 | 30.6 |
| Mean. Air temperature (°C) | 24.1 | 24.2 | 24.6 | 25.2 |
| Solar Radiation (W.m$^{-2}$) | nd | nd | nd | nd |
| Relative humidity (%) | nd | nd | nd | nd |
| Win direction (Deg) | nd | nd | nd | nd |
| Wind speed (m.s$^{-1}$) | nd | nd | nd | nd |
| Rainfall (mm) | 23. 8 | 20.0 | 32.5 | 53.34 |
| Reference Evapotranspiration (mm)¶ | 104 | 107 | 139 | 147 |

† Goyal and Gonzalez, (1989); ‡ No data.; ¶. Harmsen et al. (2004).

**TABLE 6** Cumulated $ET_c$ from V2 to R9 phenological phases for two common bean genotypes, measured by water balance methods (Lysimeter) and energy balance method (Generalized Penman-Monteith).

| Year | Genotype | Without drought stress | | With drought stress | | Reference Evapotranspiration |
|------|----------|------------|------|------------|------|------------------|
| | | Lysimetry | P-M | Lysimetry | P-M | |
| | | | | mm | | |
| 2006 | Morales | 211.0 (5.6)[†] | 172.2 | 167.3 (20.2) | 154.8 | 256 |
| 2007 | Morales | 215[‡] | 190.0 | 140 (26.6) | 151.8 | 263 |
| 2006 | SER16 | 142.0 (5.9) | 147.2 | 100.0 (6.8) | 157.6 | 256 |
| 2007 | SER16 | 152.5 (0.7) | 166.3 | 107.2 (37.7) | 137.1 | 263 |

† Parenthesis values indicated 1-SD; ‡ The other two lysimeters had plant establishment problems.

The differences in $ET_c$ (= $ET_{c,\text{without drought stress}}$ – $ET_{c,\text{with drought stress}}$) from the beginning of drought stress are presented in Fig. 3. During 2007, the drought stress was greater, and the difference in $ET_c$ was greater, with 40.2 mm for Morales, and 33.5 mm for SER 16 during R1 to R9, compared with 12.3 mm for Morales and 7.3 mm for SER 16 during 2006 for the same stage of plant development. The most critical differences were observed during: R2, R4 and R6 in 2006 and R2, R3, R4, R5, R6 and R9 stages of plant development in 2007. The common bean is most sensitive to drought stress during the preflowering and reproductive stages [15, 37].

The intermittent drought stress from the R1 to R9 growth stages induced seed-yield reduction for Morales of 33% in 2006 and 76% in 2007, and for SER 16 of 29% in 2006 and 67% in 2007, for small plots (2.0 m long, harvested at 6.0 g.kg$^{-1}$ of seed moisture). In the largest plots the yield reduction was exactly the same as in the small plots for both years for Morales, and for SER 16 was 33% in 2006 and 73% in 2007.

Without drought stress, the cumulative $ET_c$ during vegetative growth (V1 to R1-DOY 46 to 65) in 2006, was 30 mm for Morales and 21.1 mm for SER 16. During the same growth period in 2007 (DOY 28 to 54) the cumulative $ET_c$ was: 74.8 mm for Morales and 61.7 mm for SER 16. The cumulative $ET_c$ during the reproductive growth stage (R1 to R8- DOY 66 to 97) in 2006, was 118.2 mm for Morales and 108.7 mm for SER 16. During the same growth period in 2007 cumulative $ET_c$ was 103.3 mm for Morales and 92.1 mm for SER 16. During seed maturity to harvest (DOY 98 to 104), the cumulative ET in 2006, was 26.3 mm for Morales and 17.4 mm for SER 16, and for the same growth period in 2007 was 12.0 mm for Morales and 12.5 mm for SER 16.

The larger $ET_c$ rates were reached for both genotypes after pod initiation (R3), and maximum leaf area index (LAI) was registered at the R4 growth stage: Morales 4.2 m$^2$.m$^{-2}$ in 2006, and 3.0 m$^2$.m$^{-2}$ in 2007; SER 16 1.70 m$^2$.m$^{-2}$ in 2006 and 1.8 m$^2$.m$^{-2}$ in

2007. The differences in LAI are directly associated with the plant densities used in both genotypes.

The low $ET_c$ rates at the beginning of the growing season, DOYs 40 to 62 in 2006 and 28 to 46 in 2007, were associated with high surface resistances ($r_s$) as shown in Table 7. Changes in $r_s$ are also associated directly with stomatal resistance ($r_L$) and leaf area index (LAI). The low drought stress during 2006 did not generate significant changes in LAI and $r_s$; however, larger differences in LAI, $r_L$ and subsequently $r_s$ were observed in 2007, which suggest that $r_s$ is one of the most sensitive parameters controlling ET during drought stress.

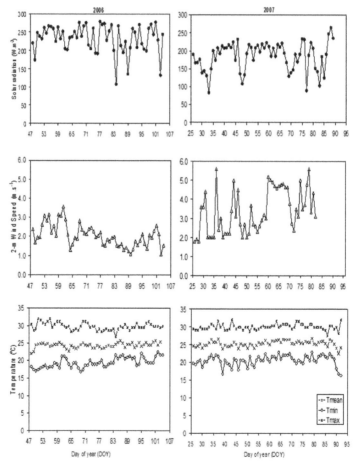

**FIGURE 3** Daily crop evapotranspiration rates for two common bean genotypes, with and without drought stress during two growing seasons: A. Morales-2006, B. Morales-2007, C. SER16-2006, and D. SER16-2007.

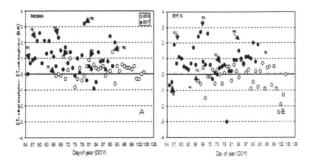

**FIGURE 4** Daily crop evapotranspiration differences with and without drought stress for two common bean genotypes, during two growing seasons, for R1 to harvest in 2006–2007: A. Morales, B. SER 16.

**TABLE 7** Surface resistance ($r_s$) distribution for two common bean genotypes, during two growing seasons, with and without drought stress conditions. The drought stress was applied starting in the R1 developmental phase.

| Water Level | Genotype | Phenologic phase | DOY[‡] | DAP[†] | $r_s$ | S.E[¶] | LAI | Phenologic phase | DOY | DAP | $r_s$ | S.E | LAI |
|---|---|---|---|---|---|---|---|---|---|---|---|---|---|
| 2006 | | | | | — s.m[-1] — | | $m^2 m^{-2}$ | 2007 | | | — s.m[-1] — | | $m^2 m^{-2}$ |
| Without drought stress | Morales | V2 | 48 | 16 | 3944.3 (651.2) | | 0.05 | V2 | 31 | 14 | 1518.2 (144.3) | | 0.10 |
| | | V4 | 56 | 24 | 1738.3 (527.7) | | 0.20 | V3 | 38 | 21 | 649.0 (69.9) | | 0.20 |
| | | V6 | 62 | 30 | 596.2 (69.7) | | 0.60 | V5 | 46 | 29 | 637.7 (177.6) | | 0.43 |
| | | R1 | 70 | 39 | 259.6 (53.6) | | 1.30 | V6 | 52 | 35 | 191.7 (26.1) | | 1.68 |
| | | R3 | 77 | 46 | 132.4 (20.2) | | 2.60 | R2 | 57 | 40 | 435.2 (18.9) | | 2.27 |
| | | R4 | 84 | 53 | 79.6 (13.2) | | 4.22 | R4 | 64 | 47 | 644.0 (6.1) | | 3.03 |
| | | R6 | 91 | 60 | 128.5 (30.5) | | 2.60 | R6 | 71 | 54 | 217.1 (25.4) | | 2.57 |
| | | R8 | 98 | 67 | 111.5 (31.5) | | 3.00 | R8 | 87 | 70 | 758.0 (473.8) | | 1.00 |
| | | R9 | 104 | 73 | 164.1 (16.8) | | 2.10 | | | | | | |
| Without drought stress | SER16 | V2 | 48 | 16 | 21020.9 (21899.9) | | 0.04 | V2 | 31 | 14 | 9100.0 (472.3) | | 0.02 |
| | | V4 | 56 | 24 | 2624.4 (298.2) | | 0.12 | V3 | 38 | 21 | 1308.0 (236.5) | | 0.10 |
| | | V6 | 62 | 30 | 1431.8 (211.5) | | 0.21 | V5 | 46 | 29 | 992.5 (402.7) | | 0.20 |
| | | R1 | 70 | 39 | 336.0 (6.8) | | 0.91 | V6 | 52 | 35 | 230.3 (32.6) | | 1.00 |
| | | R3 | 77 | 46 | 206.1 (59.5) | | 1.50 | R2 | 57 | 40 | 439.4 52.9) | | 1.27 |
| | | R4 | 84 | 53 | 192.2 (35.6) | | 1.60 | R4 | 64 | 47 | 519.4 (51.0) | | 1.77 |
| | | R6 | 91 | 60 | 184.2 (1.5) | | 1.70 | R6 | 71 | 54 | 258.2 (38.15) | | 1.67 |
| | | R8 | 98 | 67 | 252.1 (4.5) | | 1.20 | R8 | 87 | 70 | 388.8 (160.0) | | 1.43 |
| | | R9 | 104 | 73 | 581.8 (140.0) | | 0.50 | | | | | | |
| With drought stress | Morales | R1 | 70 | 39 | 221.2 (74.7) | | 1.50 | R2 | 57 | 40 | 769.0 (101.5) | | 2.00 |
| | | R3 | 77 | 46 | 108.1 (26.3) | | 3.10 | R4 | 64 | 47 | 1767.0 (410.1) | | 2.00 |
| | | R4 | 84 | 53 | 82.2 (83.5) | | 4.00 | R6 | 71 | 54 | 795.3 (61.7) | | 1.13 |
| | | R6 | 91 | 60 | 103.5 (59.5) | | 3.20 | R8 | 87 | 70 | 4800.0 (996.8) | | 0.80 |
| | | R8 | 98 | 67 | 126.2 (21.5) | | 2.60 | | | | | | |
| | | R9 | 104 | 73 | 141.9 (18.2) | | 2.30 | | | | | | |
| With drought stress | SER16 | R1 | 70 | 39 | 374.2 (36.0) | | 0.90 | R2 | 57 | 40 | 957.7 (99.5) | | 0.73 |
| | | R3 | 77 | 46 | 233.2 (12.5) | | 1.40 | R4 | 64 | 47 | 1150.0 (73.8) | | 1.20 |
| | | R4 | 84 | 53 | 257.7 (132.4) | | 1.20 | R6 | 71 | 54 | 375.1 (86.3) | | 1.00 |
| | | R6 | 91 | 60 | 182.5 (76.3) | | 1.80 | R8 | 87 | 70 | 4194.1 (539.7) | | 0.53 |
| | | R8 | 98 | 67 | 248.2 (4.4) | | 1.30 | | | | | | |
| | | R9 | 104 | 73 | 251.9 (32.7) | | 1.30 | | | | | | |

[‡] Day od the year; [†] Days after planting; [¶] Standard error.

The decreasing $ET_c$ during the DOYs: 58, 60, 61, 64 and 66 in 2007 for both water-levels, are associated with low aerodynamic resistance ($r_a$), and high surface resistance ($r_s$). The mean $r_a$ values were 53 sm$^{-1}$ for SER 16 and 54 sm$^{-1}$ for Morales without drought stress, as compared with mean $r_s$ values equal to 220 sm$^{-1}$ for SER 16 and 200 sm$^{-1}$ for Morales. The decreasing $ET_c$ is associated with high $r_s$ values for the same period (DOY 57 to DOY 64, Table 8). The low values of $r_a$ were directly related to the high wind velocities registered during that period between 10:00 am to 3:00 pm (range: 5.0 to 8.0 ms$^{-1}$), which likely induced stomatal closure. The wind speed and $r_a$ have an influence on the transpiration and have been reported by several authors [17, 18, 46]. Decreases or increases in the $r_L$ depends on the plant species [18]. Davies et al. [17] found that stomates closed markedly, resulting in increasing $r_L$ and subsequently increasing $r_s$, with abrupt increases in wind speed in prostrate plants.

## 8.3.2 CROP COEFFICIENTS

The crop coefficient curves are shown in Fig. 5. The largest differences in the $K_c$ between drought stress (open circles) and without drought stress (close circles) were observed during 2007 (Fig. 5B and 5D). The $K_c$ difference between water levels were more pronounced in Morales than in SER16 in both years.

**TABLE 8**  Length of common bean growth stages and crop coefficients ($K_c$), without drought stress. Estimated with Penman-Monteith general model, for Juana Diaz, PR.

| Genotype | Year | Bean Growth stage | Length of Stage days | Crop coefficient $K_c$ |
|---|---|---|---|---|
| Morales | 2006 | Initial | 21 | 0.25 |
| | | Crop development | 13 | 0.25 to 0.90 |
| | | Mid season | 32 | 0.90 |
| | | Late season | 6 | 0.90 to 0.50 |
| | 2007 | Initial | 18 | 0.50 |
| | | Crop development | 19 | 0.50 to 0.80 |
| | | Mid season | 35 | 0.80 |
| | | Late season | 6 | 0.80 to 0.30 |
| SER 16 | 2006 | Initial | 21 | 0.22 |
| | | Crop development | 13 | 0.22 to 0.82 |
| | | Mid season | 32 | 0.82 |
| | | Late season | 6 | 0.80 to 0.30 |
| | 2007 | Initial | 18 | 0.24 |
| | | Crop development | 19 | 0.24 to 0.80 |
| | | Mid season | 35 | 0.80 |
| | | Late season | 6 | 0.80 to 0.30 |

‡. Crops were grown during the period of January to April, which is considered to be the driest time of the year; drip irrigation was used; site elevation 28 meters above sea level, row spacing was 90 cm; plant densities for Morales and SER 16 were 13.2 plants per m$^2$ and 6.4 plants per m$^2$, respectively; registered wind velocities were greater during 2007 than during 2006.

The linearized crop coefficients ($K_c$) are shown in Table 8. SER 16 did not show differences across years, the reduction in the $K_c$ during the mid-season in 2007 was associated with low leaf area index during that year compared with the first, and differences in $r_a$ and $r_s$. An additional stress by high wind conditions during the mid-season reduced the $ET_c$ in 2007, and Morales was more susceptible. The $K_c$ values presented in this study are lower than those reported by the Irrigation and Drainage Paper-FAO 56 [3]. The large row spacing and differences in plant density (low LAI), and irrigation system (drip) help to explain lower $K_c$ values obtained in this study.

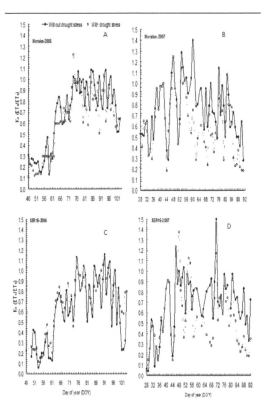

**FIGURE 5** Daily crop coefficients ($K_c$), for two common bean genotypes, with and without drought stress during two growing seasons: A. Morales-2006, B. Morales-2007, C. SER 16-2006, and D. SER 16-2007.

In Table 9, $K_c$ measured in intervals of 3 and 4 days by the drainage lysimeter are listed. The $K_{c\,mid}$ are similar to those in Table 8 estimated with the PM model. A mean value for the reproductive phase (R1 to R9) was 0.9 in 2006 and 1.0 in 2007 for Morales without drought stress and decreased to 0.8 in 2006 and to 0.7 in 2007. For the genotype SER 16, the average $K_c$ during R1 to R9 was 0.7 in 2006 and 0.6 in 2007 without drought stress; the $K_c$ decreased from 0.6 in 2006 to 0.3 in 2007 with drought stress.

The variation in crop development rates between location and year have been expressed as correlations between crop coefficients and indices such as the thermal base index, ground cover, days after emergence or planting, and growth rate [14, 25, 29, 34, 35, 39, 52]. In this study, the $K_c$ was correlated with the fraction covered by vegetation ($f_c$) calculated as the ratio between plant canopy diameter and row spacing, and with the cumulative growing degree days (Fig. 6), calculated as follows:

$$CGDD = \frac{\left[ T_{max} + T_{min} \right]}{2} - T_b \qquad (18)$$

where, $T_{max}$ is the maximum daily temperature; $T_{min}$ is the minimum daily temperature; and $T_b$ is the base temperature = 10 °C.

**TABLE 9** Crop coefficients ($K_c$) for two common bean genotypes without and with drought stress, measured in drainage lysimeters. Each observation is an average of three lysimeters values.

| Water Level | Genotype | DAP[†] | Phenologic phase | $K_c$ | S.E[¶] | DAP | Phenologic phase | $K_c$ | S.E |
|---|---|---|---|---|---|---|---|---|---|
| 2006 Trial | | | | | | 2007 Trial | | | |
| Without drought stress | Morales | 16 | V2 | 0.9 | (0.22) | 19 | V3 | 0.6 | (0.09) |
| | Morales | 21 | V3 | 1.0 | (0.22) | 26 | V4 | 0.9 | (0.13) |
| | Morales | 24 | V4 | 1.4 | (0.30) | 29 | V5 | 1.7 | (0.39) |
| | Morales | 28 | V7 | 0.6 | (0.15) | 35 | V6 | 0.8 | (0.10) |
| | Morales | 30 | V8 | 0.2 | (0.02) | 38 | R1 | 1.0 | (0.09) |
| | Morales | 35 | R1 | 1.0 | (0.75) | 40 | R2 | 0.8 | (0.02) |
| | Morales | 41 | R2 | 0.8 | (0.15) | 43 | R3 | 0.6 | (0.06) |
| | Morales | 52 | R4 | 0.9 | (0.30) | 47 | R4 | 0.9 | (0.05) |
| | Morales | 56 | R5 | 0.9 | (0.20) | 51 | R5 | 0.8 | (0.02) |
| | Morales | 59 | R6 | | | 54 | R6 | 1.6 | (0.19) |
| | Morales | 63 | R7 | 1.3 | (0.05) | 62 | R7 | 1.4 | (0.35) |
| | Morales | 66 | R8 | 0.4 | (0.13) | 70 | R8 | 1.2 | (0.28) |
| | Morales | 72 | R9 | 0.6 | (0.14) | 75 | R9 | 0.2 | (0.05) |
| | | Average[‡] | | 0.9 | | | | 1.0 | |
| With drought stress | Morales | 16 | V2 | 0.6 | (0.09) | 19 | V3 | 0.8 | (0.16) |
| | Morales | 21 | V3 | 0.9 | (0.17) | 26 | V4 | 1.0 | (0.34) |
| | Morales | 24 | V4 | 1.3 | (0.22) | 29 | V5 | 0.3 | (0.01) |
| | Morales | 28 | V7 | 0.5 | (0.10) | 35 | V6 | 1.3 | (0.28) |
| | Morales | 35 | R1 | 1.7 | (0.01) | 38 | R1 | 0.9 | (0.10) |
| | Morales | 41 | R2 | 0.2 | (0.01) | 40 | R2 | 0.1 | 0.03 |
| | Morales | 45 | R3 | 1.4 | (0.01) | 43 | R3 | | |
| | Morales | 52 | R4 | 0.8 | (0.19) | 47 | R4 | 0.3 | (0.06) |
| | Morales | 56 | R5 | 0.2 | (0.01) | 51 | R5 | 0.8 | (0.13) |
| | Morales | 59 | R6 | | (0.01) | 54 | R6 | 0.8 | (0.11) |
| | Morales | 63 | R7 | 1.3 | (0.24) | 62 | R7 | 0.9 | (0.33) |
| | Morales | 66 | R8 | 0.4 | (0.06) | 70 | R8 | 0.7 | (0.16) |
| | Morales | 72 | R9 | 0.8 | (0.15) | 75 | R9 | 0.0 | |
| | | Average | | 0.8 | | | | 0.7 | |
| Without drought stress | SER16 | 16 | V2 | 0.6 | (0.05) | 19 | V3 | 0.5 | (0.08) |
| | SER16 | 21 | V3 | 0.6 | (0.05) | 26 | V4 | 0.5 | (0.06) |
| | SER16 | 24 | V4 | ` | | 29 | V5 | 1.0 | (0.20) |
| | SER16 | 28 | V7 | 0.3 | (0.01) | 35 | V6 | 0.6 | (0.13) |
| | SER16 | 35 | R1 | 1.3 | (0.30) | 38 | R1 | 0.7 | (0.09) |
| | SER16 | 41 | R2 | 0.4 | (0.02) | 40 | R2 | 0.4 | (0.05) |
| | SER16 | 45 | R3 | 1.5 | (0.05) | 43 | R3 | 0.3 | (0.04) |
| | SER16 | 52 | R4 | 0.9 | (0.10) | 47 | R4 | 0.5 | (0.06) |
| | SER16 | 56 | R5 | 0.5 | | 51 | R5 | 0.6 | (0.07) |
| | SER16 | 59 | R6 | | | 54 | R6 | 1.1 | (0.18) |
| | SER16 | 63 | R7 | 0.8 | (0.05) | 62 | R7 | 0.7 | (0.14) |
| | SER16 | 66 | R8 | 0.3 | (0.10) | 70 | R8 | 0.6 | (0.07) |
| | SER16 | 72 | R9 | 0.4 | (0.13) | 75 | R9 | § | (0.02) |
| | | Average | | 0.7 | | | | 0.6 | |
| With drought stress | SER16 | 16 | V2 | 0.2 | (0.02) | 19 | V3 | 0.5 | (0.11) |
| | SER16 | 21 | V3 | 0.6 | (0.15) | 26 | V4 | 0.4 | (0.12) |
| | SER16 | 24 | V4 | 0.7 | (0.20) | 29 | V5 | 0.3 | (0.06) |
| | SER16 | 28 | V7 | 0.3 | (0.05) | 35 | V6 | 0.6 | (0.06) |
| | SER16 | 35 | R1 | 1.1 | (0.10) | 38 | R1 | 0.4 | (0.08) |
| | SER16 | 41 | R2 | 0.1 | (0.01) | 40 | R2 | 0.1 | (0.09) |
| | SER16 | 45 | R3 | 1.1 | (0.10) | 43 | R3 | | |
| | SER16 | 52 | R4 | 0.4 | (0.19) | 47 | R4 | 0.1 | (0.02) |
| | SER16 | 56 | R5 | 0.2 | | 51 | R5 | 0.4 | (0.05) |
| | SER16 | 59 | R6 | | | 54 | R6 | 0.3 | (0.03) |
| | SER16 | 63 | R7 | 0.8 | (0.15) | 62 | R7 | 0.7 | (0.05) |
| | SER16 | 66 | R8 | 0.3 | (0.05) | 70 | R8 | 0.3 | (0.10) |
| | SER16 | 72 | R9 | 0.4 | (0.10) | 75 | R9 | § | (0.01) |
| | | Average | | 0.6 | | | | 0.3 | |

[†] Days after planting; [¶] Standard error; [‡] Average corresponded since R1 to R9.
[§] Plants completely dry; The spaces in white, correspond with strong rainfall events, where the Kc, could no be successfully measured.

The plant density, adopted in the present study, induce differences in the seasonal trend in $f_c$ (Eqs. (19) and (20)), but not in CGDD (Eqs. (21) and (22)). A second degree polynomial equation for each genotype was fitted to relate $K_c$ with CGDD and $f_c$ (Fig. 6). For Morales with 13.6 plants.m$^{-2}$, the equations were:

$$K_c = -3x10^{-6}CGDD^2 + 0.0033CGDD - 0.053; R^2 = 0.76; p < 0.0001 \quad\quad (19)$$

$$K_c = -1.4019 f_c^2 + 2.5652 f_c - 0.2449; R^2 = 0.70; p < 0.0003 \quad\quad (20)$$

For SER 16, with 6.4 plants.m$^{-2}$ the equations were:

$$K_c = -3x10^{-6}CGDD^2 + 0.0034CGDD - 0.0515; R^2 = 0.60; p < 0.0001 \quad\quad (21)$$

$$K_c = -0.6726 f_c^2 + 1.90086 f_c - 0.2560; R^2 = 0.60; p < 0.0032 \quad\quad (22)$$

The cumulative observed and simulated evapotranspiration were 166.5 mm and 166.3 mm for Morales at 13.6 plants/m, and 143.7 mm and 146.3 mm for SER 16 at 6.4 plants/m, respectively. The comparison between observed and simulated values is presented in Fig. 7, where the relative errors were 0.17% for Morales and 1.7% for SER 16 (Eqs. (20) and (22)).

### 8.3.3 DUAL CROP COEFFICIENTS

The single crop coefficient ($K_c$) was separated into two coefficients, which represent the crop and soil participation in the evapotranspiration process, and which are used to predict the effects of specific wetting events on the $K_c$ [3]. The dual crop coefficients are especially useful in the case where the soil surface layer is dry, but the average soil water content in the root zone is adequate to sustain full plant transpiration [4]. The dual crop coefficients include the basal coefficient or transpiration coefficient ($K_{cb}$) and the soil evaporation coefficient ($K_e$).

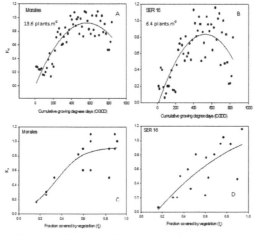

**FIGURE 6**  Crop coefficients ($K_c$) as a related to cumulative growing degree days (CGDD) and fraction covered by vegetation ($f_c$) for: A. Morales CGDD vs. $k_c$ B. SER 16 CGDD vs. $k_c$, C. Morales $f_c$ vs. $k_c$, D. SER 16 $f_c$ vs. $k_c$. The curves were fit from V1 to R9.

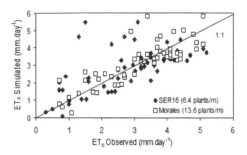

**FIGURE 7**   Observed and simulated evapotranspiration from $K_c$ models and reference ET estimated with the PM-model, and the observed ET estimated with the generalized PM-model for bean in 2006 at Juana Diaz, PR.

Figures 8 and 9 show the $K_{cb}$ distribution during the growing season for water treatment, genotype, and year. The upper limit of $K_{cb}$ in the mid-season was 0.85 for SER 16; and 0.91 for Morales without drought stress. The initial $K_{cb}$ values are lower than 0.15 for both genotypes and growing seasons, which are close to the reported values of the FAO [3], for dry bean. The mid and end values in this study were lower than the reported values.

The change from vegetative to reproductive states of development, is associated with an increase in the transpiration, due to the abrupt increase in the $K_{cb}$ observed in DOY 74 (Fig. 8), and DOY 53 (Fig. 9), that is directly associated with the increase in leaf area and transpiration surface.

During the mid-season in 2006, the drought stress treatment $K_{cb}$ values reached 0.4 for Morales (DOY's 81, 83, 84 and 91; Fig. 8D), which corresponded with the R4 to R6 stages of development, and 0.30 for SER 16 (DOY 73) corresponding with R2. Morales exhibited a higher frequency of low $K_{cb}$ values than SER 16 in the drought stressed treatments.

The low $K_{cb}$ measured in 2007, during the mid-season for both genotypes and water levels (Fig. 9), indicated low transpiration rates in an important stage of development, possible due to a "physiological stress" associated with high wind speeds and low irrigation rates, and which induced high $r_L$ as was discuss in the $ET_c$ results. SER 16, responded similarly for both water level treatments Figs. 9C and 9D.

The larger difference between $K_c$ and $K_{cb}$ in 2006 vs. 2007, can be explained in 2006 due to the water supplied in the experiment was higher than in 2007, associated with several rainfall events that kept the soil wetter for a longer time, and also due to higher wind speed in 2007 than 2006, that dismiss the transpiration rates.

The $ET_c$ was adjusted to account for water stress by using the water stress coefficient ($K_s$). This coefficient is related to the root zone depletion ($D_r$), calculated using the water balance equation:

$$D_{r,i} = D_{r,i-1} - (P - RO)_i - I_i + ET_{c,i} + DP_i \qquad (23)$$

where, $D_{r,i}$ is the root zone depletion at the end of the day i; $D_{r,i-1}$ is water content in the root zone at the end of the previous day, i–1; $(P-RO)_i$ is the difference between precipitation and runoff on the day i; $I_i$ is the irrigation depth on the day i; $ET_{c,i}$ is the crop evapotranspiration on day I; and $DP_i$ is the water loss out of the root zone by deep percolation on day i. All the units are in mm.

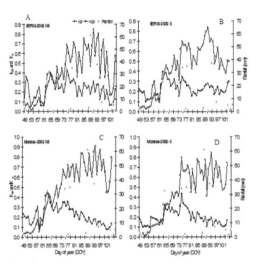

**FIGURE 8** Basal crop coefficients ($K_{cb}$) and soil evaporation coefficient ($K_e$) for two common bean genotypes-2006: A. SER 16 without stress, B. SER 16 with drought stress, C. Morales without drought stress and D. Morales with drought stress.

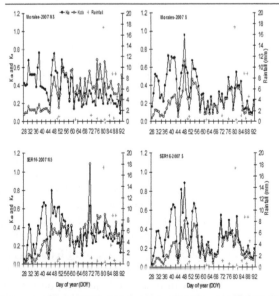

**FIGURE 9** Basal crop coefficients ($K_{cb}$) and soil evaporation coefficient ($K_e$) for two common bean genotypes-2007: A. Morales without drought stress, B. Morales with drought stress, C. SER 16 without drought stress and D. SER 16 with drought stress.

The root zone depletion associated with a $K_s = 1.0$ (i.e., no water stress), was up to 10 mm for a root depth between 0 to 20 cm, and up to 15 mm for a root depth of 0 to 40 cm (Fig. 10). Fifty percent of the transpiration reduction was reached for $D_r = 22$ mm and 25 mm in Morales and SER 16, respectively. Transpiration ceased completely ($K_s = 0$) when Dr = 37 mm and 46 mm, respectively for Morales and SER 16.

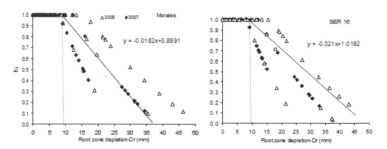

**FIGURE 10**   Water stress coefficients ($K_s$) for two common bean genotypes: A. Morales B. Morales.

## 8.4   CONCLUSIONS

In this chapter, crop evapotranspiration was estimated with the generalized Penman-Monteith model and drainage lysimeters for two common bean genotypes, with and without drought stress. The maximum ET rates for both genotypes were reached at the beginning of the reproductive phase to seed maturity, and were equivalent to 67% of the total ET for Morales and 73% for SER 16, in 2006, and 54% and 55%, respectively, in 2007. The reduction in ET in 2007 may be associated with an increase in surface resistance due to windy and drought stress conditions. The increasing surface resistance was also related to an observed decrease in the transpiration coefficient ($K_{cb}$).

The $K_{c\ mid}$ values for the well watered treatment were lower than 1.0 for both genotypes, measured by lysimeters and the PM-model, indicating relatively low water requirements for both genotypes. Both genotypes exhibited a $K_c$ reduction during drought stress of similar magnitude. The $K_c$ for nonlimited soil water conditions was well correlated with the cumulative degree day (CGDD) and with the fraction covered by vegetation ($f_c$) for both genotypes with different plant densities.

The largest differences in the ET estimations between the lysimeter and the PM-model were observed in the beginning of the crop season, which was particularly associated with low LAI, increasing $r_s$, and a decrease in ET. The changes in ET rates associated with drought stress were variable between genotypes: Morales ET in 2006 was reduced by 10% with the PM model, as compared with 0.0% by SER 16. The change in ET due to drought stress for 2007 was 20% for Morales and 18% for SER 16. Note that the two genotypes should not be compared due to differences in the plant density.

The intermittent drought stress applied from floral differentiation to harvest was stronger during 2007 than 2006, with a subsequent effect on yield components. The genotype Morales exhibited the highest reduction in evapotranspiration during critical drought stress periods (R1, R2, R5 and R6).

Values of surface resistance as a function of stomatal resistance and LAI were also derived in this study, as well as values of the crop stress coefficient ($K_s$), and critical values of root zone depletion were estimated as a $K_s$ function, for both genotypes.

The crop coefficients ($K_c$) derived in this study, are specific to the genotypes considered and the agronomic practices used, including the irrigation system. Additionally, it is important to consider that the plant density is a critical component in the $K_c$ estimation, and it is suggested that adjustments be made to the $K_c$ based on the fraction of the soil covered by vegetation ($f_c$). The specific wind conditions present during the study can have a considerable effect on the derived crop coefficients, and therefore, caution should be exercised when applying these coefficients under wind conditions, which vary from those in this study.

## 8.5 SUMMARY

The product of the single crop coefficient ($K_c$) or the dual crop coefficients ($K_{cb}$ and $K_e$) and the reference evapotranspiration ($ET_0$) is a widely used method for crop evapotranspiration estimation ($ET_c$), recommended by the Food and Agriculture Organization (FAO) of the United Nations in their Irrigation and Drainage Papers 24 and 56. $ET_c$, $K_c$, $K_{cb}$ and $K_e$ were measured for two new common bean genotypes, during two growing seasons (2006 and 2007) in southern Puerto Rico, at the University of Puerto Rico Experiment Station at Fortuna, during the driest months of the year (January–April). The genotypes (*P. vulgaris* L.) planted were: Morales, commonly grown in Puerto Rico; and SER 16, which is drought tolerant and was developed at the International Center of Tropical Agriculture (Centro Internacional de Agricultura Tropical, CIAT – Colombia); both with a type II plant architecture. Drought stress was applied for both genotypes at flowering through to maturation. $ET_c$ was measured with drainage lysimeters and estimated using the generalized Penman-Montith (PM) method with variable aerodynamic ($r_a$) and surface resistance ($r_s$). Additionally, an automatic weather station was placed in a nearby well-watered grass to estimate $ET_0$ using the PM-reference model. The linearized $K_c$ for the initial, mid, and end stages of growth for Morales were: $K_{c\,ini} = 0.25$; $K_{c\,mid} = 0.90$ and $K_{c\,end} = 0.50$, and for SER 16 were: $K_{c\,ini} = 0.22$; $K_c\,mid = 0.80$ and $K_{c\,end} = 0.30$. Morales was more adversely affected by drought stress than SER 16. The $K_c$ for both genotypes was correlated with the fraction of the soil covered by vegetation ($f_c$) and cumulative grown degree days (CGDD). The stress coefficient ($K_s$) maintained a value of 1.0 when the root zone depletion ($D_r$) was less than 10 mm within the top 0–20 cm of soil surface and less than 15 mm within the top 0–40 cm soil of the soil surface. The total average $ET_c$ for Morales without drought stress was 211 mm in the lysimeters and 172.2 mm using the PM method during 2006, and 215 mm in the lysimeters and 190 mm using the PM method in 2007. The total average $ET_c$ for SER 16 without drought stress was 142 mm in the lysimeters and 147 mm using the PM method in 2006, and 152.5 mm in the lysimeter and 166.3 mm using the PM method in 2007. The $r_s$ was a determinant variable in the $ET_c$ estimation under drought stress in both genotypes.

## KEYWORDS

- **Blaney-Criddle**
- **common bean**
- **crop coefficient**
- **drainage type lysimeter**
- **drip irrigation**
- **drought stress**
- **dry bean**
- **dual crop coefficients**
- **evapotranspiration**
- **leaf area index**
- **Penman-Monteith**
- **plant transpiration coefficient**
- **Puerto Rico**
- **references conditions**
- **soil evaporation coefficient**
- **trickle irrigation**
- **wind speed**

## REFERENCES

1. Adams, M.W., D.P. Coyne., J.H.C. Davis., P.H. Graham, and C.A. Francis. 1985. Common bean (*Phaseolus vulgaris* L.), 433–476. In: *Grain Legume Crops* by R.J. Summerfield and E.H. Roberts, Editors. Harper Collins
2. Allen, R. G. 1996. Assessing integrity of weather data for reference evapotranspiration estimation. *J. Irrig. Drain. Eng.* 122(2):97–106.
3. Allen G.R, L.S. Pereira, D. Raes, and M. Smith. 1998. Crop evapotranspiration: Guidelines for computing crop water requirements. Food and Agricultural Organization of the United Nations (FAO). Publication No. 56. Rome. 300p.
4. Allen, G.R., L.S. Pereira, M. Smith, D. Raes, and J.L Wright. 2005. FAO-56 dual crop coefficient method for estimating evaporation from soil and application extensions. *J. Irrig. Drain. Eng.* 131(1):2–13.
5. Alves, I., Perrier, A.; Pereira, L. S. 1998. Aerodynamic and Surface Resistance of Complete Cover Crops: How Good is the "Big Leaf"?. *Trans. ASAE.* 41(2):345–351.
6. Anda, A. and Löke, Z.S., 2002. Stomatal resistance investigations in maize. Proceedings of the 7th Hungarian Congress of Plant Physiology. 46(3–4):181–183.
7. Barros, L.C.G., and J.R. Hanks. 1993. Evapotranspiration and yield of bean as affected by mulch and Irrigation. *Agron. J.* 85(3): 692–697
8. Beebe, S., and B. McClafferty. 2006. Biofortified Bean. Centro Internacional de Agricultura Tropical (CIAT)-Cali, Colombia. 2p. In: www.harvestplus.org/pdfs/**bean**.pdf.
9. Beaver, J.S., P.N. Miklas, 1999. Registration "Morales" small white bean. *Crop Sci.* 39:1257.
10. Beaver, J.S. 2006. FY05 Report of the leaders of the basic grain commodity group. In.: Agricultural Experiment Station, University of Puerto Rico: http://eea.uprm.edu/decano/bienvenida.aspx.

11. Berengena, J and P. Gavilan. 2005. Reference evapotranspiration in a highly adjective semiarid environment. *J. Irri. Drain. Eng.* 132(2):147–163

12. Brian, J.B. 1991. Alfalfa ET measurements with drainage lysimeters. Pages 264–271. In: *Lysimeter For Evapotranspiration and Environmental Measurements*, Edited by Richard G. Allen, Terry A. Howell, William O. Pruitt, Ivan A. Walter and Marvin E. Jensen, ASCE.

13. Broughton, W.J., G. Hernandez., M. Blair., S. Beebe., P. Gepts., and J. Vanderleyden. 2003. Bean (*Phaseolus* spp.)-model food legumes. *Plant and Soil*. 252: 55–128.

14. Brown, P.A., C.F. Mancino., M.H Joung., T.L. Thompson., P.J Wierenga, and D.M. Kopec. 2001. Penman Monteith crop coefficients for use with desert turf system. *Crop. Sci.* 41:1197–1206.

15. Calvache, M., Reichardt, K., Bacchi. O.O.S., and Dourado-Neto, D. 1997. Deficit irrigation at different growth stages of the common bean (*Phaseolus vulgaris* L., cv. IMBABELLO). *Sci. agri. Piracicaba.* 54:1–16.

16. Caspari, H.W, S.R. Green, W.R.N. Edwards. 1993. Transpiration of well-watered and water-stressed Asian pear stress as determined by lysimetry, heat-pulse, and estimated by a Penman-Monteith model. *Agric. For. Meteorol.* 67:13–27.

17. Davis, W.J., K. Gill, and G. Halliday. 1978. The influence of wind on the behavior of stomata of photosynthetic stems of *Cyticus scoparious* (L.) Link. *Annals of Botany*. 42:1149–1154.

18. Dixon, M and J. Grace. 1984. Effect of wind on the transpiration of young trees. *Annals of Botany*. 53: 811–819.

19. Doorenbos, J. and W.O. Pruitt. 1977. Guidelines for predicting crop water requirements. Food and Agricultural Organization of the United Nations (FAO). Publication No. 24. Rome. 300p.

20. Food and Agriculture Organization. (FAO). 2002 Food Outlook, No. 2.in: www,fao.org.

21. Fritschen, J.L., and L.C. Fritschen. 2005. Bowen ratio energy balance method. In: *Micrometeorology in Agricultural Systems*, Agronomy Monogaph no, 47. American Society of Agronomy, Crop Science Society of America, Soil Science Society of America, 677 S. Soe Rd., Madison, W153711, USA: 397–405.

22. Goenaga, R., Rivera, E., Almodovar, C. 2004. Yield of papaya irrigated with fractions of Class A pan evaporation in a semiarid environment. *J. Agric. Univ. P.R.*, 88 (1–2):1:10.

23. Goyal, M.R. and E.A. Gonzalez. 1988d. Water requirements for vegetable production in Puerto Rico. ASAE Symposium on Irrigation and Drainage, July. 8p.

24. Goyal, M.R, and E.A. Gonzalez. 1989. Datos climatológicos de las subestaciones experimentales agrícolas de Puerto Rico. Universidad de Puerto RicoRecinto de Mayagüez/Colegio de Ciencias Agrícolas/Estación Experimental Agrícola. Publicación 88–70, proyecto C-411. 87p.

25. Hanson, R.B., and D.M. May. 2004. Crop coefficients for drip-irrigated processing tomatoes. In: ASAE/CSAE Annual International Meeting. Fairmont Chateau Laurier, The Westing, Government Centre, Ottawa, Ontario, Canada.12p.

26. Harmsen, E.W. 2003. Fifty years of crop evapotranspiration studies in Puerto Rico. *J. Soil Water Conserv.* 58(4):214–223.

27. Harmsen, E.W, A. Gonzalez-Pérez, and A. Winter. 2004. Re-evaluation of pan evaporation coefficients at seven location in Puerto Rico. *J. Agric. Univ. P.R.* 88(3–4):109–122.

28. Harmsen, E.W., A. Gonzalez-Pérez. 2005. Computer program for estimating crop evapotranspiration in Puerto Rico. *J. Agric. Univ. P.R.* 89(1–2):107–113.

29. Hunsaker, D.J. 1999. Basal crop coefficients and water use for early maturity cotton. *Trans. ASAE.* 42(2):927–936.

30. Jia, X., E.C. Martin, and D.C. Slack, F. 2005 Temperature adjustment for reference evapotranspiration calculation in Central Arizona. *J. Irrig. Drain. Eng.* 130(5):384–390.

31. Karam, F, R. Masaad, T. Sfeir, O. Mounzer, and Y. Rouphael. 2005. Evapotranspiration and seed yield of field grown soybean under deficit irrigation conditions. *Agrl. Water. Manage.* 75:226–244.

32. Kjelgaard, J.F.; and Stokes, C.O.2001. Evaluating Surface Resistance for Estimating Corn and Potato Evapotranspiration with the Penman-Monteith Model. *Trans. ASAE.* 44(4):797–805.

33. Lin Li, Y., Yuan Cui, J., Hui Zhang, T., and Liu Zhao, H. 2003. Measurement of evapotranspiration of irrigated spring wheat and maize in a semiarid region of north China. *Agric. Water. Manage* .61(1): 1–12.

34. Madeiros, G.A., F.B. Arruda., E. Sakai., and M. Fujiwara. 2001.The influence of the crop canopy on evapotranspiration and crop coefficients of bean (*Phaseolus vulgaris* L.). *Agric. Water. Manage*. 49:211–234.

35. Madeiros, G.A., F.B. Arruda., E. Sakai. 2004. Crop coefficient for irrigated bean derived using three reference evaporation methods. *Agric. Water. Manage*. 135:135–143.

36. Monteith, J.L, and M.H. Unsworth.1990. Principles of environmental physics. Chapman and Hall, Inc.292p.

37. Muñoz-Perea, C.G., Allen, G.R., D.T. Westermann, J.L. Wring, and S.P. Singh. 2007. Water used efficiency among dry bean landraces and cultivars in drought-stressed and nonstressed environments. *Euphytica*. 155:392–402.

38. National Agricultural Statistic Services, of the United States Department of Agriculture. (NASS). 2002.

39. Nasab, B.S., H.A. Kashkuli, and M.R. Khaledian. 2004. Estimation of crop coefficients for sugarcane (ratoon) in Haft Tappeh of Iran. In: ASAE/CSAE Annual International Meeting. Fairmont Chateau Laurier, The Westing, Government Centre, Ottawa, Ontario, Canada.10p.

40. North Dakota State University (NDSU). 1997. Dry bean production guide. Publication series No. A-1133.In: http://www.ag.ndsu.edu/pubs/plantsci/rowcrops/a1133–5.htm

41. North Dakota State University (NDSU). 2003. Dry bean production guide. Publication series No. A-1133.120p.

42. Pereira, L.S, and M.S. Adaixo. 1991. Lysimeter-based evapotranspiration research in Portugal. Pages 142–150. In: *Lysimeter For Evapotranspiration and Environmental Measurements* Edited by: Richard G. Allen, Terry A. Howell, William O. Pruitt, Ivan A. Walter and Marvin E. Jensen. ASCE.

43. Roygard K,F.R.; Alley, M.M.; and Khosla, R. 2002. No-Till Corn Yield and Water Balance in the Mid-Atlantic Coastal Plain. *Agron. J.* 94:612–623.

44. Rosenberg, J. N, B.L. Blad, and S.B, Verma. 1983. Microclimate; The Biological Environment. A Wiley-Interscience Publication, John Wiley & Sons. Edt. 495p.

45. Sheng Li, Q., Villardson, S.L., Deng, W., Jun Li, X., and Jiang, L.C. 2005. Crop water deficit estimation and irrigation scheduling in western Jilin province, North-east China. *Agric. Water. Manage*. 71(1):47–60.5

46. Smith, K.W. 1980. Importance of aerodynamic resistance to water use efficiency in three conifers under field conditions. *Plant Physiol*. 65:132–135.

47. Szeicz, G, and I.F. Long. 1969. Surface resistance of crop canopies. *Water Resour. Res*. 5:622–633.

48. U.S. Department of Agriculture (USDA). 1987. Primary characterization data of San Anton soil, Guanica country, Puerto Rico. Soil Conservation Service, National Soil Survey laboratory, Lincoln, Nebraska.

49. U.S. Department of Agriculture (USDA). 2005.Economic research services. Dry bean trade. In: http://www.ers.usda.gov/Briefing/DryBean/Trade.htm

50. Villalobos, F.J., Testi, L.; Rizzalli, R.; and Orgaz, F. 2004. Evapotranspiration and crop coefficients of irrigated garlic [*Allium sativum* L.] in a semiarid climate. Agric. Water. Manage. 64(3):233–249.

51. Wenkert, W. 1983. Water transport and balance within the plant: An overview. In: *Limitations to Efficient Water Use in Crop Production*. Edited by: Howard M. Taylor, Wayne R. Jordan and Thomas R. Sinclair. Published by: ASA-CSSA-SSSA:137–172.

52. Wright, J.L., and M.E. Jensen. 1978. Development and evaluation of evapotranspiration models for irrigation scheduling. *Trans ASAE*. 21(1):88–96.

53. Zhang, Y.; Yu, Q.: Liu, C., Jiang, J., and Zhang, X. 2004. Estimation of water wheat evapotranspiration under water stress with two semi empirical approaches. *Agron J.* 96:159–168.

# CHAPTER 9

# WATER REQUIREMENTS FOR PAPAYA ON A MOLLISOL SOIL

RICARDO GOENAGA, EDMUNDO RIVERA, and CARLOS ALMODOVAR

## CONTENTS

*Modified and printed with permission from: "*Ricardo Goenaga, Edmundo Rivera, and Carlos Almodovar. Yield of papaya irrigated with fractions of Class A pan evaporation in a semiarid environment. J. Agric. Univ. P.R., January/April 2004, 88(1–2):1–10.*" This research was carried out cooperatively between the USDA-Agricultural Research Service and the Agricultural Experiment Station of the University of Puerto Rico. The authors acknowledge Roberto Bravo and Tomas Soto for their assistance in carrying out the experiment. The names in this chapter are used only to provide specific information. Mention of a trade name or manufacturer does not constitute a warranty of materials by the USDA-ARS or the Agricultural Experiment Station of the University of Puerto Rico, nor is this mention a statement of preference over other materials.

## 9.1   INTRODUCTION

Papaya is a tropical herbaceous dicotyledonous plant producing melon-like fruits, which are rich in vitamins A and C. Total world production of papaya is estimated at 5.6 billion kg produced on about 338,000 hectares. The largest producers are Brazil and Mexico. Mean worldwide yield of papaya in 2002 was 16,544 kg/ha, with Mexico and Costa Rica topping all countries at 40,155 kg/ha and 38,667 kg/ha, respectively [3]. Increased popularity of papaya during the last 10 years has resulted in a 255% increase in world exports.

Papaya can be commercially produced in a wide range of agro-environments. The most essential requirement, particularly at early growth stages, is good soil drainage to prevent the development of root rot caused by soil-borne pathogens. Average marketable yield of five papaya hybrids and an open-pollinated cultivar was almost doubled when plants were grown on an Oxisol (123,333 kg/ha) versus a heavy-clay Ultisol (65,339 kg/ha). This yield difference was attributed in part to the effect of excess rain and less drainage at the Ultisol site [6].

Papaya can also be grown in a semiarid environment with irrigation. Drip irrigation technology permits the efficient use of water and can help maximize the utilization of semiarid lands for agricultural production. This technology is particularly suited to widely spaced crops such as papaya. There is very little information regarding optimum water requirements for papaya, particularly under semiarid conditions. It has been suggested that the water needs of papaya are ideally supplied by 100 mm of evenly distributed rainfall each month [7]. This amount is seldom encountered in the wet-and-dry climate in which papaya is normally grown, a climate characterized by erratic rainfall patterns and prolonged dry periods, or in semiarid environments where evaporation may be three times greater than rainfall.

This chapter evaluates the optimum water requirement for papaya grown under semiarid conditions under drip irrigation and evaluates how yield and fruit quality traits are affected by various levels of irrigation.

## 9.2   MATERIALS AND METHODS

### 9.2.1   EXPERIMENTAL DESIGN

An experiment was conducted from 1996 to 1998 at the Fortuna Agricultural Experiment Station of the University of Puerto Rico in the semiarid agricultural zone of Puerto Rico. The San Anton soil is a well-drained Mollisol (fine-loamy, mixed, iso-hyperthermic Cumulic Haplustoll) with pH of 7.5, bulk density 1.4 g/cm$^3$, and 1.7% organic carbon in the first 14 cm of soil. The 28-year mean annual rainfall is 917 mm and annual Class A pan evaporation is 2,149 mm. Mean monthly maximum and minimum temperatures are 31.2 and 20.8°C. Figure 1 shows total monthly rainfall and evaporation during the experimental period, and Table 1 shows average monthly irrigation supplied to plants.

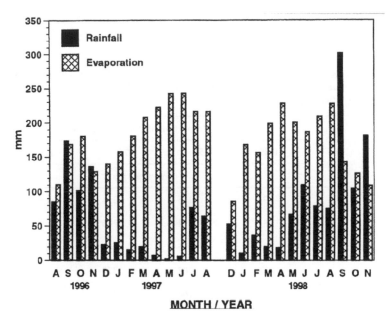

**FIGURE 1**   Total monthly rainfall and Class A pan evaporation during growth of papaya at the Fortuna Agricultural Experiment Station, Juana Diaz – Puerto Rico, USA.

Seed of papaya cv Red Lady (Known You Seed Co., Taiwan) was germinated in Styrofoam seedling trays with open bottom cells containing Pro-Mix BX. Tray cells were 38 mm deep and 13 mm on each side. About 50 days after germination, seedlings with four to six leaves and a total dry weight of about 0.6 g were transplanted to the field 15 August 1996 and 16 December 1998 at a spacing of 1.8x1.8m. Five treatments representing different moisture regimes were arranged in a randomized complete block design with four replications. There were two rows per plot, each with 10 experimental plants and surrounded by 3.7-m alleys, with two guard plants at the end of each row to prevent over-lapping of the irrigation treatments.

## 9.2.2   MICRO IRRIGATION MANAGEMENT

Two days prior to transplanting, silver polyethylene mulch (1.2 m wide x 0.15 mm thick) and drip lateral lines were installed by using a mechanical mulch applicator. Drip irrigation lines were placed on the surface along the center of each row. The equation by Young and Wu [12] was used to calculate the amount of irrigation applied to plants. The equation assumes that the evapotranspiration (ET) of a papaya plant is equal to the evaporation from a body of water with a free surface equal to the plant area as determined by a Class A pan evaporimeter. In this research, the equation was modified to include a pan coefficient (Kp) value of 0.70 and a modified average crop coefficient (Kc) of 0.42 [2] to obtain a theoretical value of potential evapotranspiration (ETc). Class A pan factors (proportion of pan evaporation), ranging from 0.25 for treatment one to 1.25 for treatment five in increments of 0.25, were used to obtain fractions of ETc. A pan factor of 1.0 means that the water applied to the plants of that treatment replaced that lost through calculated evapotranspiration; this amount was therefore, considered the theoretical optimum.

**TABLE 1**   Average monthly irrigation applied to papaya plants subjected to five levels of irrigation as determined by pan factor (proportional to Class A pan evaporation) during a two-year period, 1996–1998.

| Month | Proportion of Pan Evaporation | | | | |
|---|---|---|---|---|---|
| | **0.25** | **0.50** | **0.75** | **1.0** | **1.25** |
| | **Liters per plant** | | | | |
| January | 79 | 155 | 234 | 318 | 392 |
| February | 68 | 134 | 200 | 268 | 336 |
| March | 88 | 173 | 261 | 347 | 436 |
| April | 101 | 202 | 277 | 369 | 509 |
| May | 85 | 164 | 247 | 339 | 426 |
| June | 96 | 194 | 288 | 388 | 484 |
| July | 76 | 151 | 226 | 304 | 381 |
| August | 79 | 169 | 252 | 338 | 424 |
| September | 16 | 29 | 45 | 60 | 76 |
| October | 18 | 26 | 31 | 39 | 45 |
| November | 31 | 42 | 53 | 63 | 74 |
| December | 66 | 129 | 192 | 255 | 313 |
| Total | 803 | 1,568 | 2,306 | 3,088 | 3,896 |
| Average | 66.9 | 130.7 | 192.1 | 257.3 | 324.7 |

The plants were subjected to the five moisture treatments starting about two months after transplanting. The amount of water applied varied weekly, depending on Class A pan evaporation and rainfall. The previous week's evaporation and rainfall data were used to determine the irrigation needs for the following week. Irrigation was supplied three times during the following week on alternate days. No irrigation was provided when the total rainfall exceeded 19 mm per week as this amount was sufficient to wet the soil area that was exposed as a result of breaking small sections of the mulch to make the planting hole at transplanting time. Water for irrigation was obtained from a supply well which draws water from the unconfined alluvial aquifer underlying the study area. Submain lines equipped with volumetric metering valves to monitor the water from the main line were provided for each treatment. Lateral lines equipped with built-in 4 Lph-emitters spaced 61 cm apart branched out from the submains along the inner side of each plant row and about 21 cm from the plant stem.

### 9.2.3 FERTILIZATION

At transplanting, each plant received 11 g P provided as triple superphosphate. Throughout the experimental period, fertilization through the drip system was provided weekly at the rate of 10.5 kg/ha of N and 13.0 kg/ha of K; urea and potassium nitrate were the nutrient sources. Weekly fertilization also included 0.26 and 0.08 kg/ha of Zn and Fe, respectively, supplied in their EDTA chelate forms, and 0.29 kg/ha Mn supplied as DTPA chelate. Insects and weeds were controlled by following the recommended cultural practices [10].

### 9.2.4 CROP PERFORMANCE

The first harvest of fruits was at about six months after transplanting. Fruits were harvested at color break when they started to show a tinge of yellow at the apical end of the fruit. At each harvest, number and weight of marketable and nonmarketable (deformed) fruits were determined. Representative fruits totaling about 25% of those harvested were used to determine fruit length. Brix readings were also taken with a sugar refractometer when the fruits ripened, about five days after harvest. Analysis of variance and best-fit curves were determined by using the ANOVA and GLM procedures, respectively, of the SAS program package [9]. Only coefficients with P≤0.05 were considered statistically significant.

### 9.3 RESULTS AND DISCUSSION

Differences among irrigation treatments and years were highly significant (P≤0.01) for most of the response variables that were studied (analysis of variance not shown). An exception was the brix value, which was not affected by the irrigation treatments. The "treatment x year interaction" was not significant; therefore, data were averaged over years.

Total Class A pan evaporation was 2.5 times greater than the amount of total rainfall recorded during the 25-month experimental period. In 16 of the 25 months, evaporation was twice the amount of rainfall (Fig. 1). This indicates that large soil-water deficits would have occurred without irrigation. Overall, less irrigation was required during September through November; and more irrigation was needed during January through June (Table 1).

Total number of fruits produced was linearly related to the amount of water applied (i.e., pan factor). This response was mainly the result of a significant increase in the number of marketable fruits with increments in pan factor treatment (Fig. 2). The total number of fruits increased from 53,731 to 74,891 kg/ha for pan factor treatments 0.25 and 1.25, respectively. In both treatment extremes, marketable fruits represented 78% of the total fruits produced.

Increments in pan factor treatment significantly increased fruit yield. The highest marketable fruit weight of 75,907 kg/ha was obtained with the application of irrigation according to a pan factor of 1.25 (Fig. 3). This weight was 33, 27, 17, and 8%

higher than that obtained in plants irrigated with a pan factor of 0.25, 0.50, 0.75, and 1.0, respectively. The increase in marketable fruit weight with increments in pan factor treatment was the result of significant increases in fruit number and fruit length (Figs. 2 and 4). Similar responses have been obtained with other fruit crops [4, 5]. Fruits from plants irrigated with a pan factor treatment of 1.25 were almost 13% longer than those irrigated with a pan factor of 0.25. Water deficits occurring during reproductive stages are known to cause yield loss.

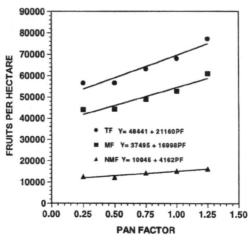

**FIGURE 2**   Number of total (TF), marketable (MF) and nonmarketable (NMF) papaya fruits as influenced by irrigation based on pan factors of 0.25 to 1.25.

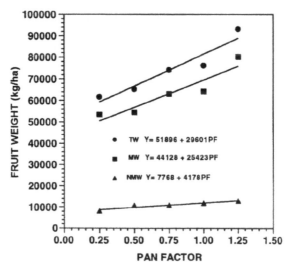

**FIGURE 3**   Relationship between irrigation based on pan factors and total (TW), marketable (MW), and nonmarketable (NMW) fruit weight of papaya.

In this study, the significantly lower number of marketable fruits and yield in plants irrigated with low pan factor treatments may have been the result of inhibition of flower development, failure of fertilization or reduction of both vegetative source and reproductive sink activity [8, 11].

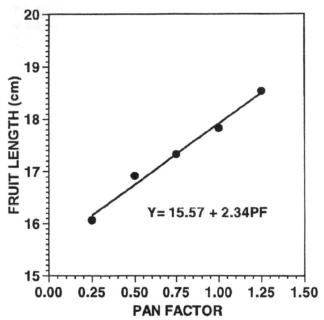

**FIGURE 4**  Relationship between irrigation based on pan factors and length of papaya fruits.

Brix (sweetness) values were not significantly different among five irrigation treatments and averaged 11.7. This value is higher than that obtained by Goenaga et al. [6] when the same variety was grown in different agro-environments.

The results reported in this study demonstrate that papaya should be irrigated by using a pan factor of at least 1.25. The use of a lower pan factor results in significant reduction in yield. Profitable papaya yields can be attained in the semiarid agricultural zone of Puerto Rico if proper irrigation practices are followed. Assuming that a grower can achieve the highest marketable fruit yield in this study and that farm gate prices are $9.00 per box of 18.14 kg, then a gross income of $37,660 per hectare can be attained. Updated procedures recommending the use of the Penman-Monteith method for calculating crop evapotranspiration were published by FAO after this study was completed [1]. Therefore, it is recommended that this method may be used in future studies on irrigation requirements of papaya and that, if necessary, refinements be made to our recommendation.

## 9.4  SUMMARY

There is a scarcity of information regarding the optimum water requirement for papaya (*Carica papaya*) grown under semiarid conditions with drip irrigation in the tropics. A two-year study was conducted to determine water requirement, yield, and fruit quality traits of papaya cv Red Lady subjected to five levels of irrigation. The irrigation treatments were based on Class A pan factors that ranged from 0.25 to 1.25 in increments of 0.25. Drip irrigation was supplied three times a week on alternate days. Results showed significant effects of irrigation treatments on number of fruits, yield and fruit length. Irrigation treatments did not have a significant effect on brix (sweetness). Highest marketable fruit weight (75,907 kg/ha) was obtained from plants irrigated according to a pan factor of 1.25. It was concluded that papaya grown under semiarid conditions should be irrigated according to a pan factor of not less than 1.25.

## KEYWORDS

- Brix value
- Carica papaya
- Class A pan factor
- crop coefficient, Kc
- drip irrigation
- DTPA
- EDTA
- evapotranspiration
- fruit length
- fruit quality
- fruit weight
- fruit yield
- gross income
- Irrigation
- irrigation recommendations
- marketable fruit weight
- marketable yield
- micro irrigation
- Mollisol soil
- nonmarketable fruit
- pan coefficient, Kp
- Papaya
- potential evapotranspiration
- semiarid land
- Yield

## REFRENCES

1. Allen,R.G.,L.S.Pereira,D.Raes and M. Smith,1998.*Crop Evapotranspiration: Guidelines for Computing Crop Water Requirements.* Irrigation Drainage Paper 56. FAO, Rome.
2. Doorenbos, J. and W. 0. Pruitt, 1977. *Guidelines for Predicting Crop Water Requirements.* Irrigation Drainage Paper 24, FAO, Rome.
3. Food and Agriculture Organization, 2003. FAOSTAT statistics database: Agriculture. Available at http://apps.fao.org/cgi-bin/nph-db.pl?subset=agriculture (accessed March 2003).
4. Goenaga, R. and H. Irizarry, 1995. Yield performance of banana irrigated with fractions of Class A pan evaporation in a semiarid environment. *Agron. J.*,87:172–176.
5. Goenaga, R., H. Irizarry and E. Gonzalez, 1993. Water requirement of plantains *(Musa acuminata x Musa balbisiana* AAB) grown under semiarid conditions. *Trop. Agric.* (Trinidad), 70:3–7.
6. Goenaga, R., H. Irizarry and E. Rivera-Amador, 2001. Yield and fruit quality of papaya cultivars grown at two locations in Puerto Rico. *J. Agric. Univ. P.R.,* 85:127–134.
7. Nakasone, H.Y. and R.E. Paull, 1998. *Tropical Fruits.* CAB International, New York, NY.
8. Pugnaire, F. I., L. Surano and J. Pandos, 1999. Constraints by water stress on plant growth. *In:* M. Pessarakli (ed.). *Handbook of Plant and Crop Stress.* Marcel Dekker, Inc., New York, NY. pp. 271–283.
9. SAS Institute, 1987. *SAS/STAT Guide for Personal Computers.* Version 6. SAS Institute, Cary, NC.
10. Toro, E. E., 1993. Papaya Cultivation (Spanish). Agricultural Extension Service, College of Agricultural Sciences of University of Puerto Rico.
11. Westgate, M. E., 1994. Seed formation in maize during drought. *In:* K. J. Boote, J. M. Bennett, T. R. Sinclair and G. M. Paulsen (eds.). *Physiology and Determination of Crop Yield.* American Society of Agronomy, Madison, Wisconsin. pp. 361–364.
12. Young, S. C. H. and I. P. Wu, 1981. Final report of the banana drip irrigation studies at the Waimanolo Experiment Station. pp. 51–69. *In: Proc. Annual Meeting Hawaii Banana Industry Association.* 13th Conf. Res. Ext. Service 021. College of Tropical Agric. and Human Resources, Univ. of Hawaii, Honolulu.

# CHAPTER 10

# WATER REQUIREMENTS FOR TANIER (*XANTHOSOMA* SPP.)

RICARDO GOENAGA

## CONTENTS

*Modified and reprinted with permission from, "Ricardo Goenaga, Growth, nutrient uptake and yield of tanier (Xanthosoma spp.) grown under semiarid conditions. J. Agric. Univ. P.R. July and October 1994, 78(3–4):87–98." The author acknowledges the field assistance of Roberto Bravo, Vidal Marti, Roberto Luciano and Tomas Miranda during the course of this investigation.* The names in this chapter are used only to provide specific information. Mention of a trade name or manufacturer does not constitute a warranty of materials by the USDA-ARS, nor is this mention a statement of preference over other materials.

## 10.1  INTRODUCTION

Tanier (Xanthosoma spp.) is a herbaceous perennial root crop that serves as an important food staple in some subtropical and virtually all tropical regions. In Puerto Rico, tanier ranks second in economic importance among root crops [14]. The major production area is the mountain region where periodic droughts often reduce yields. The demand for a year-round supply of high quality cormels, high farm-gate prices, and the availability of arable land with an irrigation infrastructure has generated interest among farmers and government agencies to shift tanier production from the highlands to the fertile but semiarid lowlands previously used for sugarcane production. In a normal year, evaporation in this agricultural zone may be three times greater than rainfall.

Micro irrigation technology allows efficient use of water and may maximize the utilization of semiarid land for agricultural production. There is little information regarding water requirements for optimum tanier production in Puerto Rico or elsewhere in the tropics, particularly under semiarid conditions. It has been suggested that the water needs of tanier are ideally supplied by 140 to 200 mm of evenly distributed rainfall throughout the year [13]. The importance of adequate soil moisture for tanier growth was demonstrated by Lugo et al. [10], who found that tanier marketable yields were significantly higher in plots that received frequent irrigation. Under field conditions without irrigation, corms and cormels of tanier plants accumulated most of their dry matter at 16 to 20 weeks after planting, but maximal leaf area was attained at 14 weeks after planting and then declined rapidly [9]. Maximum yield potential was not attained because of the rapid decline in leaf area during the corm and cormel bulking period.

Experiments conducted in Puerto Rico have demonstrated that tanier yields as high as 34,000 kg/ha can be attained when the crop is grown with micro irrigation and intensive management [5, 6]. Under rainfed conditions in Africa and many Caribbean countries, tanier yields are very low, ranging from 1,270 to 5,440 kg/ha [11]. These results illustrate the importance of tanier irrigation and suggest that the potential for commercial production of tanier depends on the implementation of management practices that prevent stress conditions during tanier growth, particularly if the goal is to increase food production for the growing population in the tropics.

In this chapter, the author evaluated the growth, nutrient uptake and yield performance of tanier grown in fertile but semiarid lands and subjected to various irrigation regimes. The study also formed part of a continued effort to collect growth analysis data to develop the SUBSTOR-Aroid model [6, 16].

## 10.2  MATERIALS AND METHODS

### 10.2.1  EXPERIMENTAL DESIGN

A field experiment was conducted at the Fortuna Agriculture Experiment Station of the University of Puerto Rico at Juana Diaz. The soil is a well-drained Mollisol (fine-loamy, mixed, isohyperthermic Cumulic Haplustoll) with pH

6.8, bulk density of 1.56 g/cm³, organic carbon of 1.7%; and exchangeable bases, 23.2 cmol(+)/kg of soil. Soil nitrate and ammonium at the 0 to 15 cm depth were 11.2 and 12.9 μg/g of soil, respectively.

Plants of cultivar Blanca were established in a screenhouse from excised corm buds with fresh weights of about 16 g (2.7 g dry weight) and planted in plastic trays containing Pro-Mix growing medium. On 11 September 1991, plants at the one-leaf stage were transplanted to the field and arranged in a split-plot design with five replications. Each replication contained four main plots representing different moisture regimes. The main plots were split to accommodate nine biomass harvests. Each subplot contained 20 plants spaced 0.91 *x* 0.46 m apart, the inner six of which were sampled for biomass production. Twelve plants per plot were harvested from each treatment to calculate final yield. An alley of 2.7 m separated the main plots to prevent overlapping of water between treatments. The experiment was surrounded by two rows of guard plants.

### 10.2.2   MICRO IRRIGATION MANAGEMENT

The amount of irrigation applied to plants was calculated with the equation by Young and Wu [18]. The equation assumes that the evapotranspiration of a tanier plant is equal to the evaporation from a body of water with a free surface equal to the plant area as determined by a Class A pan evaporimeter. In this study, the equation was modified to include a pan coefficient (Kp) value of 0.70 and a crop coefficient (Kc) value of 0.87 to obtain a theoretical value of potential evapotranspiration [3]. Class A pan factors, which ranged from 0.33 for treatment one (increasing by 0.33 per treatment) to 1.32 for treatment 4, were used to obtain fractions of the potential evapotranspiration. A pan factor of 0.33 means that the irrigation applied to the plants would replace 33% of the "water lost through evapotranspiration (WLET)."

Plants were subjected to one of the four moisture regimes 7 weeks after field planting. The amount of water applied varied weekly depending on Class A pan evaporation and rainfall. Evaporation and rainfall data of each week were used to determine the irrigation needs for the next week. Irrigation was supplied three times per week on alternate days, and no irrigation was provided when the total rainfall exceeded 20 mm per week.

The water source was a well-fed reservoir. Submain lines equipped with volumetric metering valves to monitor the water from the main line were provided for each treatment. Lateral lines equipped with 8-L emitters spaced 45.7 cm apart branched out from the submains along each plant row.

### 10.2.3   FERTIGATION

At planting, each plant received 3.5 g of granular P provided as triple superphosphate. Throughout the experimental period, fertigation was provided monthly at the rate of 3.7 and 8.0 kg/ha of N and K, respectively, with potassium nitrate and urea as the nutrient sources. Fertilization also included 0.45

and 0.13 kg/ha of Zn and Fe, respectively, supplied in their EDTA chelate forms and 0.48 kg/ha of Mn supplied as DTPA chelate.

## 10.2.4   CROP PERFORMANCE

Samples for plant and biomass measurements were collected at 49, 91, 133, 181, 223, 278, 322, 364 and 398 days after planting (DAP), except for treatments 1 and 2, in which the last biomass harvest was made at 364 DAP. At each harvest, leaves of plants were cut at the midrib-petiole intersection. Plants in the subplots were harvested by digging an area of 0.42 m² around each plant to a depth of 30.5 cm. Because of the labor and carefulness required during the root sampling operation, only two plants were used for that purpose. Plants were then pulled from the soil, washed, and separated into petioles, corms, cormels, roots and suckers. Samples were dried to constant weight at 70 °C for dry matter determination. The dry samples were ground to pass a 1.0-mil mesh screen and analyzed for N, P, K, Ca, Mg and Zn. Nitrogen was determined by the micro Kjeldahl procedure [12], P by the molybdovanadophosphoric acid method [8] and K, Ca, Mg and Zn by atomic absorption spectrophotometry [2].

Analyzes of variance and best fit curves were determined with the ANOVA and GLM procedures, respectively, of the SAS program package [15]. Only coefficients significant at $P \leq 0.05$ were retained in the models.

## 10.3   RESULTS AND DISCUSSION

### 10.3.1   DRY MATTER ACCUMULATION

Total class A pan evaporation was two times greater than the amount of total rainfall recorded during the experimental period (Table 1). Under those conditions large soil-water deficits would have existed without irrigation. The average evaporation and rainfall data collected during the same period for over 20 years were 2,482 mm and 1,212 mm, respectively [7]. These figures are similar to those obtained during the experimental period, which can therefore, be considered as a normal one.

**TABLE 1**   Monthly evaporation and rainfall registered at the Fortuna Substation and the amount of water applied to four irrigation treatments throughout the tanier-growing cycle, 1991–1992.

| Month | Evaporation mm | Rainfall mm | Irrigation supplied, liters/plant | | | |
|-------|----------------|-------------|-----------|------|------|------|
| | | | Pan factor | | | |
| | | | 0.33 | 0.66 | 0.99 | 1.32 |
| Sep | 183.6 | 55.1 | — | — | — | — |
| Oct | 182.6 | 45.5 | 18.0 | 36.0 | 54.0 | 72.0 |
| Nov | 153.2 | 82.0 | 3.4 | 6.8 | 10.2 | 13.6 |
| Dec | 134.1 | 5.1 | 7.4 | 14.8 | 22.2 | 29.6 |
| Jan | 134.6 | 228.3 | 4.0 | 8.0 | 12.0 | 16.0 |
| Feb | 160.0 | 20.3 | 11.1 | 22.2 | 33.3 | 44.4 |

**TABLE 1**  *(Continued)*

| Month | Evaporation mm | Rainfall mm | Irrigation supplied, liters/plant | | | |
|---|---|---|---|---|---|---|
| | | | Pan factor | | | |
| | | | 0.33 | 0.66 | 0.99 | 1.32 |
| Mar | 196.8 | 18.5 | 16.2 | 32.4 | 48.6 | 64.8 |
| Apr | 184.7 | 78.0 | 4.8 | 9.6 | 14.4 | 19.2 |
| May | 169.2 | 471.9 | 7.8 | 15.6 | 23.4 | 31.2 |
| Jun | 202.4 | 33.5 | 9.0 | 18.0 | 27.0 | 36.0 |
| Jul | 232.2 | 3.3 | 20.1 | 40.2 | 60.3 | 80.4 |
| Aug | 207.5 | 85.1 | 11.3 | 22.6 | 33.9 | 45.2 |
| Sep | 177.5 | 99.1 | 2.9 | 5.8 | 8.7 | 11.6 |
| Oct | 160.3 | 171.2 | 0 | 0 | 0 | 0 |
| Total | 2478.7 | 1396.9 | 116.0 | 232.0 | 348.0 | 464.0 |
| Average | 177.0 | 99.8 | 8.9 | 17.8 | 26.8 | 35.7 |

Note: Initiation of irrigation treatments. Previously, the experimental area was irrigated with sprinkler irrigation until the establishment of the tanier crop.

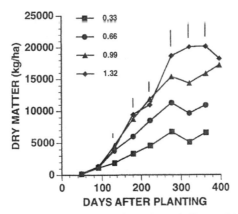

**FIGURE 1**   Accumulation of total dry matter in tanier as influenced by plant age and irrigated with fractions of evapotranspiration. The vertical bars are significant Fisher's lsd values at P = 0.05; absence of bars denotes lack of significant differences.

There were no significant differences in total dry matter production among treatments during the first 91 DAP (Fig. 1). This growth period was character- ized by low rates of total dry matter production with leaves and petioles ac- counting for over 70% (of the total dry matter produced (Figs. 2A-B). Between 91 and 223 DAP, plants that received 99 and 132% (i.e., pan factors of 0.99 and 1.32) of the water lost through evapotranspiration (WLET) exhibited simi- lar amounts of total dry matter (Fig. 1). During this period, roots and corms from most treatments showed high rates of dry matter accumulation whereas growth of cormels was severely affected in plants receiving the low irrigation treatments (Figs. 2C-E). By 223 DAP, the average production of total dry matter from plants replenished with 99% WLET was 26% greater than that from those replenished with 669% WLET; total dry matter from plants replenished with 132% WLET was 60% greater than that produced by plants replenished with 33% WLET.

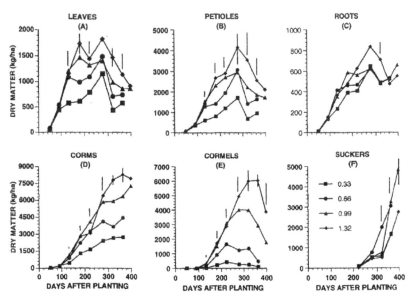

**FIGURE 2**    Accumulation of dry matter in tanier as influenced by plant age and irrigated with fractions of evapotranspiration. All symbols in Figs. A to E are same as in Fig. 1-F above. The vertical bars are significant Fisher's lsd values at P = 0.05; absence of bars denotes lack of significant differences.

**TABLE 2**    Average percentage nutrient concentration in leaf laminas of tanier plant with fractions of evapotranspiration.

| Nutrient | Treatment | Days after Planting, Dap | | | | | | | | |
|---|---|---|---|---|---|---|---|---|---|---|
| | | 49 | 91 | 133 | 181 | 223 | 278 | 322 | 364 | 398 |
| | | Nutrient concentration, % | | | | | | | | |
| N | 0.33 | 4.65 | 4.72 | 4.75 | 4.06 | 3.82 | 3.80 | 3.49 | 3.66 | — |
| | 0.66 | 4.50 | 4.62 | 4.20 | 3.67 | 3.45 | 3.63 | 3.35 | 3.29 | — |
| | 0.99 | 4.31 | 4.62 | 4.16 | 3.71 | 3.37 | 3.40 | 3.19 | 3.26 | 3.55 |
| | 1.32 | 4.57 | 4.53 | 4.09 | 3.40 | 3.20 | 3.35 | 2.97 | 3.25 | 3.48 |
| P | 0.33 | 0.35 | 0.37 | 0.37 | 0.38 | 0.31 | 0.42 | 0.39 | 0.36 | — |
| | 0.66 | 0.34 | 0.38 | 0.43 | 0.40 | 0.35 | 0.43 | 0.39 | 0.35 | — |
| | 0.99 | 0.32 | 0.38 | 0.41 | 0.40 | 0.35 | 0.41 | 0.40 | 0.37 | 0.46 |
| | 1.32 | 0.34 | 0.37 | 0.44 | 0.40 | 0.35 | 0.43 | 0.39 | 0.44 | 0.51 |

**TABLE 2**   *(Continued)*

| Nutrient | Treatment | Days after Planting, Dap | | | | | | | | |
|---|---|---|---|---|---|---|---|---|---|---|
| | | 49 | 91 | 133 | 181 | 223 | 278 | 322 | 364 | 398 |
| | | Nutrient concentration, % | | | | | | | | |
| K | 0.33 | 4.55 | 4.60 | 4.53 | 4.55 | 4.28 | 4.92 | 3.84 | 3.83 | — |
| | 0.66 | 5.41 | 4.66 | 4.51 | 4.22 | 4.24 | 4.84 | 4.01 | 3.78 | — |
| | 0.99 | 5.24 | 4.52 | 4.61 | 4.52 | 4.09 | 4.85 | 3.95 | 3.74 | 3.96 |
| | 1.32 | 4.77 | 4.52 | 4.50 | 4.40 | 3.99 | 4.62 | 4.15 | 3.84 | 4.08 |
| Ca | 0.33 | 1.92 | 1.91 | 1.52 | 1.14 | 1.74 | 1.77 | 1.86 | 1.44 | — |
| | 0.66 | 1.97 | 1.72 | 1.46 | 1.24 | 1.78 | 1.90 | 1.76 | 1.85 | — |
| | 0.99 | 2.16 | 1.91 | 1.82 | 1.34 | 1.88 | 1.85 | 1.97 | 2.06 | 1.84 |
| | 1.32 | 1.95 | 1.89 | 1.69 | 1.39 | 1.99 | 1.99 | 2.07 | 1.97 | 1.83 |
| Mn | 0.33 | 0.29 | 0.30 | 0.23 | 0.27 | 0.28 | 0.30 | 0.37 | 0.27 | — |
| | 0.66 | 0.30 | 0.29 | 0.22 | 0.25 | 0.28 | 0.32 | 0.34 | 0.33 | — |
| | 0.99 | 0.31 | 0.28 | 0.23 | 0.22 | 0.26 | 0.32 | 0.36 | 0.36 | 0.33 |
| | 1.32 | 0.29 | 0.29 | 0.22 | 0.20 | 0.24 | 0.30 | 0.35 | 0.33 | 0.36 |
| Zn | 0.33 | 0.35 | 0.46 | 0.35 | 0.38 | 0.46 | 0.46 | 0.53 | 0.48 | — |
| | 0.66 | 0.32 | 0.48 | 0.37 | 0.36 | 0.47 | 0.50 | 0.48 | 0.49 | — |
| | 0.99 | 0.28 | 0.51 | 0.36 | 0.36 | 0.48 | 0.48 | 0.48 | 0.54 | 0.64 |
| | 1.32 | 0.31 | 0.04 | 0.32 | 0.39 | 0.37 | 0.50 | 0.47 | 0.52 | 0.58 |

Cormel development was initiated at 91 DAP regardless of the irrigation treatment imposed on plants. This finding indicates that the initiation of cormel development is insensitive to water stress conditions. The importance of proper irrigation during the cormel bulking period was manifested by the low rates of dry matter accumulation in cormels from plants replenished with 33 and 66% WLET (Fig. 2E). By 223 DAP, cormel dry matter in plants receiving 33 and 66% WLET was 86 and 45% less than in those that received 99% WLET. Results published elsewhere [4] showed that leaf area indices were considerably lower in plants subjected to 33 and 66% WLET. As a result, the potential contribution of leaf assimilates toward the cormels was reduced, resulting in a drastic decline of cormel dry matter in these treatments. Cormel dry matter production was also influenced by water stress conditions, which caused cormel sprouting and consequent sucker development (Fig. 2E). By 364 DAP, suckers were more than 20% of the total dry matter in plants replenished with 33 to 99% WLET, but only 8% in those receiving 132% WLET.

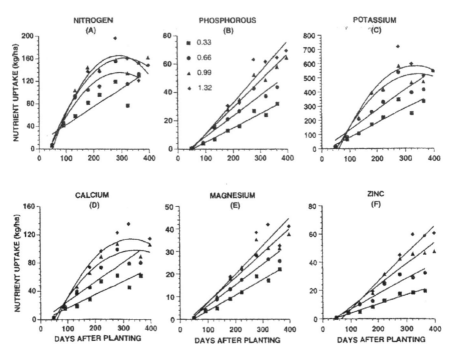

**FIGURE 3**   Nutrient content in tanier cultivars as influenced by plant age and irrigated with fractions of evapotranspiration. All symbols in Figs. A, C to F are same as in Fig. 3B.

### 10.3.2   NUTRIENT UPTAKE

Figure 3A-F shows that nutrient uptake increased steadily during most of the growing season. The amount of N, P, K, Ca, Mg and Zn taken up by plants replenished with 99 and 132% WLET was similar, whereas the content of these nutrients in plants replenished with 33 and 66% WLET was considerably lower (Figs. 3A-F). Plants from the latter treatments exhibited lower nutrient contents because of their lower production of dry matter rather than by a reduced concentration of nutrients in the tissue (Figs. 2A-F and Table 2).

Maximum uptake values for plants replenished with 132% WLET were 165 kg/ha of N, 76 kg/ha of P, 582 kg/ha of K, 114 kg/ha of Ca, 44 kg/ha of Mg and 65 kg/ha of Zn. Similar studies [5] conducted in an Oxisol showed higher values for N, Mg and Zn; similar values for P and Ca and lower values for K. In that study, a total of 185 of N and 232 of K kg/ha, were supplied from fertilizer sources. The fact that only 49.4 of N and 105.8 of K kg/ha of N and K, were provided from fertilizer sources in this study indicates that soils in this agricultural zone are highly fertile.

**TABLE 3** Effects of irrigation treatments and age of a crop on the marketable cormels [no./ha] and marketable yield [Kg/ha] of tanier.

| Pan factor treatment | | Days after planting (DOP), crop age | | | | | |
|---|---|---|---|---|---|---|---|
| | | 181 | 223 | 278 | 322 | 364 | 398 |
| 0.33 | Number of cormels**, No./ha | 0 | 797 | 1,594 | 0 | 0 | — |
| 0.66 | | 0 | 12,753 | 15,941 | 10,361 | 3,985 | — |
| 0.99 | | 7,970 | 38,258 | 54,996 | 45,431 | 33,476 | 25,505 |
| 1.32 | | 11,956 | 33,476 | 64,560 | 70,139 | 74,124 | 56,590 |
| LSD* | | 8,161 | 12,094 | 15,288 | 19,370 | 11,086 | 35,407 |
| 0.33 | Tanier yield** Kg/ha | 0 | 114 | 342 | 0 | 0 | — |
| 0.66 | | 0 | 2,083 | 2,880 | 1,815 | 606 | — |
| 0.99 | | 1,128 | 7,160 | 11,423 | 10,990 | 7,773 | 5,368 |
| 1.32 | | 1,790 | 6,058 | 14,827 | 17,885 | 19,479 | 12,172 |
| LSD* | | 1,197 | 2,540 | 3,798 | 4,402 | 2,943 | 6,699 |

* Fisher's (protected) LSD, significant at P = 0.05.
** Cormels were considered marketable when these attained a weight of 130 grams or more.

## 10.3.3 YIELD PERFORMANCE

Large significant differences for cormel yield and number of cormels were observed among irrigation treatments (Table 3). Cormel yield and cormel number in plants replenished with 33 and 66% WLET were extremely low. The yield of plants that received 99% WLET was considerably higher than that of plants replenished with 33 and 66% WLET, but significantly lower than that from plants that received 132% WLET. Maximum cormel yields of 19,479 kg/ha were obtained at 364 DAP in plants replenished with 132% WLET. The yields obtained in this investigation were higher than those obtained at the same location [10] and in other studies conducted in the humid mountain region of Puerto Rico [1, 17]. However, maximum yields in other studies [5] with the same tanier cultivar grown in an Oxisol soil under micro irrigation were 16% higher than those obtained in this study.

The results reported in this study demonstrate that profitable tanier yields can be attained in the semiarid agricultural zone of Puerto Rico if proper irrigation and cultural practices are followed. Assuming that a grower can achieve the highest yields obtained in this study and farm gate prices are $0.88/Kg of commercial cormels, then gross income of $17,140 per hectare can be attained.

This study also supports the concept that the potential for profitable tanier production in Puerto Rico depends on the intensive management of the crop through the use of modern and efficient agrotechnology. It should be noted, however, that for tanier production in Puerto Rico, alternate sites to the semi-arid zone should also be considered because higher yields have been obtained at other agricultural zones with less dependency on irrigation [5].

## 10.4 SUMMARY

There is little information regarding optimum water requirement for tanier grown under semiarid conditions with irrigation. A study was conducted to determine the growth, nutrient uptake and yield performance of tanier plants irrigated with the equivalent of fractions of evapotranspiration. The irrigation regimes were based on class A pan factors ranging from 0.33 to 1.32 with increments of 0.33. Tanier plants grown under field conditions were harvested for biomass production about every 6 weeks during the growing season. At each harvest, plants were separated into various plant parts to determine dry matter accumulation, N, P, K, Ca, Mg, and Zn uptake and yield. During the first 278 days after planting, plants replenished with 99 and 132% of the water lost through evapotranspiration (WLET) exhibited similar total dry matter content; however, their dry matter content was significantly greater than that in plants supplied with 33 and 66% WLET. The amount of N, P, K, Ca, Mg, and Zn taken up by plants replenished with 99 and 132% WLET was similar, whereas the content of these nutrients in plants replenished with 33 and 66% WLET was considerably lower. The yield of plants replenished with 99% WLET was considerably greater than that of plants supplied with 33 and 66% WLET, but significantly lower than that from plants receiving 132% WLET. Maximum cormel yields of 19,479 kg/ha were obtained from plants replenished with 132% WLET.

## KEYWORDS

- **Class A pan evaporation**
- **cormel**
- **cormel bulking**
- **cormel development**
- **cormel sprouting**
- **crop coefficient**
- **crop growth**
- **crop yield**
- **drip irrigation**
- **dry matter accumulation**
- **dry matter production**

- **DTPA**
- **EDTA**
- **evapotranspiration**
- **highlands**
- **irrigation**
- **irrigation recommendations**
- **leaf area**
- **marketable cormels**
- **marketable yield**
- **Mollisol soil**
- **nutrient concentration**
- **nutrient uptake**
- **pan coefficient**
- **semiarid lands**
- **sucker development**
- **suckers**
- **Tanier**
- **water lost through evapotranspiration, WLET**
- ***Xanthosoma* spp.**
- **yield performance**

## REFERENCES

1. Abruña, F., J. Vicente-Chandler, E. Boncta and S, Silva, 1967. Experiments on tanier production with conservation in Puerto Rico's mountain region. *J. Agric. Univ. P.R.* 51(2):167–175.
2. Chapman, H.D. and P.F. Pratt, 1961. Methods of Analysis for Soils. Plants, and Water. Division of Agricultural Sciences, University of California, Riverside, CA.
3. Doorenbos, J., and W.O. Pruitt, 1977. Guidelines for Predicting Crop Water Requirements; Irrigation and Drainage Paper 24, FAO United Nations, Rome. Italy.
4. Goenaga, R. 1994. Partitioning of dry matter in tanier (*Xanthosoma* spp.) irrigated with fractions of evapotranspiration. Annals of Botany, 73:251–261.
5. Goenaga, R. and U. Chardon, 1993. Nutrient uptake, growth and yield performance of three tanier *(Xanthosoma* spp.) cultivars grown under intensive management. *J. Agric. Univ. P.R.* 77(1–2):1–10.
6. Goenaga, R., U. Singh, F.H. Beinroth and H. Prasad, 1991. SUBSTOR-Aroid: A model in the making. *Agrotechnology* Transfer, 14:1–4.
7. Goyal, M.R. and E.A. Gonzalez, 1989. Climatological Data Collected in the Agricultural Experiment Stations of Puerto Rico. Publication 88–70. Agricultural Experiment Station, College of Agricultural Sciences, Mayaguez, Puerto Rico (Spanish).

8.  IBSNAT, 1987. Field and laboratory methods for IBSNAT. Technical Report 2. Department of Agronomy and Soil Science, College of Tropical Agriculture and Human Resources, University of Hawaii, Honolulu, Hawaii

9.  Igbokwe, M.C., 1983. Growth and development of *Colocasia* and *Xcmthosoma* spp. under upland conditions. *In:* E.R. Terry, E.V. Doku. O.B Arene and N.M. Mahungu (eds.), Tropical Root Crops: Production and Cases in Africa. Proc. Second Triennial Symp. of the Int. Soc. for Trop. Root Crops, Douala, Cameroon.

10. Lugo, W.L, H. Lugo-Mercado, J. Badillo, A Beale, M. Santiago and L. Rivera, 1987. Response of tanier to different water regimes. *In:* B.R. Cooper and R. Patterson (eds.), Twenty-Sixth Annual Meeting of the Caribbean Food Crops Society. Saint John, Antigua.

11. Lyonga, S.N. and S. Nzietchueng. 1986. Cocoyam and the African Food Crisis. *In:* E.R. Terry, M.O. Akoroda, and O.B. Arene (eds.), Tropical Root Crops: Root Crops and the African Food Crisis. Proc. Third Triennial Symp. of the Int. Soc. For Root Crops. Owerri, Nigeria.

12. McKenzie, H.A and H.S. Wallace, 1954. The Kjeldahl determination of nitrogen: a critical study of digestion conditions-temperature, catalyst, and oxidizing. *Australian Journal of Chemistry*, 7(1):55–70.

13. National Academy of Sciences, 1978. Underexploited tropical plants with promizing economic value. Washington, D.C.

14. Ortiz-Lopez, J., 1990, Farinaceous Crops. Pages 114–164. In: Situation and Perspectives of Agricultural Commodities in Puerto Rico during 1988–89. Agricultural Experiment Station, College of Agricultural Sciences. Mayaguez, Puerto Rico (Spanish).

15. SAS Institute, Inc. 1987. SAS/STAT Guide for Personal Computers. Raleigh, North Carolina.

16. Singh, U., G.Y. Tsuji, R. Goenaga and H.K Prasad, 1992. Modeling growth and development of taro and tanier. Pages 45–56. *In:* U. Singh, (ed.) Proc. Workshop on Taro and Tanier Modeling, College of Tropical Agriculture and Human Resources. Honolulu, Hawaii.

17. Vicente-Chandler, J., H. Irizarry, and S. Silva, 1982. Nutrient uptake by taniers as related to stage of growth and effect of age on yields of the Morada variety. *J. Agric. Univ. P.R.* 66(1):1–10.

18. Young S. and Wu I.P., 1981. Proc. XIII Annual Meeting. Pages 51–69. Conference Research Extension Series 021, College of Tropical Agriculture and Human Resources, University of Hawaii, Honolulu, Hawaii.

# CHAPTER 11

# WATER REQUIREMENTS FOR TANIER (*XANTHOSOMA* SPP.) ON A MOLLISOL SOIL

RICARDO GOENAGA

## CONTENTS

*Modified and reprinted with permission from: "Ricardo Goenaga. Partitioning of Dry Matter in Tanier (Xanthosoma spp.) Irrigated with Fractions of Evapotranspiration. Annals of Botany, 1994. 73: 257–261."* <www.oxfordjournals.org>, © Copyright 1994, Annals of Botany Company. Oxford University Press. The author acknowledges the field assistance by Roberto Bravo, Vidal Marti, Roberto Luciano and Tomas Miranda during the course of this experiment.

## 11.1    INTRODUCTION

Tanier (Xanthosoma spp.) is a herbaceous perennial root crop that serves as an important food staple for inhabitants in some subtropical and virtually all tropical regions. Morphologically, the plant is characterized by a subterranean stem or corm enclosed by dry, scale like leaves. Secondary corms or cormels represent the edible portion of the plant whereas the main corm is used for replanting. The above-ground portion of the plant consists of four to five sagittated leaves. Because of the low research priority given to this crop in the past, the National Academy of Sciences [10] has classified it as a neglected food crop with promizing economic potential.

Not enough information is available on: The water requirements of tanier and how soil moisture deficits affect the accumulation and partitioning of dry matter among various parts of the plant. Studies with greenhouse lysimeters showed that increasing the depth of the water table from 15 to 45 cm resulted in almost a two-fold increase in total cormel yield [12]. Under field conditions without irrigation, corms and cormels of tanier plants accumulated most of their dry matter at 16 to 20 weeks after planting, but maximal leaf area was attained at 14 weeks after planting and then declined rapidly [7]. The rapid decline in leaf area during the corm and cormel bulking period prevented the realization of maximum yield potential.

Under rainfed situations in the Caribbean and Pacific Basins as well as in Africa, yield for aroids (*Xanthosoma* and *Colocasia*) is very low, ranging from 1270 to 5440 kg/ha [8]. Experiments conducted with aroids in Puerto Rico and Hawaii have shown potential yields ranging from 34,000 to 45,000 kg/ha when grown with drip irrigation under intensive management [3, 5]. These results illustrate that much of the potential for aroids lies in commercial rather than subsistence production particularly if the goal is to increase food production for the growing population in the tropics.

Although some studies have shown the adverse effects of water stress on tanier yield, yet no field studies have been reported on the detailed analysis of growth and dry matter accumulation in relation to various levels of supplied irrigation.

This chapter discusses the research results on a continued effort to collect growth analysis data to develop the "SUBSTOR Aroid model" [5, 132], which will serve to accelerate transfer of agrotechnologies for aroid production in the tropics. The author determined the dry matter accumulation and partitioning to various parts of the plant during growth and development of tanier plants subjected to various irrigation regimes.

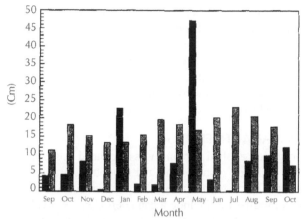

**FIGURE 1**   Mean monthly class A pan evaporation (▧) and rainfall (■) during September 1991 through October 1992, at Fortuna Agricultural Experiment Station, Puerto Rico.

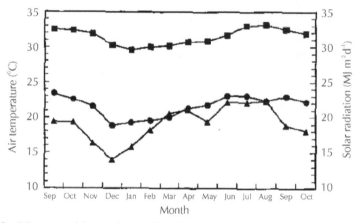

**FIGURE 2**   Mean monthly maximum (■) and minimum (●) temperatures, and mean solar radiation (▲) from September 1991 through October 1992 at the Fortuna Agricultural Research Station, Puerto Rico.

## 11.2   MATERIALS AND METHODS

### 11.2.1   EXPERIMENTAL DESIGN

A field experiment was conducted at the Fortuna Agricultural Research Station of the University of Puerto Rico. The soil is a well-drained Mollisol (fine-loamy, mixed, iso-hyperthermic Cumulic Haplustoll) with pH of 6·8; bulk density of 1·56 g.cm$^{-3}$; organic carbon of 1.7%; and exchangeable bases of 23·2 cmol(+) per kg of soil. Soil nitrate and ammonium at the 0–15 cm depth were 11·2 and 12·9 µg/g of soil, respectively. Figures 1 and 2 show mean monthly rainfall, Class A pan evaporation, maximum and minimum air temperatures, and mean solar radiation during the experimental period.

Plants of cultivar Blanca were established in a screenhouse from excised corm buds with fresh weights of about 16 g (2·7 g of dry weight) and planted in plastic trays containing Pro-Mix growing medium. On 11 September 1991, plants at the one-leaf stage were transplanted to the field and arranged in a split plot design with five replications. Each replication contained four main plots representing different moisture regimes. The main plots were split to accommodate nine biomass harvests. Each subplot contained 20 plants spaced 0·91 x 0.46 m apart from which the inner six plants were sampled. An alley of 2·7 m separated the main plots to prevent overlapping of water between treatments. The experiment was surrounded by two rows of guard plants.

## 11.2.2  WATER MANAGEMENT

The amount of irrigation applied to plants was calculated using the equation of Young and Wu [15]. The equation assumes that the evapotranspiration of a tanier plant is equal to the plant stand as determined by a Class A pan evaporimeter. In this research, their equation was modified to include a pan coefficient $(K_p)$ value of 0·70 and crop coefficient $(K_c)$ value of 0·87 [1] to obtain a theoretical value of potential evapotranspiration. Class A pan factors ranged from 0·33 for treatment one (increasing by 0·33 per treatment) to 1·32 for treatment four were used to obtain fractions of the potential evapotranspiration. A pan factor of 0·33 means that the water applied to the plants would replace 33% of that lost through evapotranspiration.

Plants were subjected to one of the four moisture regimes seven weeks after field planting. The amount of water applied varied weekly depending on Class A pan evaporation and rainfall. Evaporation and rainfall data of the previous week were used to determine the irrigation needs for the following week. Irrigation was supplied three times the following week on alternate days, and no irrigation was provided when the total rainfall exceeded 20 mm per week.

The water source was a well-fed reservoir. Each treatment was equipped with Bermad© volumetric metering valves in the submain line to monitor the water from the main line. Lateral lines equipped with eight one-Lph emitters spaced 45·7 cm apart branched out from the submains along each plant row.

## 11.2.3  FERTIGATION

At planting, each plant received 3·5 g of granular P provided as triple superphosphate. Throughout the experimental period, fertigation was provided monthly @ of 4·0 kg/ha of N and 8·2 kg/ha of K, respectively, using potassium nitrate and urea as the nutrient sources. Zinc, manganese, and iron were supplied through the drip system at monthly rates of 0.45, 0·48, and 0·13 kg/ha, respectively.

## 11.2.4  CROP PERFORMANCE

Plants and biomass measurements were collected at 49, 91, 133, 181, 223, 278, 322, 364, and 398 days after planting (DAP) except for treatments 1 and 2 in which the last biomass harvest was made on 364 DAP. At each harvest, leaves of plants were cut at the midrib-petiole intersection and brought to the laboratory for leaf area determination using a LI-COR-3000A area meter. Each of the six plants in the subplots was harvested by digging an area of 0.42 m² around each plant and to a depth of 30·5 cm. Due to the intensive labor required during the root sampling operation, only two of the six plants were used for this purpose. Plants were then pulled from the soil, washed, and separated into petioles, corms, cormels, root, and suckers. Samples were dried to constant weight at 70°C for dry matter determination. Best-fit curves were determined using the *General Linear Model* (GLM) procedures of the Statistical Analysis System (SAS) program package [11]. Only coefficients significant at P ≤0·05 were retained in the models.

## 11.3   RESULTS AND DISCUSSION

### 11.3.1   DRY MATTER ACCUMULATION

The rate of accumulation of total dry matter indicated the presence of three growth stages (GSl, GS2, GS3) in the growth cycle of tanier plants. These stages varied in magnitude but were similar in duration regardless of the irrigation treatment (Fig. 3A). GSl was characterized by low rates of total dry matter production during the first 90 DAP. No significant differences were observed in total dry matter produced among treatments.

Following GSl, a period of rapid growth (GS2) occurred in which total dry matter increased almost linearly until about 278 DAP as a result of increased rates in dry matter accumulation in all plant components (Fig. 3A to 3F). During this stage, plants that received 99 and 132% of the water lost through evapotranspiration (WLET) exhibited similar total dry matter content, which was significantly greater than that in all other treatments (Fig. 3A). During GS2, the dry matter content of leaves, petioles, roots, corms, and cormels was not statistically different in plants that were replenished with 99 and 132% WLET (Fig. 3B to 3F).

Although cormel development was initiated by 90 DAP in all treatments, the rates of cormel dry matter accumulation indicated that the duration of the cormel - bulking period varied. These rates increased until 223 DAP in plants replenished with 33 and 66% WLET, whereas they increased until 278 and 322 DAP in plants replenished with 99 and 132% WLET, respectively.

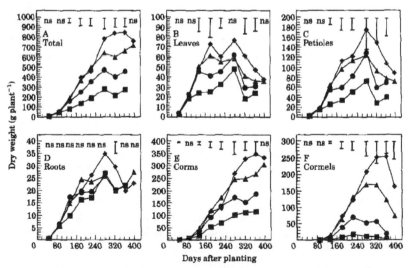

**FIGURE 3**   Effects of irrigation treatments and plant age on dry weight of different parts of tanier plant. The vertical bars are significant LSD values at P ≤0·05, subjected to: 0·33 WLET (■); 0·66 WLET (●); 0·99 WLET (▲); and 1·32 WLET (♦).

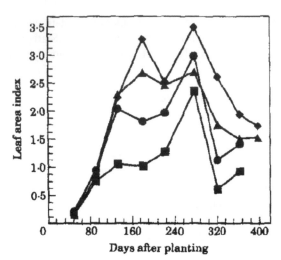

**FIGURE 4**   Leaf area index [LAI] of plants throughout their growth cycle, subjected to: 0·33 WLET (■); 0·66 WLET (●); 0·99 WLET (▲); and 1·32 WLET (♦).

Maximum leaf area indices (LAI) were obtained at 278 DAP and then declined sharply in all treatments (Fig. 4). Water stress conditions prevented the development of higher LAIs in plants replenished with 33 and 66% WLET. As a result, the potential contribution of leaf assimilates toward the storage organs (cormels), particularly during the cormel bulking period, was reduced and resulted in a drastic decline of cormel dry matter yield in these treatments. These results support previous studies [2] that showed positive correlations between LAI and yield of taniers. Although cormels represented a major sink for assimilates during GS2, it is noteworthy that corm dry matter production increased until late in the growing season in all treatments (Fig. 3E). Similar results with other tanier cultivars grown under optimum soil moisture conditions were obtained by Goenaga and Singh [6]. These results are of particular importance because, in contrast to other aroids (e.g., *Colocasia* spp.) tanier corms are usually not edible and they may compete for assimilates with the cormels. Spence [14] suggested a structural relationship between leaf production and corm growth and proposed that a reduction in leaf production might reduce corm growth and benefit cormel yield. However, a reduction in the rate of leaf production may only be advantageous if leaf longevity could be increased in order to maintain optimum leaf area.

The third growth stage (GS3) commenced after 278 DAP and was characterized by a leveling off in total dry matter accumulation and a sharp decline in the dry matter content of leaves, petioles, roots, and cormels (Fig. 3A, 3D, and 3F). GS3 was also characterized by rapid cormel sprouting and consequent sucker development, which renders the cormels unmarketable. At this point, a new biological cycle is initiated at GS1 for each sprouted cormel is considered a corm of a new plant.

Shoot (leaves and petioles) development of suckers was initiated at 223 DAP in all treatments, however, the rate of dry matter accumulation in sucker shoots varied among treatments. From 223 to 364 DAP, the rate of dry matter accumulation in sucker shoots from plants replenished with 33% WLET was 0·20 g/day. From 223 to 398 DAP, shoots of suckers replenished with 132% WLET accumulated only 0·03 g of dry matter per day (data not shown). This difference in dry matter accumulation among treatment extremes was the result of a reduced number of suckers in plants that received more irrigation.

## 11.3.2   DRY MATTER PARTITIONING

Ratios of dry matter partitioning to leaves, petioles, roots, corms, cormels, and suckers as a fraction of total plant dry matter are presented in Fig. 5. Up to 91 DAP, leaves and petioles accounted for over 70% of the total dry matter in plants. This response is expected during early growth as plants become autotrophic and less dependent on stored assimilates from the planting material for growth. As plants matured, the partitioning ratio decreased significantly in leaves, petioles, and roots but increased in corms, cormels, and eventually in suckers (Fig. 5). It is noteworthy that the dry matter partitioning to roots remained considerably higher throughout the experimental period in plants replenished with 33% WLET (Fig. 5C). This response was the result of significantly lower production of dry matter in all parts except roots of plants receiving this treatment (Fig. 3A to 3F).

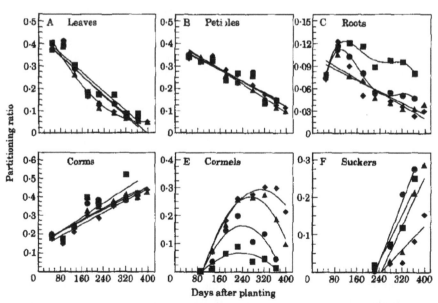

**FIGURE 5**   Effects of days after planting [DAP] on dry matter partitioning to different parts of tanier plant, subjected to various levels of irrigation: 0·33 WLET (■); 0·66 WLET (●); 0·99 WLET (▲); and 1·32 WLET (◆).

Dry matter partitioning to cormels increased linearly between 91 and 223 DAP in plants replenished with 99 and 132% WLET (Fig. 5E). At 223 DAP, the partitioning ratio in both treatments was about 0·27. This response contrasted with that observed in plants replenished with 33 and 66% WLET in which the partitioning ratios were considerably smaller at 223 DAP. The decline in the partitioning ratios in all treatments after 223 DAP was mainly the result of cormel sprouting and consequent sucker development.

In contrast to cormels, the partitioning of dry matter to corms increased linearly throughout the growing season, and the ratios were very similar within each biomass harvest (Fig. 5D). It is noteworthy that partitioning of dry matter to corms from each treatment continued after LAIs had peaked (Figs. 4 and 5D). Several investigators [9, 14] have suggested that bulking of storage organs during periods of declining LAIs may occur from assimilates of current photosynthesis moving preferentially into these organs or assimilates being translocated from senescing leaves. It is most probable that both mechanisms are taking place, with older senescing leaves contributing assimilates to the storage organs (corms and cormels) and products of current photosynthesis being translocated to new leaves and petioles as well as storage organs.

The partitioning of dry matter to suckers was greater in plants replenished with 33 and 66% WLET than in those receiving more heavily irrigated treatments (Fig. 5F). By 364 DAP, the partitioning ratio ranged from 0·19 to 0·28 in plants supplied with WLET values between 33 to 99%, whereas suckers from plants replenished with 132% WLET accounted for only 9% of the total dry matter. Although leaves and petioles were well developed in suckers after 223 DAP in all treatments, most of their dry matter was associated with the sucker corms (previously cormels). Higher partitioning ratios in the less-irrigated treatments (33 and 66% WLET) were the result of increased rates of cormel sprouting and, consequently, a greater number of suckers being formed. In plants replenished with 33% WLET, the number of suckers per plant ranged from 0–4 at 223 DAP to 2–4 at 364 DAP. In plants replenished with 132% WLET, this number ranged from 0·03 at 223 DAP to 1·2 at 398 DAP.

This study demonstrated that commercial production of tanier requires abundant irrigation, which should replenish at least 100% of the evapotranspirational demand of the crop in order to attain proper growth and development. This is in contrast with other crops with high water requirements in which acceptable yields can be attained by replenishing 75% WLET [4]. Inadequate irrigation management in taniers will cause a severe reduction in growth rates and will promote early cormel sprouting, rendering the cormels unmarketable and causing a reduction in economic yields.

## 11.4  SUMMARY

The accumulation and partitioning of dry matter was determined in tanier plants irrigated with fractions of the water lost through evapotranspiration (WLET) in an effort to establish growth analysis data from which a tanier growth simulation model could be developed. The irrigation regimes were based on Class A pan factors ranging from 0·33 to 1·32 with increments of 0·33. Tanier plants were planted in the field and harvested for biomass production about every 6 weeks during the growing season. At each harvest, plants were separated into various plant parts, and their dry matter content was determined. The first 90 dafter planting (DAP) were characterized by low rates of dry matter accumulation, with only leaves and petioles showing substantial growth. A grand growth period followed in which leaves, petioles, and roots rapidly accumulated dry matter until 278 DAP. During this period, plants that received 0·99 and 1·32 WLET exhibited similar total dry matter content, and this was significantly greater than in plants supplied with 0·33 and 0·66 WLET. Cormel dry matter content peaked at 29% of total plant dry matter by 322 DAP in plants replenished with 1·32 WLET. Partitioning of dry matter to cormels in other treatments was significantly reduced. Partitioning of dry matter to corms increased linearly throughout the growing season in all treatments. Dry matter partitioning to suckers and the number of suckers formed from plants replenished with 0·33 and 0·66 WLET was greater than in the more irrigated treatments.

## KEYWORDS

- biomass
- Class A pan evaporation
- cormels
- corms
- crop coefficient, Kc
- crop growth
- days after planting, DAP
- drip irrigation
- dry matter partition
- evapotranspiration
- irrigation
- leaf area index, LAI
- micro irrigation
- pan coefficient, Kp
- Puerto Rico
- semiarid land
- tanier
- USDA
- water lost through evapotranspiration, WLET
- water management
- *Xanthosoma* spp.

## REFRENCES

1. Doorenbos, J. and Pruitt W.O. 1977. *Guidelines for Predicting Crop Water Requirements*. Irrigation and Drainage Paper 24. Food and Agriculture Organization, United Nations, Rome, Italy.
2. Enyi, B.A.C. 1968. Growth of cocoyam (*Xanthosoma sagittifolium*). Indian Journal of Agricultural Science, 38: 627–633.

3. Goenaga R., and Chardon, U. 1993. Nutrient uptake, growth and yield performance of three tanier (Xanthosoma spp.) cultivars grown under intensive management. Journal of Agriculture of the University of Puerto Rico, 77(1–2): 1–10.

4. Goenaga R., Irizarry, H., and Gonazalez, E. 1993. Water requirement of plantains (*Musa acuminata x Musa balbisiana* AAB) grown under semiarid conditions. Tropical Agriculture, 70 (1): 3–7.

5. Goenaga, R., Singh, U., Beinroth, F.H., and Prasad, H. 1991. SUBSTOR-Aroid: A model in the making. Agrotechnology Transfer, 14: 1–4.

6. Goenaga, R., and Singh, U. 1992. Accumulation and partition of dry matter in tanier (*Xanthosoma spp.*) In: Singh U. (ed.) *Proceedings of the Workshop on Taro and Tanier Modeling.* Honolulu, Hawaii: College of Tropical Agriculture and Human Resources, 37–43.

7. Igbokwe, M.C., 1983. Growth and development of *Colocasia* and *Xanthosoma* spp. under upland conditions. In: Terry E.R., Doku E.V., Arene O.B. and Mahungu N.M. (eds.) *Tropical Root Crops: Production and Uses in Africa.* Dovala, Cameroon: Proceedings of the Second Triennial Symposium of the International Society for Tropical Root Crops, 172–174.

8. Lyonga, S.N., and Nzietchueng, S. 1986. Cocoyam and the Africa food crisis. In: Terry E.R., Doku E.V., Arene O.B. and Mahungu N.M. (eds.) *Root Crops and the African Food Crisis.* Owerri, Nigeria: Proceedings of the Third Triennial Symposium of the International Society for Tropical Root Crops, 84–87.

9. Milthorpe F.L., 1967. Some physiological principles determining the yield of root crops. In: Charles W.B., lton E.F., Haynes P.H., Leslie K.A. (eds.) *Proceedings International Symposium Tropical Root Crops*, St. Augustine, Trinidad: University of the West Indies, 1–19.

10. National Academy of Sciences. 1978. *Underexploited Tropical Plants with Promizing Economic Value.* Washington, D.C.: National Academy of Sciences.

11. SAS Institute Inc., 1987. SAS/STAT *Guide for Personal Computers.* Raleigh, North Carolina: SAS.

12. Silva, S., and Irizarry, H. 1980. Effect of depth of water table on yields of tanier. Journal of Agriculture of the University of Puerto Rico, 64 (2): 241–242.

13. Singh, U., Tsuji, G.Y., Goenaga, R., and Prasad, H.K. 1992. Modeling growth and development of taro and tanier. In: Singh U. (ed.) *Proceedings of the Workshop on Taro and Tanier Modeling.* Honolulu, Hawaii: College of Tropical Agriculture and Human Resources, 45–56.

14. Spence, J.A. 1970. Growth and development of tanier (*Xanthosoma* spp.). In: Plucknett D.L. (ed.) *Proceedings International Symposium Tropical Root Crops*. Honolulu, Hawaii: College of Tropical Agriculture, 47–52.

15. Young, S. and Wu, I.P. 1981. *Proceedings of XIII Annual Meeting.* Conference Research Extension Series 021, Honolulu, Hawaii: College of Tropical Agriculture and Human Resources, University of Hawaii, 51–69.

# CHAPTER 12

# WATER REQUIREMENTS FOR BANANA ON A MOLLISOL SOIL

RICARDO GOENAGA and HEBER IRIZARRY

## CONTENTS

*Modified and reprinted with permission from: *"Ricardo Goenaga and Heber Irizarry. Yield of Banana Irrigated with Fractions of Class A Pan Evaporation in a Semiarid Environment. Agron. J. 87:172–176 (1995)."* All rights reserved. © American Society of Agronomy https://www.agronomy.org/. Authors acknowledge the support of R. Bravo, V. Marti, and R. Luciano for their field assistance.

## 12.1  INTRODUCTION

Total world production of banana is estimated at 41.4 billion kg, produced on about 4 million hectares (ha). In many tropical regions, banana is grown either in wet-and-dry climates characterized by erratic rainfall patterns and prolonged dry periods, or in fertile but semiarid lands under irrigation [2, 4, 5]. Depending on the prevailing climatic conditions and method of measurement, estimates of the annual evapotranspiration (ET) of banana range from 1200 to 2690 mm [8]. The high evaporative demand in semiarid environments, combined with the large transpiring surface area and shallow root system of banana, makes it susceptible to lodging and water deficits. Consequently, banana plants require irrigation during dry periods to prevent reductions in yield and fruit quality [7].

Semiarid regions comprise a large percentage of the world's arable land [3]. Drip irrigation technology permits the efficient use of water and can help maximize the utilization of semiarid lands for agricultural production. This technology is particularly suited to widely spaced crops such as banana. There is little information regarding optimum water requirement for banana in the tropics, particularly under semiarid conditions. In addition, most irrigation studies have emphasized the impact of irrigation treatments on yield (i.e., bunch weight) but have disregarded the effect of water supply on yield components such as fruit size, number of hands (fruit clusters) per bunch, and average hand weight.

This study chapter evaluates the optimum water requirement for banana on a Mollisol soil grown under semiarid conditions under drip irrigation and evaluates how banana performance is affected by various levels of irrigation.

## 12.2  MATERIALS AND METHODS

### 12.2.1  EXPERIMENTAL DESIGN

An experiment was conducted from 1990 to 1993 at the Fortuna Agricultural Research Station of the University of Puerto Rico (18°2' N, 66°31' W; elevation 21 m) in the semiarid agricultural zone of Puerto Rico. The San Anton soil is a well-drained Mollisol (fine-loamy, mixed, isohyperthermic Cumulic Haplustoll) with pH of 7.5, bulk density 1.4 g.cm$^{-3}$, and 1.7% of organic carbon in the first 14 cm of soil. The 28-year mean annual rainfall is 917 mm and Class A pan evaporation is 2149 mm. Mean monthly maximum and minimum temperatures are 31.2 and 20.8 °C. Total monthly rainfall and evaporation during the experimental period are shown in Fig. 1, and average monthly irrigation supplied to plants is in Table 1.

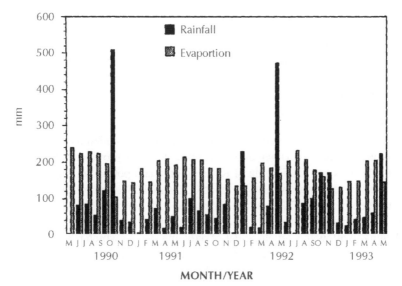

**FIGURE 1** Total monthly rainfall and Class A pan evaporation during the growth cycle of a plant crop and two ratoon crops of banana at the Fortuna Agricultural Research Station, PR.

**TABLE 1** Three-year average monthly irrigation supplied to banana plants subjected to five levels of irrigation based pan factor (proportional to Class A pan evaporation).

| Month | Pan factor | | | | |
|---|---|---|---|---|---|
| | **0.25** | **0.50** | **0.75** | **1.0** | **1.25** |
| | Liters per plant | | | | |
| January | 77 | 154 | 231 | 308 | 385 |
| February | 78 | 156 | 234 | 312 | 390 |
| March | 133 | 266 | 399 | 532 | 665 |
| April | 87 | 174 | 261 | 348 | 435 |
| May | 97 | 194 | 291 | 388 | 485 |
| June | 84 | 168 | 252 | 336 | 420 |
| July | 159 | 318 | 477 | 636 | 795 |
| August | 165 | 330 | 495 | 660 | 825 |
| September | 85 | 170 | 255 | 340 | 425 |
| October | 36 | 72 | 108 | 144 | 180 |
| November | 33 | 66 | 99 | 132 | 165 |
| December | 73 | 146 | 219 | 292 | 365 |
| Total | 1107 | 2214 | 3321 | 4428 | 5535 |
| Average | 92.2 | 184.5 | 276.7 | 369.0 | 461.2 |

Corms of *Grande Naine* banana were planted at a 1.8x1.8 m spacing (equivalent to 1990 plants/ha) on 10 May 1990. Five treatments representing different moisture regimes were arranged in a randomized complete block design with four replications. There were two rows per plot, each with eight experimental plants and surrounded by alleys of 3.7 m, with two guard plants at the end of each row to prevent overlapping of the irrigation treatments.

## 12.2.2   FERTILIZATION

At planting, each plant received 11 grams of granular P provided as triple superphosphate. Throughout the experimental period, fertigation was provided weekly at the rate of 10.2 kg/ha of N and 28.5 kg/ha of K, using urea and potassium nitrate as nutrient sources. Weekly fertilizations also included 0.26 and 0.08 kg/ha of Zn and Fe, respectively, supplied in their EDTA chelate forms and 0.29 kg/ha of Mn supplied as DTPA chelate. A desuckering program in the plant crop (PC) was implemented ≈5 months after planting to allow the development of only one sucker, which represented the first ratoon crop (Rl). Similarly, only one sucker was allowed to develop from Rl plants in order to establish the second ratoon crop (R2).

## 12.2.3   MICRO IRRIGATION MANAGEMENT

The equation of Young and Wu [12] was used to calculate the amount of irrigation applied to plants. The equation assumes that the ET of a banana plant is equal to the evaporation from a body of water with a free surface equal to the plant area as determined by a Class A pan evaporimeter. In this study, the equation was modified to include a pan coefficient ($K_p$) of 0.70 and a modified average crop coefficient ($K_c$) of 0.88 [1] to obtain a theoretical value of potential evapotranspiration (PE). Class A pan factors ranging from 0.25 for Treatment 1 to 1.25 for Treatment 5 in increments of 0.25, were used to obtain fractions of PE. A pan factor of 1.0 means that the water applied to the plants of that treatment replaced that lost through calculated evapotranspiration. Therefore, this was considered the theoretical optimum.

The plants were subjected to the five moisture treatments starting ≈2.5 months after planting. The amount of water applied varied weekly, depending on Class A pan evaporation and rainfall. The previous week's evaporation and rainfall data were used to determine the irrigation needs for the following week. Irrigation was supplied three times during the following week on alternate days, and no irrigation was provided when the total rainfall was>19 mm per week. The water source was a well-fed reservoir. Submain lines equipped with volumetric metering valves to monitor the water from the main line were provided for each treatment. Lateral lines equipped with built-in 4 Lph emitters spaced 61 cm apart branched out from the submains along the inner side of each plant row and ≈21 cm from the pseudostems.

## 12.2.4   CROP PERFORMANCE

At flowering, the number of functional leaves was recorded. Two weeks later, the male flower bud and the false hands were removed from the immature bunches. Immediately, the bunches were bagged with blue plastic sleeves. The number of days to flower was calculated as the time interval between planting and flowering (bunch-shooting) in the plant crop, and the interval between harvest of the previous crop and flowering of the next in the ratoon crops. Banana bunches were harvested when the fruits were in the mature-green stage, ≈110 days after flowering.

At harvest, outer length and diameter were measured in three inner and three outer fruits from the middle section of the third upper and last hands in the bunch. These measurements were pooled to obtain an average for each hand. Values for bunch weight and yield per area were obtained after subtracting the rachis weight from the total bunch weight. Analyzes of variance and best fit curves were determined using the ANOVA and GLM procedures, respectively, of the SAS program package [10]. Only coefficients significant at P≤0.05 were retained in the models.

## 12.3   RESULTS AND DISCUSSION

Differences among irrigation treatments and crops were highly significant (P≤0.01) for all the response variables that were studied (analysis of variance not shown). The "treatment x crop interaction" was highly significant (P ≤0.01), except for fruit length in the bunch last hand and number of leaves at flowering. Therefore, results are reported for each treatment-crop combination.

Total Class A pan evaporation doubled the amount of rainfall recorded during the experimental period. In 22 out of 37 months, evaporation/rainfall ratios were ≥3.0 (Fig. 1). This indicates that large soil-water deficits would have existed without irrigation. Less irrigation was required during the months of October through December and more irrigation was required in March, July, and August (Table 1).

Bunch weight was linearly related to the amount of water applied (i.e., pan factor) in the R1 and R2 crops (Fig. 2). The greatest response to irrigation was obtained in the R2 crop, which produced an average maximum bunch weight of 43.3 kg when irrigated using a pan factor of 1.25. This bunch weight represents increases of 91 and 23% over those obtained for PC and R1, respectively, when irrigated using the same pan factor. The increase in bunch weight in plants that received irrigation from the two highest pan factor treatments can be attributed largely to a greater number of marketable hands per bunch (Fig. 2). Bunches harvested from PC, Rl, and R2 plants that were irrigated with a pan factor of 1.25 had 25, 54, and 40% more hands, respectively, than when irrigated with a pan factor of 0.25 (Fig. 2). Similar improvements in PC bunch weight and hands per bunch were obtained by Hedge and Srinivas [4] when the evaporation replenishment was increased from 20 to 120%. In their study, however, bunch weight and number of hands from R1 bunches were considerably smaller than obtained in this study. As a result of the increase in the number of hands per bunch with increments in pan factor, the number of fruits per bunch also increased. The number of fruits per bunch between treatment extremes (pan factors 0.25 and 1.25) ranged from 109 to 133 fruits in PC, 111 to 193 fruits in Rl, and 133 to 207 in R2 (data not shown).

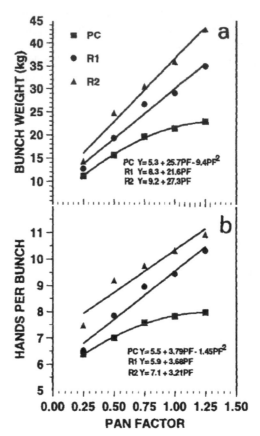

**FIGURE 2** Effects of irrigation treatments based on pan factors on the banana performance in Mollisol soil: (a) Bunch weight (Top); and (b) Hands per bunch (Bottom).

The weight of the third-upper and last hand in the bunch also increased with pan factor increments (Fig. 3). This response was more pronounced in R2, where an increase in pan factor from 0.25 to 1.25 resulted in a third-upper hand weight gain of 2.616 kg, compared with gains of 1.923 kg in R1 and 1.248 kg in PC. The same pan factor increment caused a less pronounced effect in the last hand, with weight gains of only 582 grams in PC, 634 grams in Rl, and 875 grams in R2.

Increments in pan factor treatment resulted in significant increases in length and diameter of fruits of the bunch third-upper and last hands (Fig. 3). Third-hand fruits in PC, Rl, and R2 that received irrigation according to a pan factor of 1.25 were 20, 21, and 32% longer, respectively, than when the crops were irrigated using a pan factor of 0.25. Similar trends of smaller magnitude were measured of fruits in the last hand of PC, Rl, and R2. Similarly, increasing the amount of irrigation resulted in an increased diameter for fruits in the third-upper and last hands (Fig. 3). The greatest increase in fruit diameter (8.5 mm) was observed in the third-upper hand of R2 when the pan factor was incremented from 0.25 to 1.25.

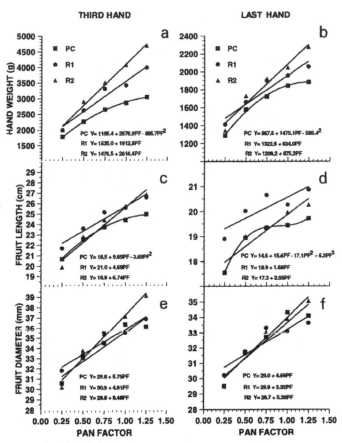

**FIGURE 3** Relationship between five irrigation treatments based on pan factor and hand weight (Figs. 3a, 3b), fruit length (Figs. 3c, 3d), and fruit diameter (Figs. 3e, and 3f) in the third-upper and last hands of the banana bunch.

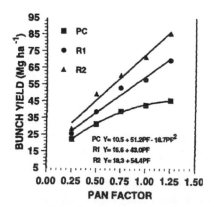

**FIGURE 4** Relationship between irrigation treatments based on pan factors and banana bunch yield for plant crop (PC) and first ratoon crop (R1) and second ratoon crop (R2).

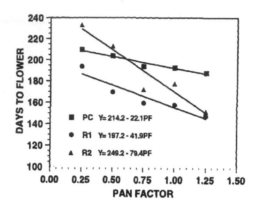

**FIGURE 5**   Relationship between irrigation treatments based on pan factors and days to flower in the banana plant crop (PC) and first ratoon crop (R1) and second ratoon crop (R2).

The number of functional leaves present at flowering is an important physiological trait for proper banana fruit filling [11]. Increments in pan factor caused significant (P ≤ 0.01) increases in the number of functional leaves present at flowering in this study (data not shown). The average number of functional leaves present at flowering was 14.4 and 15.1, respectively, for pan factors 0.25 and 0.50. Robinson et al. [9] indicated that retention of eight leaves at flowering is sufficient to avoid significant reductions in yield and fruit size. Thus, the smaller fruit length and diameter values obtained from PC, Rl, and R2 subjected to the pan factor increments of 0.25 and 0.50 cannot be attributed to a reduced leaf area that might have hindered translocation of photosynthate to fruits in these treatments. This suggests that fruit growth in those treatments was restricted due to drought stress that reduced the rate of cell expansion.

The highest marketable yield of 86.3 Mg/ha was obtained from R2 and with a pan factor of 1.25 (Fig. 4). This yield represented an increase of 41 and 16 Mg/ha over PC and Rl, respectively, when they were subjected to the same pan factor treatment. After harvest of the plant crop, banana yields tend to increase in successive ratoon crops [6]. In this study, the yield average was 36.1 Mg/ha in PC, 48.8 Mg/ha in Rl, and 59.1 Mg/ha in R2. With each increment in pan factor, marketable yield increased by 10.8 Mg/ha, in Rl and 13.6 Mg/ha in R2. This linear effect was not observed in PC, where a maximum yield gain of 9.3 Mg/ha was obtained when irrigation according to the pan factor was increased from 0.25 to 0.50. Thereafter, yield gains were significantly reduced with each pan factor increment (Fig. 4).

Increments in pan factor resulted in a significant (P ≤ 0.01) reduction in the number of days to flower and consequently, the planting-to-harvest cycle was shortened in plants that received irrigation corresponding to the higher pan factors (Fig. 5). The Rl and R2 plants irrigated according to a pan factor of 1.25 flowered 42 and 79 days earlier, respectively, than those irrigated according to a pan factor of 0.25. Range in days to flower between the 0.25 and 1.25 pan factors in PC was only 22 days. This response may have been the result of abnormally high rainfall during the period prior to flowering (Sept. through Oct. 1990; Fig. 1), which probably allowed PC plants irrigated according to the lowest pan factor to partially recover from drought stress conditions.

There was a significant (P ≤ 0.05) treatment and crop effect on the number of days from flowering to harvest; however, the "treatment x crop interaction" was not significant (data not shown). Plant crop, Rl, and R2 plants irrigated according to a pan factor of 0.25 required 110, 111, and 113 days from flowering to harvest, respectively. When irrigated using a pan factor of 1.25, the number of days from flowering to harvest was 105 for PC, 104 for Rl, and 110 for R2. Although the pan factor treatment affected the number of days to flower (Fig. 5), the flowering-to harvest period appeared to be fixed, regardless of the irrigation treatment.

From this investigation, authors conclude that banana should be irrigated using a pan factor of ≥1.0. The use of a lower pan factor results in significant reductions in yield and fruit quality, particularly in ratoon crops,

## 12.4   SUMMARY

There is a scarcity of information regarding the optimum water requirement for banana (*Musa acuminata* Colla, AAA group) grown under semiarid conditions with drip irrigation in the tropics. A three year study was conducted on a fine-loamy, mixed, isohyperthermic Cumulic Haplustoll to determine water requirement, yield, and fruit-quality traits of the plant crop (PC) and two ratoon crops (Rl and R2) of 'Grande Naine' banana subjected to five levels of irrigation. The irrigation treatments were based on Class A pan factors of 0.25 to 1.25 in increments of 0.25. Drip irrigation was supplied three times a week on alternate days. Results showed significant (P ≤ 0.01) irrigation treatment and crop effects for all yield components, fruit length and diameter, days to flower, and days from flowering to harvest. Highest marketable yield (86.3 Mg/ha) was obtained from the R2 crop with water application according to a pan factor of 1.25. Plant crop and R1 plants irrigated using the same pan factor yielded 45.3 and 70.3 Mg/ha, respectively. Increasing the pan factor from 0.25 to 1.25 resulted in weight gains of the third-upper hand of 70% in PC, 90% in Rl, and 122% in R2. Irrigation according to increasing pan factors resulted in significant increases on the number of hands per bunch and the length and diameter of fruits in the third-upper and last hands in the bunch. It can be concluded that, to attain high yields, banana grown under semiarid conditions should be irrigated with a pan factor of not less than 1.0.

## KEYWORDS

- banana
- bunch weight
- bunch yield
- class A pan evaporation
- corms
- crop coefficient, Kc
- days to flower
- drip irrigation

- evapotranspiration
- first ratoon
- fruit diameter
- fruit length
- hand weight
- hands per bunch
- irrigation
- mean hand weight
- micro irrigation
- Mollisol soil
- *Musa* spp.
- pan coefficient, Kp
- pan factor
- plant crop
- Puerto Rico
- second ratoon
- semiarid land

## REFERENCES

1. Doorenbos, J., and W.O. Pruitt. 1977. *Guidelines For Predicting Crop Water Requirements.* Irrigation Drainage Paper 24. FAO, Rome.
2. Ghavami, M. 1974. Irrigation of Valery bananas in Honduras. *Trop. Agric.*, 51:443–446.
3. Grove, A.T. 1985. The arid environment. 9–18. In G.E. Wickens et al. (ed.) *Plants for Arid Lands.* Proc. Kew Int. Conf. Economic Plants for Arid Lands, Kew - England. 23–27 July 1984. Unwin Hyman - London.
4. Hedge, D.M., and K. Srinivas. 1990. Growth, productivity and water use of banana under drip and basin irrigation in relation to evaporation replenishment. *Indian J. Agron.*, 35:106–112.
5. Hill, T.R., R.J. Bissell, and J.R. Burt. 1992. Yield, plant characteristics, and relative tolerance to bunch loss of four banana varieties (Musa AAA Group, Cavendish subgroup) in the semiarid subtropics of Western Australia. *Aust. J. Exp. Agric.*, 32:237–240.
6. Irizarry, H., E. Rivera, I. Beauchamp de Caloni, and R. Guadalupe. 1989. Performance of elite banana (Musa acuminata, AAA) cultivars in four locations of Puerto Rico. *J. Agric. Univ. P.R.*, 73:209–221.
7. Norman, M.J.T., C.J. Pearson, and P.G.E. Searle. 1984. Bananas (Musa spp.). Pages 271–285. In Norman et al. (ed.) *The Ecology of Tropical Food Crops.* Cambridge Univ. Press, Cambridge.
8. Robinson, J.C., and A.J. Alberts. 1989. Seasonal variations in the crop water-use coefficient of banana (cultivar 'Williams') in the subtropics. *Sci. Hortic.*, 40:215–225.
9. Robinson, J.C., T. Anderson, and K. Eckstein. 1992. The influence of functional leaf removal at flower emergence on components and photosynthetic compensation of banana. *J. Hortic. Sci.*, 67:403–410.
10. SAS Institute. 1987. *SAS/STAT Guide for Personal Computers.* Version 6 ed. SAS Jnst., Cary, NC.

11. Soto, M. 1985. Bananas: Cultivation and Trading (In Spanish), 19–97. LIL Press, Tibas, Costa Rica.
12. Young, S., and I.P. Wu. 1981. Final report of the banana drip irrigation studies at the Waimanolo Experiment Station. 51–69. *Proc. Annual Meeting of Hawaii Banana Ind. Assoc.,* 13th. Conf. Res. Ext. Ser. 021. College of Tropical Agric. Human Resources, Univ. of Hawaii, Honolulu.

# CHAPTER 13

# WATER REQUIREMENTS FOR BANANA ON AN OXISOL

RICARDO GOENAGA and HEBER IRIZARRY

## CONTENTS

*Modified and reprinted with permission from: "*Ricardo Goenaga and Heber Irizarry. Yield and Quality of Banana Irrigated with Fractions of Class A Pan Evaporation on an Oxisol. Agron. J. 92:1008–1012 (2000).*" All rights reserved. © American Society of Agronomy https://www.agronomy.org/. Authors acknowledge Roberto Bravo for his field assistance. The names in this chapter are used only to provide specific information. Mention of a trade name or manufacturer does not constitute a warranty of materials by the USDA-ARS, nor is this mention a statement of preference over other materials.

## 13.1 INTRODUCTION

Total world production of banana in 1998 was estimated at 5.7 x $10^{10}$ kg [4]. While most of the global banana production is for local consumption, bananas are the world's second most important traded fruit after citrus and, along with rubber, cocoa, sugar and coffee, one of the five major tropical products entering into world trade [8].

The banana plant is a tropical herbaceous evergreen, which has no natural dormant phase; it has a high leaf area index and a very shallow root system [10]. These factors make the crop extremely susceptible to water shortage. Consequently, banana plants require irrigation during dry periods to prevent reductions in yield and fruit quality.

Depending on the prevailing climatic conditions, estimates of the annual ET of banana plants range from 1200 to 2690 mm [9]. Water requirements of drip-irrigated banana grown under semiarid conditions on a Mollisol or on an Ultisol under transient dry periods were determined by Goenaga and Irizarry [5, 6]. Using Class A pan factors of 0.25 to 1.25, they found that all yield components for the plant crop and two ratoon crops were significantly improved with an increase in water applied. Young et al. [15] reported similar results when banana was irrigated according to pan factor of 0.2 to 1.8.

Little is known about water requirements of banana grown on an Oxisol or about possible differences in water requirements among banana cultivars. A local selection of cultivar 'Johnson' is thought by some growers to be more tolerant to water deficits than '*Grande Naine*,' the most common cultivar used in Puerto Rico and many other tropical regions. In fields planted to 'Grande Naine' and 'Johnson' authors have observed plants of the latter with more vigorous growth and better fruit quality during dry periods.

This chapter evaluates the optimum water requirement for banana cv's 'Johnson' and 'Grande Naine' grown on an Oxisol; and how yield, fruit size, and other bunch and plant traits were affected by various levels of irrigation.

## 13.2 MATERIALS AND METHODS

### 13.2.1 EXPERIMENTAL DESIGN

An experiment was conducted from 1995 to 1998 at the research farm of the USDA-ARS Tropical Agriculture Research Station, Isabela, Puerto Rico. The Coto soil is a well-drained Oxisol (very-fine, kaolinitic, isohyperthermic, Typic Hapludox) with pH of 6.1, bulk density of 1.4 g $cm^{-3}$, 2.0% organic carbon, and 8.3 $cmol_c kg^{-1}$ of exchangeable bases in the first 14 cm of soil. The 23-year mean annual rainfall is 1649 mm and Class A pan evaporation is 1672 mm. Mean monthly maximum and minimum air temperatures are 29.8 and 19.9°C [7]. Total monthly rain and pan evaporation during the experimental period are shown in Fig. 1. The average monthly irrigation supplied to plants is shown in Table 1.

Sword suckers of 'Grande Naine' and 'Johnson' banana were planted at a 1.8- by 1.8-m spacing (equivalent to 1990 plants $ha^{-1}$) in a split-plot design with four replications. Each replication contained five irrigation treatments (main plot) that were split to accommodate both banana varieties. There were two rows per main plot, each with eight experimental plants per variety and surrounded by alleys of 3.7 m, with two guard plants at the end of each row to prevent overlapping of the irrigation treatments.

**TABLE 1** Average monthly irrigation applied to banana plants subjected to five levels of irrigation as determined by pan factor (proportional to Class A pan evaporation) during a three-year period, 1995–1998.

| Month | Irrigation supplied, as proportion of pan evaporation | | | | |
|---|---|---|---|---|---|
|  | 0.25 | 0.50 | 0.75 | 1.0 | 1.25 |
|  | ————————— L plant [1] ————————— | | | | |
| January | 25 | 50 | 75 | 100 | 125 |
| February | 46 | 92 | 138 | 184 | 230 |
| March | 95 | 191 | 287 | 383 | 479 |
| April | 107 | 214 | 321 | 428 | 535 |
| May | 65 | 130 | 196 | 261 | 326 |
| June | 38 | 76 | 115 | 153 | 191 |
| July | 27 | 55 | 82 | 110 | 137 |
| August | 33 | 66 | 100 | 133 | 166 |
| September | 46 | 93 | 139 | 186 | 232 |
| October | 46 | 93 | 140 | 187 | 234 |
| November | 32 | 65 | 97 | 130 | 162 |
| December | 43 | 86 | 129 | 172 | 215 |
| Total | 603 | 1211 | 1819 | 2427 | 3032 |
| Avg. | 50.2 | 101.0 | 151.6 | 202.2 | 252.7 |

## 13.2.2 FERTILIZATION

At planting, each plant received 11 grams of granular P provided as triple superphosphate. Throughout the experimental period, fertilization through the drip system with potassium nitrate was provided weekly at the rate of 3.6 kg.ha$^{-1}$ of N and 12.4 kg.ha$^{-1}$ of K.

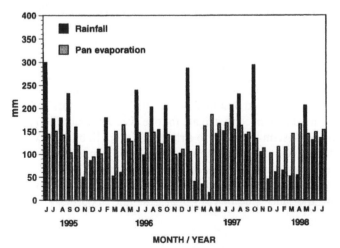

**FIGURE 1** Total monthly rainfall and Class A pan evaporation during the growth cycle of a plant crop and two ratoon crops of banana at the USDA-ARS Research Station, Puerto Rico.

### 13.2.3 CULTURAL PRACTICES

A desuckering program in the plant crop (PC) was implemented during the fifth month after planting to allow the development of only one sucker, which represented the first ratoon crop (R1). Similarly, only one sucker was allowed to develop from R1 plants in order to establish the second ratoon crop (R2). On 10 September 1996, Hurricane Hortense completely destroyed R1 plants, which at the time were almost two months away from harvest. Suckers from R1 were not affected by the hurricane. Therefore, a new sucker (R2) was selected from R1 plants and the experiment was continued until a third ratoon crop (R3) was harvested. Yellow sigatoka, nematodes, soil-borne insects and weeds were controlled following recommended cultural practices [1].

### 13.2.4 MICRO IRRIGATION MANAGEMENT

The equation of Young and Wu [16] was used to determine the amount of irrigation applied to plants:

$$AI = 10.03 \times APE \times PA \times PF \tag{1}$$

where, AI (liters.plant$^{-1}$.d$^{-1}$) is applied irrigation, 10.03 is a constant (100,289 L.ha. cm$^{-1}$/9996 m$^2$.ha$^{-1}$), APE (cm.d$^{-1}$) is the average daily class A pan evaporation, PA (m$^2$) is the plant area and PF is pan factor treatments (0.25, 0.50, 0.75, 1.0, 1.25).

However, since evaporimeter data cannot be correlated to crop water use directly [14], AI values were multiplied by a pan coefficient (Kp) of 0.70 and an average crop coefficient (Kc) of 0.88 [3] to obtain an AI value equivalent to theoretical potential evapotranspiration (PE). The use of pan factors in the equation, which ranged from 0.25 for treatment one to 1.25 for treatment five, in increments of 0.25. Therefore, this allowed to replenish plants with fractions of water lost through PE. For example, a pan factor of 1.0 means that the water applied to the plants of that treatment replaced that lost through PE; this was hence considered the theoretical optimum.

The plants were subjected to the five moisture treatments starting on 14 August 1995. The amount of water applied varied weekly, depending on Class A pan evaporation and rain which were recorded daily from a weather station located near the experimental site. The previous week's evaporation and rain data were used to determine the irrigation needs for the following week. Following commercial practices, irrigation was supplied three times during the following week on alternate days, and no irrigation was provided when the total rain was>19 mm per week. From tensiometer readings, the authors determined that this amount of rain keeps this soil sufficiently wet (10–15 kPa) to avoid the need to irrigate for one week (unpublished results).

A surface drip system was used to irrigate the crop. Submain PVC lines equipped with volumetric metering valves to monitor the water from the main line were provided for each treatment. Lateral drip lines (Drip In Irrigation Co., Madera, CA) equipped with in-line 4 L.h$^{-1}$ emitters spaced 0.61 m apart branched out from each treatment submain along the inner side of each plant row and about 0.21 m from the pseudostems.

### 13.2.5 CROP PERFORMANCE

At flowering and harvest, the number of functional leaves per plant was recorded. About two weeks after flowering, the male flower bud and the false hands were removed from the immature bunches. Immediately, the bunches were bagged with blue polyethylene sleeves. Banana bunches were harvested when the fruits were about three-quarters full, about 110 days after flowering. At harvest, the bunches were weighed and the number of hands counted and then cut from the rachis. The outer length and diameter were measured in three inner and three outer fruits from the middle section of the third-upper and last hands in the bunch. These measurements were pooled to obtain an average for each hand. The weight of these hands was also recorded. Values for bunch weight and yield per area were obtained after subtracting the rachis weight from the total bunch weight.

### 13.2.6 DATA ANALYSIS

Analyzes of variance and best-fit curves were determined using the ANOVA and GLM procedures, respectively, of the SAS program package [12]. The GLM Solution Option was used in cases in which significance was found for treatment and crop effects but not for the "treatment x crop interaction" (Victor Chew, personal communication). Only coefficients significant at $P<0.05$ were retained in the models.

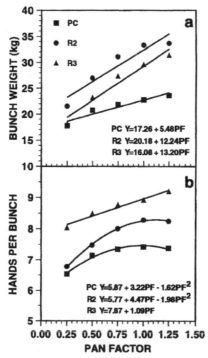

**FIGURE 2** Bunch weight (a) and hands per bunch (b) of a banana plant crop (PC) and two ratoon crops (R2 and R3) as influenced by irrigation based on proportion of pan evaporation (pan factor).

## 13.3   RESULTS AND DISCUSSION

Cultivar and the 'treatment x cultivar' interaction were not significant (P<0.05). Therefore, data were averaged over cultivars. Irrigation treatments and crops showed significant effects (P<0.01) on bunch weight and yield, number of hands per bunch, weight and fruit diameter of the third and last hands, length of fruits in the third hand, and, number of functional leaves at flowering and harvest (analysis of variance not shown).

Total Class A pan evaporation (5162 mm) was similar to the amount of total rainfall (5277 mm) recorded during the 38-month experimental period (Fig. 1). Although this may suggest that plants were never exposed to soil-water deficits, yet it is noteworthy that 30% of the total rain recorded during the experimental period fell during the months of June and September of 1995; June 1996; and January, August, and October of 1997. In 20 of 38 months, rain was lessthan Class A pan evaporation indicating that soil-water deficits would have existed without irrigation. More irrigation was required during the months of March, April and May (Table 1). The water requirement of banana plants in this study was twice that for plants subjected to similar treatments on an Ultisol [6] and about half that for plants grown on a Mollisol in a semiarid environment [5].

Bunch weight was linearly related to the amount of water applied (i.e., pan factor) in the plant crop and R2 and R3 crops (Fig. 2). The greatest response to irrigation was obtained in the R2 crop, which produced an average maximum bunch weight of 35.5 kg when irrigated using a pan factor of 1.25. This bunch weight represents increases of 9 and 47% over those obtained for R3 and PC, respectively, when irrigated using the same pan factor. Similar experiments [5, 6] have also shown greater bunch weight in the R2 crop of banana grown under semiarid conditions with drip irrigation or in humid high elevations with supplemental irrigation. Bunches harvested from PC, R2, and R3 plants that were irrigated with a pan factor of 1.25 had 12, 22, and 13% more hands, respectively, than when irrigated with a pan factor of 0.25 (Fig. 2). Bunches harvested from R2 plants had fewer hands than those of R3 (Fig. 2). Therefore, the increase in bunch weight in R2 plants can be attributed to an increase in individual fruit size and weight (Fig. 3).

Fruit diameter and length in the third-upper hand and fruit diameter in the last hand significantly increased with increments in pan factor treatment (Fig. 3). This response was probably responsible for the significant bunch weight increases in plants of R2 (Fig. 2). Third-hand fruits in PC, R2, and R3 that received irrigation according to a pan factor of 1.25 were 5, 10, and 12% thicker, respectively, than when the crops were irrigated using a pan factor of 0.25.

Similar trends of smaller magnitude were measured on fruits in the last hand of PC, R2, and R3. Increasing the amount of irrigation resulted in an increase in length for fruits in the third-upper hand but not in the last hand (Fig. 3). The weight of the third-upper and last hand in the bunch also increased with pan factor increments (Fig. 3). This response was more pronounced in R2 and R3 where an increase in pan factor from 0.25 to 1.25 resulted in a third-upper hand weight gain of 1284 and 1429 g, respectively, compared to only a gain of 594 g in PC. The same pan factor increment caused a less pronounced effect in the last hand, with weight gains of only 190 g in PC, 356 g in R2, and 596 g in R3.

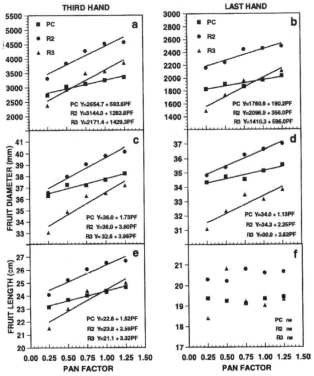

**FIGURE 3** Relationship between irrigation pan factor and hand weight (a,b), fruit diameter (c,d), and fruit length (e,f) in the third-upper and last hands of the banana bunch as influenced by irrigation based on proportion of pan evaporation (pan factor).

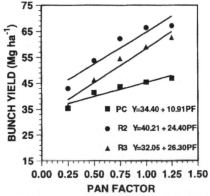

**FIGURE 4** Relationship between irrigation based on proportion of pan evaporation (pan factor) and bunch yield in the banana plant crop (PC) and two ratoon crops (R2 and R3).

The number of functional leaves present at flowering and at harvest is an important trait for proper banana fruit filling [11, 13]. Increments in pan factor caused significant (P<0.01) increases in the number of functional leaves present at flowering in the study (data not shown). The average number of functional leaves at flowering was 14, 14.5, 15.0, 15.3, and 15.5, respectively, for pan factors 0.25 to 1.25. The average number of functional leaves at harvest (data not shown) ranged from 8.9 to 9.7 for pan factors 0.25 to 1.25. There should be a minimum of 12 functional leaves at flowering, and nine at harvest to achieve maximum bunch filling in banana [11]. Thus, the smaller fruit length and diameter values obtained from PC, R2 and R3 plants subjected to the lower pan factor treatments (0.25 and 0.50) cannot be attributed to a reduced leaf area that might have hindered translocation of photosynthate to fruits in these treatments. This suggests that fruit growth in those treatments was restricted due to drought stress that reduced the rate of cell expansion.

Increments in pan factor treatment significantly increased bunch yield in PC, R2 and R3. The highest marketable yield of 70.7 Mg ha$^{-1}$ was obtained from R2 and the application of irrigation according to a pan factor of 1.25 (Fig. 4). This yield represented an increase of 23 and 6 Mg.ha$^{-1}$ over PC and R3, respectively, when they were subjected to the same pan factor treatment. Even though rain is an important component of the annual water requirement for banana grown in this region, PC, R2, and R3 plants irrigated with a pan factor of 1.25 had a 29, 53, and 68% higher bunch yield than those irrigated with a pan factor of 0.25 (Fig. 4). These results confirm that a banana plantation requires large quantities of water for maximum productivity [10, 11].

## 13.4   CONCLUSIONS

From this investigation, it is concluded that banana grown on an Oxisol should be irrigated using a pan factor of 1.0 or more. The use of a lower pan factor may reduce bunch yield significantly and affect fruit quality. Similar results were obtained by Goenaga and Irizarry [5, 6] with banana grown under semiarid conditions on a Mollisol or on an Ultisol under transient drought periods. Updated procedures recommending the use of the Penman-Monteith method for calculating crop evapotranspiration were published by FAO after this study was completed [2]. Therefore, it is recommended that this method is used in future studies on irrigation requirements of banana and refinements be made to our recommendation presented in this study if necessary. No significant differences in yield and fruit quality traits were observed between the two cultivars used in this study. Future studies should be directed to the screening of a larger number of cultivars in an effort to identify materials with some drought tolerance.

## 13.5   SUMMARY

There is a scarcity of information regarding the optimum water requirement for banana (*Musa acuminata* Colla, AAA group) grown with supplemental drip irrigation on an Oxisol. A three-year study was conducted on a very-fine, kaolinitic, isohyperthermic Typic Hapludox to determine water requirement, yield, and fruit-quality traits of the plant crop (PC) and two ratoon crops (R2 and R3) of 'Grande Naine' and 'Johnson' banana subjected to five levels of irrigation. The irrigation treatments were based on Class A pan factors of 0.25 to 1.25 in increments of 0.25. Drip irrigation was supplied three times a week on alternate days. Results showed significant (P<0.001) irrigation treatment and crop effects for all yield components, fruit length and diameter, number of leaves at flowering and harvest and number of hands per bunch. Cultivar and the treatment by cultivar interaction were not significant (P<0.05). The highest marketable yield (70.7 Mg ha$^{-1}$) was obtained from the R2 crop with water application according to a pan factor of 1.25. Plant crop and R3 plants irrigated using the same pan factor yielded 48 and 65 Mg.ha$^{-1}$, respectively. Increasing the pan factors from 0.25 to 1.25 resulted in weight gains of the third-upper hand of 594 g in PC, 1284 g in R2, and 1429 g in R3. It was concluded that banana grown on an Oxisol should be drip irrigated with a pan factor of 1.0 or more three times a week.

## KEYWORDS

- banana
- bunch weight
- bunch yield
- class A pan evaporation
- crop coefficient, Kc
- days to flower
- drip irrigation
- drip irrigation
- evapotranspiration
- first ratoon
- fruit diameter
- fruit length
- gross sales
- hand weight
- hands per bunch
- mean hand weight
- *Musa* spp.
- Oxisol soil
- pan coefficient, Kp
- pan factor
- plant crop
- Puerto Rico
- second ratoon

## REFERENCES

1. Agricultural Experiment Station. 1995. Technological package for the production of plantains and bananas. Agricultural Experiment Station Publication 97. College of Agricultural Sciences, University of Puerto Rico, Mayaguez.
2. Allen, R.G., Pereira, L.S., Raes, D. and Smith, M. 1998. *Crop Evapotranspiration: Guidelines for Computing Crop Water Requirements.* Irrigation Drainage Paper 56, FAO, Rome.
3. Doorenbos, J., and W. O. Pruitt. 1977. *Guidelines for Predicting Crop Water Requirements.* Irrigation Drainage Paper 24, FAO, Rome.
4. FAO (Food and Agriculture Organization). 1999. FAOSTAT statistics database – Agriculture [on line]: Available at http://apps.fao.org/cgi-bim/mph-db.pl?subset=agriculture (verified 15 Sep. 1999).
5. Goenaga, R., and H. Irizarry. 1995. Yield performance of banana irrigated with fractions of Class A pan evaporation in a semiarid environment. *Agronomy Journal* 87:172–176.
6. Goenaga, R., and H. Irizarry. 1998. Yield of banana grown with supplemental drip irrigation on an Ultisol. *Experimental Agriculture* 34:439–448.
7. Goyal, M.R., and E.A. González. 1989. Climatological data collected in the Agricultural Experiment Stations of Puerto Rico. Agricultural Experiment Station Publication 88–70. College of Agricultural Sciences, University of Puerto Rico, Mayaguez.
8. Hallam, D. 1995. The world banana economy. p. 509–533. In: S. Gowen (ed.) *Bananas and Plantains.* Chapman and Hall, London, UK.
9. Robinson, J.C., and A.J. Alberts. 1989. Seasonal variations in the crop water-use coefficient of banana (cultivar 'William') in the subtropics. Scientia Horticulturae, 40:215–225.
10. Robinson, J.C. 1995. Systems of cultivation and management. p. 15–65. In: S. Gowen (ed.). *Bananas and Plantains.* Chapman and Hall, London, UK.
11. Robinson, J.C. 1996. *Bananas and Plantains.* CAB International, Wallingford, UK.
12. SAS Institute. 1987. *SAS/STAT Guide for Personal Computers.* Version 6 edition. SAS Institute, Cary, N.C.
13. Soto, M. 1985. Bananas: Cultivation and trading. p. 19–97. (In Spanish.) LIL Press, Tibas, Costa Rica.
14. Van der Gulik, T. 1999. *British Columbia Trickle Irrigation Manual,* pages 209–221. British Columbia Ministry of Agriculture and Irrigation Industry Association of British Columbia, Abbotsford, B.C., Canada.
15. Young, S.C.H., T.W. Sammis, and I. P. Wu. 1985. Banana yield as affected by deficit irrigation and pattern of lateral layouts. Trans. ASAE, 28(2):507–510.
16. Young, S.C.H., and I. P. Wu. 1981. Final report of the banana drip irrigation studies at the Waimanolo Experiment Station. Pages 51–69. *Proceedings Annual Meeting Hawaii Banana Industry Association,* 13th Conference Research Extension Service 021. College Tropical Agriculture and Human Resources, University of Hawaii, Honolulu.

# CHAPTER 14

# WATER REQUIREMENTS FOR PLANTAINS ON A MOLLISOL SOIL

RICARDO GOENAGA, HEBER IRIZARRY, and ELADIO GONZALEZ

## CONTENTS

*Modified and reprinted with permission from: "*Ricardo Goenaga, Heber Irizarry and Eladio Gonzalez. Water requirements for plantains (Musa acuminata x Musa balbisiana AAB) grown under semiarid conditions. Tropical Agriculture (Trinidad), January 1993, 70(1): 3–7.*" © 1993 Tropical Agriculture (Trinidad), <Tropical.Agri@sta.uwi.edu>. This research was conducted cooperatively between the USDA-Agricultural Research Service and the Agricultural Experiment Station of the University of Puerto Rico. Authors acknowledge the field assistance by Roberto Bravo. The names in this chapter are used only to provide specific information. Mention of a trade name or manufacturer does not constitute a warranty of materials by the USDA-ARS or the Agricultural Experiment Station of the University of Puerto Rico, nor is this mention a statement of preference over other materials.

## 14.1   INTRODUCTION

Plantain is a tropical rhizomatous perennial plant closely related to bananas (Musa spp.). The starchy fruits are known as "cooking bananas" and are an important staple food in Africa, Central and South America, Oceania, South-East Asia, and the Caribbean Basin [6]. In Puerto Rico, plantain is an important cash crop with an annual farm value of $30.4 million [2]. The main area of production is in the humid mountain region where periodic droughts can reduce yields and affect fruit quality. The demand for a year-round supply of high quality fruits, high farm prices, and the availability of arable land with an irrigation infrastructure have contributed to the shifting of plantain production from the highland to the fertile but semiarid lowlands previously used for sugar-cane production. In a normal year, evaporation in this agricultural zone may be three times greater than rainfall.

Micro irrigation technology allows efficient use of water and can help maximize the utilization of semiarid lands for agricultural production. Not enough information is available regarding optimum water requirement for plantains either in Puerto Rico or elsewhere in the tropics particularly under semiarid conditions. It has been suggested that the water needs of plantains are ideally supplied by 100 mm of evenly distributed rainfall each month and that a serious shortage is experienced when there are less than 50 mm in any month [8]. Plantains have a shallow root system and a very large transpiring surface [5] making the crop susceptible to lodging and water deficit.

Plantain bunch weight and consequently fruit weight often responds to an increasing amount or frequency of irrigation [1] presumably because transpiration (and thus photosynthesis) is sensitive to available soil water. The importance of fruit weight in plantain production is critical since in most tropical areas, plantains are marketed by fruit units which must weigh 270 g or more to be considered marketable.

In this chapter, authors discuss the research results to determine the optimum water requirement for intensively managed plantains grown in semiarid conditions under drip irrigation, and to examine the effect of five irrigation regimes on fruit size.

## 14.2   MATERIALS AND METHODS

### 14.2.1   EXPERIMENTAL DESIGN

An experiment was conducted at the Fortuna Agricultural Research Station of the University of Puerto Rico (18°2' N, 66°31' W; elevation 21 m) in the semiarid agricultural zone of Puerto Rico. The mean annual rainfall and Class A pan evaporation were 866 mm and 2149 mm, respectively. Mean monthly maximum and minimum temperatures were 31 and 21 °C. The soil is Mollisol (Cumulic Haplustoll) with pH of 7.5, bulk density 1.4 Mg.m$^{-3}$, 1.7% of organic carbon, and exchangeable bases of 25 cmol(+) per kg of soil.

Corms of the horn-type Mari Congo cultivar were planted on 14 January, 1988 in a randomized complete block design with four replications. Each replication contained five plots representing different moisture regimes. There were two rows of six plants per plot spaced 1.8 x 1.8 m of which eight were used to collect data on plant, bunch, and fruit variables. An alley of 3.7 m separated the plots to prevent overlapping of water between treatments. The experiment was surrounded by two rows of guard plants.

## 14.2.2   WATER MANAGEMENT

The equation by Young and Wu [9] was used to calculate the amount of irrigation applied to plants. The equation assumes that the ET of a banana plant is equal to the evaporation from a body of water with a free surface equal to the plant area as determined by a Class A pan evaporimeter. In this study, the equation was modified to include a pan coefficient ($K_p$) of 0.70 and a modified average crop coefficient ($K_c$) of 0.88 [3] to obtain a theoretical value of potential evapotranspiration (PE). Class A pan factors ranging from 0.25 for treatment 1 to 1.25 for treatment 5 in increments of 0.25, were used to obtain fractions of PE. A pan factor of 1.0 means that the water applied to the plants of that treatment replaced that lost through calculated evapotranspiration. Therefore, this was considered the theoretical optimum.

The plants were subjected to the five moisture treatments 2.5 months after planting. The amount of water applied varied weekly, depending on Class A pan evaporation and rainfall. The previous week's evaporation and rainfall data were used to determine the irrigation needs for the following week. Irrigation was supplied three times during the following week on alternate days, and no irrigation was provided when the total rainfall was >20 mm per week.

The water source was a well-fed reservoir. Submain lines equipped with volumetric metering valves to monitor the water from the main line were provided for each treatment. Lateral lines branched out from the submains along the inner side of each of the two rows of plants and 46 cm from the base of the pseudostems. Two in-line 8 Lph emitters were spaced 61 cm between each end.

## 14.2.3   FERTIGATION

At planting, each plant received 11 grams of granular P provided as triple superphosphate. Throughout the experiment, fertigation was provided weekly at the rate of 8.6 kg/ha of N and 1.4 kg/ha of K with urea and potassium nitrate serving as sources for these nutrients.

## 14.2.4   CROP PERFORMANCE

Recommended cultural practices were followed regarding sucker control and pesticide and herbicide applications. The bunches were harvested when the fruits reached the maturity green stage, about 120 days after flowering.

A desuckering program was implemented five months after planting to allow the development of only one sucker per stump for the subsequent establishment of the first ratoon crop. In contrast to the mother crop, ratoon suckers developed a poor radical system, which caused plant lodging halfway through their growth cycle. Consequently the experiment was terminated at this point (about six months after harvesting the mother crop). Root and corm weevil were not the limiting factors for the problem as is usually suggested. Apparently, there are other factors associated with the poor development of ratoon crops in plantain, a condition commonly encountered and often referred to as "plantain decline" in the tropics.

Best-fit curves were determined using the General Linear Model (GLM) procedures of Statistical Analysis System (SAS) program package [7]. Only coefficients at $P \leq 0.01$ were retained in the models.

## 14.3 RESULTS AND DISCUSSION

Total class A pan evaporation was three times greater than the amount of rainfall recorded during the experimental period (Table 1). This indicated that large soil-water deficits would have existed without irrigation. The average annual evaporation and rainfall data collected for over 20 years at the study location were 2149 mm and 866 mm, respectively [4]. These figures are similar to those obtained during the experiment and, therefore, 1988 can be considered as representative of a normal year.

Days to flower (bunch-shooting) was one of the crop parameters most affected by the irrigation treatments (Fig. 1). Flowering and consequently the planting to harvest cycle were substantially delayed in plants that received the lower levels of moisture. Plants replenished with 25 and 50% (pan factors of 0.25 and 0.50) of the water lost through evapotranspiration flowered at about 318 and 296 days after planting (DAP), respectively. Those in the remaining treatments flowered between 277 and 282 DAP.

The number of leaves at harvest ranged between 11.6 and 14.4 for pan factors 0.25 and 1.25, respectively. Although this parameter was statistically significant, it exceeded the minimum of 10 leaves required at bunch-shooting to fill the fruits. The total number of hands per bunch was linearly related to the pan factor and varied between 7.1 and 8.1 for the lowest and highest pan factor treatments, respectively. Consequently, the number of fruits per bunch was 45 for pan factor of 0.25, and 52 for pan factor of 1.25 (Fig. 2A).

**TABLE 1** Monthly evaporation and rainfall registered at the Fortuna Agricultural Experiment Station in Juana Diaz – Puerto Rico; and the amount of water applied in five treatment throughout the plantain growing cycle (1988–1989).

| Month | Class A pan evaporation | Rainfall | Irrigation applied, Liters per plant | | | | |
|---|---|---|---|---|---|---|---|
| | | | Pan factor | | | | |
| | mm | mm | 0.25 | 0.50 | 0.75 | 1.00 | 1.25 |
| Jan | 146.3 | 11.7 | — | — | — | — | — |
| Feb | 168.1 | 4.1 | — | — | — | — | — |
| Mar | 191.8 | 21.8 | — | — | — | — | — |
| Apr[1] | 193 | 62.0 | 143 | 286 | 429 | 572 | 715 |
| May | 216.9 | 26.7 | 82 | 164 | 246 | 328 | 410 |
| Jun | 195.3 | 45.0 | 67 | 134 | 201 | 268 | 335 |
| Jul | 205.7 | 92.5 | 69 | 138 | 207 | 276 | 345 |
| Aug | 194.6 | 316.5 | 77 | 154 | 231 | 308 | 385 |

**TABLE 1**   *(Continued)*

| Month | Class A pan evaporation | Rainfall | Irrigation applied, Liters per plant | | | | |
|-------|------------------------|----------|------|------|------|------|------|
|       |                        |          | Pan factor | | | | |
|       | mm                     | mm       | 0.25 | 0.50 | 0.75 | 1.00 | 1.25 |
| Sep     | 164.6   | 54.6  | 65  | 130  | 195  | 260  | 325  |
| Oct     | 171.7   | 75.9  | 124 | 248  | 372  | 496  | 620  |
| Nov     | 146.3   | 116.8 | 32  | 64   | 96   | 128  | 160  |
| Dec     | 140.2   | 8.9   | 102 | 204  | 306  | 408  | 510  |
| Jan     | 150.1   | 11.2  | 107 | 214  | 321  | 428  | 535  |
| Feb     | 121.9   | 9.4   | 93  | 186  | 279  | 372  | 465  |
| **Total** | **2406.5** | **857.1** | **961** | **1922** | **2883** | **3844** | **4805** |
| **Average** | —   | —     | **87.4** | **174.7** | **262.1** | **349.4** | **436.4** |

[1] Initiation of irrigation treatment. Previously, the experimental plot as sprinkler-irrigated until the establishment of the plantain crop.

Bunch weight was linearly related to the amount of water applied (i.e., pan factor; Fig. 2B). The regression analysis for pan factor values and bunch weight showed that the maximum bunch weight (17.02 k g per bunch) was obtained with a of pan factor 1.25. This response represents an increase of 70.7, 45.0, 26.1, and 11.5% as compared to plants that received pan factor treatments of 0.25, 0.50, 0.75, and 1.0, respectively.

Maximum bunch mean fruit weight (313.2 g) was attained with the application of a pan factor between 1.0 and 1.25 (Fig. 2C). Only plants that received a pan factor of 0.75 or more (i.e., average monthly irrigation of 262 L) yielded bunches containing marketable fruits with a mean weight of 270 grams. This mean fruit weight is the minimum threshold to qualify the horn-type plantains as marketable.

The weight of the bunch third upper hand increased linearly with pan factor increments (Fig. 2D). The average fruit weight in this hand was 198, 250, 314, 296, and 311 grams for pan factors of 0.25, 0.50, 0.75, 1.00 and 1.25, respectively. Since the first three upper hands of typical "Mari Congo" bunches contain over 50% of the total fruits, any irrigation regime exceeding a pan factor of 0.75 should result in the production of high-grade marketable fruits that will surpass the 270-g market criteria.

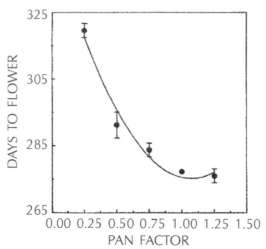

**FIGURE 1**   Relationship between days to flower and irrigation pan factors. Vertical bars represent the SE.

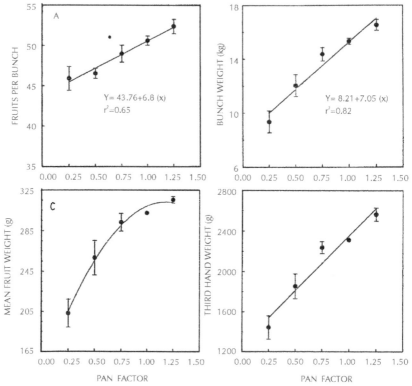

**FIGURE 2**   Effects of irrigation pan factor on the: Number of fruits per bunch (Fig. 2A); bunch weight (Fig. 2B); bench mean fruit weight (Fig. 2C); and weight of bunch third upper hand (Fig. 2D).

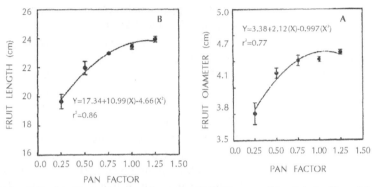

**FIGURE 3** Effects of irrigation pan factor on fruit diameter (Fig. 3A, Left) and fruit length (Fig. 3B, Right) in the third hand of upper plantain bunch.

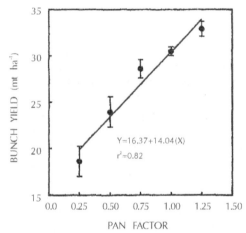

**FIGURE 4** Effects of irrigation pan factor on the plantain bunch yield (metric tons per ha).

**TABLE 2** Effect of pan factor treatments on mean fruit weight and yield of plantains.

| Pan factor treatment | Average monthly Irrigation (Table 1) | Mean fruit weight | Number of fruits | Fruit yield |
|---|---|---|---|---|
| Fraction | Liters per plant | grams | per ha | Tons per ha |
| 0.25 | 87.4 | 205.0 | 90,464 | 19.7 |
| 0.50 | 174.7 | 255.5 | 93,851 | 23.3 |
| 0.75 | 262.1 | 290.2 | 97,239 | 26.9 |
| 1.00 | 349.4 | 309.3 | 100,626 | 30.5 |
| 1.25 | 436.8 | 312.6 | 104,013 | 33.9 |

Fruit diameter and length in the third upper hand were significantly affected by the increments in irrigation regimes (Figs. 3A and 3B). Regression-predicted values showed that maximum fruit diameter (4.5 cm) and length (23.8 cm) were attained with the use of a pan factor treatment between 1.00 and 1.25.

Highest marketable plantain yields (33.9 tons per ha) were obtained with the application of a pan factor treatment of 1.25 (Fig. 4). This represented an increase of 14.2, 10.6, 7.0 and 3.4 tons per ha for pan factor values of 0.25, 0.50, 0.75, and 1.00, respectively. Table 2 summarizes the importance of scheduled irrigation on plantain fruit size (mean fruit weight) and yield per ha. Considering the market demand for large fruits at premium prices, it is recommended that intensively managed plantains grown in the semiarid lowlands of Puerto Rico should be irrigated with not less than a pan factor of 0.75. The irrigation based on pan factors between 1.00 and 1.25 is recommended if the farm gate price for premium fruits is sufficiently high as to compensate for the additional irrigation costs. These amounts represent an average of 262 to 437 L of monthly irrigation per plant (Table 1).

## 14.4   SUMMARY

An experiment was conducted to determine the optimum water requirement of drip-irrigated plantains (*Musa acuminata x Musa balbisiana* AAB) grown under semiarid conditions. The irrigation regimes were based on Class A pan factors ranging from 0.25 to 1.25 with increments of 0.25. Plants with irrigation based on a pan factor of 0.75 or above averaged 280 days from planting to flowering. Flowering in plants that received irrigation based on pan factors of 0.25 and 0.50 was substantially delayed. All yield components were significantly affected by the amount of water applied. Only fruits from treatments with irrigation based on a pan factor of over 0.75 exceeded the average weight of 270 g, the minimum accepted for marketable plantains. Highest marketable yield (33.9 tons per ha) was obtained with the application of a pan factor treatment of 1.25, an increase of 14.2, 10.6, 7.0, and 3.4 tons per ha over pan factor treatments 0.25, 0.50, 0.75, and 1.0, respectively. Over 50% of the fruits harvested in pan factor treatments 0.25 and 0.50 were classified as nonmarketable.

## KEYWORDS

- banana
- bunch weight
- bunch yield
- class A pan evaporation
- crop coefficient, Kc
- days to flower
- drip irrigation
- evapotranspiration
- fruit weight
- fruit diameter

- fruit length
- fruits per bunch
- hands per bunch
- irrigation
- mean fruit weight
- micro irrigation
- Mollisol soil
- Musa spp.
- pan coefficient, Kp
- pan factor
- plant crop
- plantain
- Puerto Rico
- semiarid land
- third hand weight
- USDA
- water management

## REFERENCES

1. Abruña, F., Vicente-Chandler, J., and Silva, S. 1980. Evapotranspiration with plantains and the effect of frequency of irrigation on yields. 1980. J. Agric. Univ. P.R., 64(2): 204–210.
2. Antoni, M. 1990. Plantains and bananas, In: Perspectives of Agricultural Commodities in Puerto Rico (eds. Antoni, M., Cortes, M., Gonzalez, G., and Velez, S.), Agricultural Experiment Station, Univ. Puerto Rico, Rio Piedras, Puerto Rico, pp. 14–28.
3. Doorenbos, J., and Pruitt, W.O. 1977. *Guidelines for predicting crop and water requirements*, Irrigation and Drainage Paper 24, FAO, United Nations, Rome, Italy.
4. Goyal, M. R. and E. A. Gonzalez, 1989. Climatological data collected in the agricultural experiment stations of Puerto Rico. Publication 88–70. Agricultural Experiment Station, College of Agricultural Sciences, Mayaguez, Puerto Rico (In Spanish).
5. Irizarry, H., Vicente-Chandler, J. and Silva, S. (1981) Root distribution of plantains growing on five soil types, J. Agri. Univ. P.R. 65 (1): 29–43.
6. Plucknett, D.L. (1978) Tolerance of some tropical root crops and starch-producing tree crops to suboptimal land conditions, in: Crop Tolerance to Suboptimal Land Conditions (ed. Jung, G.A.), Am. Soc. Agron., Madison, Wisconsin. pp. 125–144.
7. SAS Institute, Inc. (1987) SAS/STA T Guide for Personal Computers, Raleigh, North Carolina.
8. Tai, E.A. (1977) Banana, in: Ecophysiology of Tropical Crops (eds. Alvim, P. de T. and Kozlowski, T.T.), Academic Press, New York, N.Y., pp. 441–460.
9. Young, S. and 1-Pai Wu, 1981. Proceedings of XIII Annual Meeting. Pages 51–69. Conference Research Extension Series 021, College of Tropical Agriculture and Human Resources, Univ. Hawaii.

# CHAPTER 15

# DRIP IRRIGATION MANAGEMENT: PLANTAIN AND BANANA

RICARDO GOENAGA, HEBER IRIZARRY, BRUCE COLEMAN, and
EULALIO ORTIZ

## CONTENTS

*Modified and reprinted with permission from, *"Ricardo Goenaga, Heber Irizarry, Bruce Coleman and Eulalio Ortiz. Drip irrigation recommendations for plantain and banana on the semiarid southern coast of Puerto Rico. J. Agric. Univ. P.R. January and April 1995, 79(1–2):13–27."* The authors acknowledge the field assistance of Roberto Bravo and Vidal Marti, during the course of this investigation.* The names in this chapter are used only to provide specific information. Mention of a trade name or manufacturer does not constitute a warranty of materials by the USDA-ARS, nor is this mention a statement of preference over other materials.

## 15.1 INTRODUCTION

Plantain *(Musa acuminata* x *Musa balbisiana,* AAB) and banana *(Musa acuminata* AAA) are important cash crops in Puerto Rico with a combined annual farm value of $52.6 million [11]. Both crops are grown mainly in the mountain region where cyclical droughts reduce yields and affect fruit quality. The demand for a year-round supply of high quality fruits, the high farm prices, and the availability of arable land with an irrigation infrastructure have contributed to shifting plantain and banana production from the highlands to the fertile, but semiarid, lowlands previously used for sugar cane production. Efficient use of irrigation for plantain and banana production in this agricultural zone is imperative because of the combination of high evaporation and low rainfall.

Drip irrigation technology allows efficient use of water and can help maximize the utilization of semiarid lands for agricultural production. There is a scarcity of information regarding optimum water requirement for plantain and banana in the tropics, particularly under semiarid conditions. Both crops are known to require large quantities of well-distributed water in order to attain maximum productivity and fruit quality as well as to ensure adequate sucker development [1, 5, 14].

Other investigations [3, 8, 9, 12] have been conducted to examine the yield response of plantain and banana subjected to various irrigation regimes. However, few have reported irrigation recommendations that are readily available to growers.

Goyal and Gonzalez [7] used a modified Blaney-Criddle model to estimate the drip irrigation requirement for plantain grown in the semiarid zone of Puerto Rico. Their results indicated that a total of 149.1 cm/plant of drip irrigation would be required yearly for proper growth and development of plantains. However, the authors stressed the fact that their results were not substantiated by field studies.

Lahav and Kalmar [10] conducted field studies in the northern coastal plain of Israel to study the response of drip-irrigated banana subjected to various irrigation regimes based on class A evaporation factors. Their results showed higher yields (67.9 t/ha) when plants received irrigation corresponding to a constant evaporation factor of 1.0 throughout the growing season and equivalent to 11,630 cubic meters per hectare. However, the authors did not present monthly rates of applied irrigation to demonstrate changes due to seasonal patterns.

This chapter discusses the research results to determine how marketable yield and other crop traits of plantain and banana grown in semiarid conditions under drip irrigation are influenced by five levels of irrigation based on "Class A pan evaporation." To provide practical irrigation recommendations to growers, authors made projections on crop productivity, gross sales, and on irrigation expenses incurred in the operation of a 20-hectare planting of drip irrigated plantain or banana.

## 15.2 MATERIALS AND METHODS

### 15.2.1 EXPERIMENTAL DESIGN

Field studies with plantain were conducted in 1988–1989 [5] and with bananas during 1990–1992 at the Fortuna Agricultural Research Station, latitude 18° 2'N and longitude 66° 31' W, in the semiarid agricultural zone of Puerto Rico. The soil is a Mollisol (Cumulic Haplustolls) with good drainage, a pH of 7.5, a bulk density of 1.4 Mg/m³, 1.7% organic carbon, and 25 cmol(+) per kg of exchangeable bases.

Corms of the horn-type Mari Congo plantain cultivar or Grand Nain banana were planted in a randomized complete block design with four replications. Each replication contained five plots representing different moisture regimes. Each plantain plot was 14.6 m long by 1.8 m wide and contained two eight-plant rows. Banana plots were 21.9 m long by 1.8 m wide, and each of the two rows per plot contained 12 plants. Plots were separated by alleys of 3.7 m to prevent overlapping of the irrigation treatments. The experiments were surrounded by two rows of guard plants. The population density was about 1,990 plants per hectare. Data on bunch and fruit variables were collected from 10 of the 16 plantain plants per plot, and on banana from 16 of the 24 plants per plot. Data collected for plantain are representative of the plant crop only. For banana, data were collected for the plant and first ratoon crops.

### 15.2.2 MICRO IRRIGATION MANAGEMENT

In both experiments, the equation used by Young and Wu [15] was used to calculate the amount of irrigation applied to plants. The equation assumes that the evapotranspiration of a banana or plantain mat is equal to the evaporation from a body of water with a free surface equal to the mat area as determined by a class A pan evaporimeter. In this study, the equation was modified to include a pan coefficient (Kp) value of 0.70 and a modified average crop coefficient (Kc) of 0.88 [4] to obtain a theoretical value of potential evapotranspiration. Class A pan factors (proportion of pan evaporation), which ranged from 0.25 for treatment one to 1.25 for treatment five in 0.25 increments, were used to obtain fractions of the potential evapotranspiration. A pan factor of 1.0 means that the water applied to the plants of that treatment replaced that lost through evapotranspiration and hence was considered the theoretical optimum.

The plants were subjected to the five moisture regimes two and a half months after planting. The amount of water applied varied weekly depending on class A pan evaporation and rainfall. The previous week's evaporation and rainfall data were used to determine the irrigation needs for the following week. Irrigation was supplied three times the following week on alternate days, and no irrigation was provided when the total rainfall exceeded 19 mm per week.

In both experiments, submain lines were equipped with volumetric metering valves to monitor the water provided to each treatment. Lateral lines branched out from the submains along the inner side of each of the two rows of plants and about 46 cm apart from the pseudostem base. Two in-line 8 L/h emitters were spaced 21 cm from the pseudostems, and two feeder lines with 4 L/h emitters provided a uniform supply of water around each plantain plant. In the banana experiment, laterals were equipped with built-in 4 L/h emitters spaced 61 cm apart along the line and feeder lines were not used.

Rainfall and evaporation data collected from 1988 to 1992 were used to calculate the amount of applied irrigation during the field plantings. However, final drip irrigation recommendations for these crops were calculated by also including historical weather data collected at the research station from 1982 to 1987, and assuming that both crops had been grown at the test site during an 11-year period (i.e., 1982 to 1992; see Table I).

### 15.2.3  FERTIGATION

At planting, each plantain and banana plant received 11 g of granular P provided as triple superphosphate. Plantain plants were fertigated weekly at the rate of 8.6 and 14.4 kg/ha of N and K, respectively, with urea and potassium nitrate serving as sources of these nutrients. Banana plants were fertigated weekly at rates of 10.2 kg/ha of N and 28.5 kg/ha of K using potassium nitrate. Banana weekly fertigations also included 0.26 and 0.08 kg/ha of Zn and Fe supplied in the EDTA chelate forms and 0.29 kg/ha of Mn supplied as DTPA chelate. Because of adequate rainfall, when irrigation was not necessary, weekly fertigation was postponed and rates were doubled the following week.

### 15.2.4  CULTURAL PRACTICES

Recommended cultural practices [2] were followed regarding bunch and sucker management and pesticide and herbicide applications. Plantain bunches were harvested when the fruits reached the mature green stage. The banana bunches were harvested when the fruits were three-fourths full, about 120 days after flowering. Collected data were analyzed for an analysis of variance using the ANOVA procedure of the SAS program package [13]. All results for differences among means were considered to be significant at the 0.05 probability level or lower.

**TABLE 1**  Average and standard deviation for monthly evaporation and rainfall registered at the Fortuna Agricultural Station from 1982 to 1992.

| Month | Class A pan Evaporation, $E_{pan}$ | Rainfall, R | Difference, $E_{pan}$ - R |
|-------|-------------------------------------|-------------|----------------------------|
|       | mm                                  | mm          | mm                         |
| Jan   | 150.9 ± 15.9                        | 38.2 ± 65.1 | 112.7                      |
| Feb   | 154.5 ± 9.3                         | 24.0 ± 15.1 | 130.5                      |
| Mar   | 192.5 ± 14.2                        | 45.4 ± 51.6 | 147.1                      |

**TABLE 1** *(Continued)*

| Month | Class A pan Evaporation, $E_{pan}$ | Rainfall, R | Difference, $E_{pan}$ - R |
|---|---|---|---|
| | mm | mm | mm |
| Apr | 193.5 ± 13.5 | 58.4 ± 51.4 | 135.1 |
| May | 179.9 ± 33.4 | 144.3 ± 151.2 | 35.6 |
| Jun | 195.7 ± 15.6 | 49.8 ± 56.4 | 145.9 |
| Jul | 207.9 ± 22.9 | 56.7 ± 37.4 | 151.2 |
| Aug | 209.7 ± 11.5 | 89.3 ± 83.1 | 120.4 |
| Sep | 175. ± 18.9 | 109.5 ± 73.9 | 65.6 |
| Oct | 155.0 ± 23.2 | 192.8 ± 220.0 | −(37.8) |
| Nov | 134.1 ± 14.1 | 148.6 ± 108.0 | −(14.5) |
| Dec | 137.6 ± 14.0 | 16.7 ± 15.5 | 120.9 |
| Total | 2086.4 | 973.7 | — |
| Average | 173.9 ± 31.1 | 81.1 ± 106.2 | — |

### 15.2.5 PROJECTIONS FOR ENERGY COSTS FOR PLANTAIN AND BANANA CULTIVATION

Projections for energy costs (Table 4) on a 20-hectare farm were based on the following assumptions:

1. Farm was divided into four 5-ha sections, each of which received irrigation separately;
2. Use of a 40 brake horsepower motor pump with an efficiency of 75%;
2. A total dynamic head of 61 m discharges 1,977.7 L/min of water to each 5-ha section;
3. Pumping time to each 5-ha section on a given day was: 1.7, 2.3, and 2.8 h for pan factors of 0.75, 1.0, and 1.25, respectively;
4. Under the 711 agricultural tariff provided by the Electric Energy Authority, the cost per kilowatt-hour is 0.054 cents. Therefore, the total kilowatt-hour consumption per year was estimated at 31,619.99; 42,160.08; and 52,699.98 for pan factors 0.75, 1.0, and 1.25, respectively;
5. Energy costs also included fuel adjustment and minimum charge costs of 0.033 cents and $120.00, respectively. The cost of well water (Table 4) was based on a charge by the Department of Natural Resources of $25.00 per 3,785,000 L of water.

Table 5 provides a detailed economic analysis of all the operational costs and net income generated in the establishment and management of one hectare of bananas. The data were obtained from growers, federal and state government agencies, private financial institutions, and distributors of agricultural products. Projections of operational costs included the following assumptions:

1. Salary of a labor @ of $3.50/hour;

2. A plant population density of 1,990 plants per ha ;

3. Grower participation of the Puerto Rico Department of Agriculture's incentive program for the application of nematicide.

Because of the seasonal and often large market-price fluctuations for plantains, a similar economic analysis for this crop was not made. However, if the operational costs involved in the establishment and management of a plantain farm are known, the gross sales reported in this manuscript for drip-irrigated plantains may be helpful to estimate the net income at prevailing market conditions.

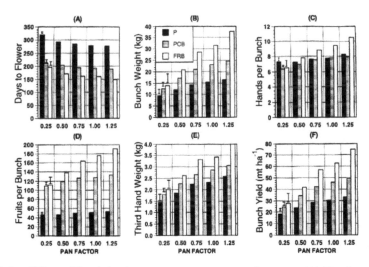

**FIGURE 1**    Effect of five irrigation treatments on crop performance: (A) Number of days to flower; (B) Bunch weight; (C) Hands per bunch; (D) Fruits per bunch; (E) Third hand weight; and (F) Bunch yield. (Note: P = Plantain; PCB = Plant crop; FRB = First ratoon bananas. Vertical bars represent Waller-Duncan values (P<0.05) to compare differences between irrigation treatments within a crop.)

## 15.3   RESULTS AND DISCUSSION

### 15.3.1   FIELD STUDIES

For the 11 years of weather data used in this study, total Class A pan evaporation at the Fortuna Agricultural Station was two times greater than the amount of total rainfall (Table 1). This ratio is similar to that obtained using the available historical data collected for more than 20 years at the same location, which_ show annual evaporation and rainfall values of 2,149 and 917 mm, respectively [7]. This finding indicates that the study period was representative of the normal climate conditions that prevail in this agricultural zone.

Increases in pan factor treatments caused a significant reduction in the number of days required to flower (bunch shooting), and consequently the planting to harvest cycle was shortened in both plantain and banana (Fig. 1A). Plantain, plant crop banana (PCB) and first ratoon banana (FRB) replenished with 100% of the water lost through evapotranspiration (pan factor 1.0) flowered at about 277, 193 and 352 days after planting, respectively. As compared to plants replenished with a pan factor of 0.25, plantains, PCB and FRB replenished with a pan factor of 1.0 flowered 43, 17 and 36 days earlier, respectively. Days to flowering in PCB ranged by only 22 days in plants subjected to treatment extremes (i.e., pan factors 0.25 and 1.25). This response contrasted with that observed for plantain and FRB and was probably the result of abnormally high rainfall (totaling 711 mm during the last four months of 1990), which probably allowed drought-stressed plants from the low pan factor treatments to develop faster.

Maximum plantain and PCB bunch weights (16.5 and 24.5 kg, respectively) were obtained with the use of pan factor 1.25. Although statistical differences in plantain and PCB bunch weights were not observed between pan factors 1.25 and 1.0, significant differences existed between pan factors 1.25 and 0.75 (Fig. 1B). The greatest response to irrigation treatments was obtained for FRB plants, which produced maximum bunch weights of 37.6 kg, using a pan factor of 1.25 (Fig. 1B). This response represented an increase of 172, 80, 24, and 19% as compared to FRB plants that received pan factor treatments of 0.25, 0.50, 0.75, and 1.0, respectively.

The total number of hands per bunch ranged from 7.2 to 8.3 in plantains subjected to the five irrigation regimes (Fig. 1C). The number of hands per bunch in plantain plants irrigated with a pan factor of 1.25 was significantly greater than those produced in the remaining irrigation treatments (Fig. 1C). The number of hands per bunch in PCB and FRB plants significantly increased with increases in pan factor treatments. This response was more evident in FRB plants, which produced four more hands per bunch with a pan factor of 1.25 than with 0.25 (Fig. 1C).

As a consequence of the increase in the number of hands per bunch with increments in pan factor treatments, the number of fruits per bunch increased from 111 to 190 fruits in FRB plants receiving pan factor treatments of 0.25 and 1.25, respectively (Fig. 1D). The average number of fruits per bunch in PCB plants subjected to the higher three pan factor treatments was 128, whereas it was 155 in FRB plants. In plantain, the number of fruits per bunch significantly increased from 45 to 52 in plants irrigated with pan factors of 0.25 and 1.25, respectively (Fig. 1D). There were no significant differences in the number of fruits per bunch in irrigated plantain plants when pan factors ranged from 0.75 to 1.25.

The first three upper hands of plantain bunches contain about 50% of the total fruits and about 40–45% in banana bunches. Therefore, the weight of the third upper hand is often used to represent an average weight of all the hands in a bunch. Pan factor treatments significantly affected the weight of the bunch's third upper hand in plantain, PCB, and FRB plants (Fig. 1E). In plantain, however, there were no significant differences in the weight of the bunch's third upper hand from plants irrigated with pan factors of 0.75 and 1.0. However, the differences between 0.75 and 1.25 were significant. This response greatly contrasted with that observed in PCB and FRB plants, in which practically each pan factor increment resulted in a significant increase in the weight of the third upper hand (Fig. 1E).

The highest yields for plantain, PCB and FRB plants were 33, 49, and 75 t/ha, respectively, and were obtained with the application of a pan factor treatment of 1.25 (Fig. 1F). Although statistical differences in plantain and PCB yields were not observed between pan factors 1.25 and 1.0, yet both crops showed that yields obtained with a pan factor of 1.25 were significantly greater than those obtained with a pan factor of 0.75 (Fig. IF). Yields of FRB plants irrigated with a pan factor of 1.25 were significantly greater than those obtained in the remaining pan factors.

### 15.3.2 IRRIGATION RECOMMENDATIONS AND ECONOMIC ANALYSIS

The results presented in this study demonstrate the importance of irrigation to obtain adequate yields in plantain and banana grown in the semiarid zone of Puerto Rico. On the basis of our field studies and historical weather data, Table 2 presents the monthly amounts of irrigation that would be applied to plantain and banana subjected to five irrigation regimes in the semiarid zone of Puerto Rico.

Less irrigation was required during the months of September through November whereas March, April, July and August demanded higher quantities of irrigation (Table 2). October and November were the only months in which rainfall exceeded Class A pan evaporation. In most of the remaining months of the year, pan evaporation exceeded rainfall by over 100 mm (Table 1).

**TABLE 2** Monthly amount of water applied (Liters per plant) to plantains or bananas subjected to five irrigation treatments on the semiarid southern coast of Puerto Rico.

| Month | Amount of irrigation applied each month, Liters per plant | | | | |
|-------|------|------|------|------|------|
| | Pan factor | | | | |
| | **0.25** | **0.50** | **0.75** | **1.00** | **1.35** |
| Jan | 97 | 194 | 291 | 388 | 485 |
| Feb | 98 | 195 | 293 | 391 | 489 |
| Mar | 120 | 240 | 361 | 481 | 601 |
| Apr | 103 | 206 | 309 | 412 | 515 |
| May | 75 | 150 | 225 | 300 | 375 |
| Jun | 99 | 199 | 298 | 397 | 497 |
| Jul | 118 | 236 | 354 | 473 | 591 |
| Aug | 109 | 218 | 328 | 437 | 546 |

**TABLE 2** *(Continued)*

| Month | Amount of irrigation applied each month, Liters per plant | | | | |
| --- | --- | --- | --- | --- | --- |
| | Pan factor | | | | |
| | 0.25 | 0.50 | 0.75 | 1.00 | 1.35 |
| Sep | 65 | 131 | 197 | 263 | 328 |
| Oct | 59 | 118 | 178 | 237 | 296 |
| Nov | 39 | 77 | 116 | 154 | 193 |
| Dec | 88 | 175 | 263 | 351 | 439 |
| Total | 1,070 | 2,139 | 3,213 | 4,284 | 5,355 |
| Average | 89 | 178 | 268 | 337 | 446 |

Based on plantain and banana irrigation requirements (Table 2), and the yield and fruit quality data obtained from our field studies (Figs. 1B–F), we estimated the yearly gross sales and irrigation costs incurred by a grower operating a 20-hectare planting irrigated with pan factors of 1.25, 1.0, 0.75. Irrigation with lower pan factors was detrimental to yield and fruit quality; and therefore, these were not considered in our analyzes.

Increasing the amount of applied irrigation from pan factor 0.75 to 1.25 resulted in gross sales increases of $40,482 for PCB and $108,054 for FRB (Table 3). However, the increase in the amount of supplied irrigation resulted in an additional energy and water cost of only $2,388.52 (Table 4). It is noteworthy that increasing the pan factor from 0.75 to 1.25 increased the number of fruit boxes by 6,747 in PCB. However, the same irrigation increment in FRB resulted in 18,009 additional boxes (Table 3).

There were no significant differences in the number of plantain fruits from plants irrigated with pan factors ranging from 0.75 to 1.25 (Fig. 1; Table 3). Nevertheless, variables such as bunch weight, weight of the third upper hand, number of hands, and total yield were significantly greater in plants irrigated with a pan factor of 1.25 than in those irrigated with a pan factor of 0.75 (Fig. 1). Moreover, the average plantain fruit weight from plants irrigated with a pan factor of 0.75 and 1.25 was 290 and 316 g, respectively (data not shown). These findings indicate that plantain irrigation with a pan factor of 1.25 results in an improvement in fruit quality and probably greater gross sales since plantains are marketed by fruit units, which must weigh 270 grams or more to be considered marketable.

**TABLE 3** Effects of three levels of irrigation based on Class A pan evaporation factors on the estimated yield and gross sales of plantain or banana: 20-hectare farm on the southern coast of Puerto Rico.

| Pan factor | Bunch yield[1] | No. of boxes of fruits[2] | Gross sales[2] |
|---|---|---|---|
| | Kg | No. | US $ |
| **Plant crop banana** | | | |
| 0.75 | 757,800 | 41,775 | 250,650 |
| 1.00 | 828,000 | 45,645 | 273,870 |
| 1.25 | 880,200 | 48,522 | 291,132 |
| Waller (0.05) | 60,010 | 3,308 | — |
| **First ratoon banana** | | | |
| 0.75 | 1,022,400 | 56,362 | 338,172 |
| 1.00 | 1,129,950 | 62,290 | 373,740 |
| 1.25 | 1,349,100 | 74,371 | 446,226 |
| Waller (0.05) | 145,524 | 8,022 | — |
| **Plantain** | | | |
| **Pan factor** | | **Fruits[1]** No./20 ha | **Gross sales[2]** US $ |
| 0.75 | | 1,657,458 | 198,895 |
| 1.00 | | 1,712,653 | 205,521 |
| 1.25 | | 1,771,953 | 212,634 |
| Waller (0.05) | | 115,097 | — |

[1] Values reflect a 1.5% and 10% yield reduction for plantain and bananas, respectively, due to losses caused by wind damage, nonproductive plants, and other factors that may reduce production in a commercial plantation. The mean number of fruits per plantain bunch for the three pan factors was 50.6 with an average of 304 g per fruit. The average number of hands per bunch was 7.8 for PCB and 9.6 for FRB, respectively. The average hand weight was 2.9 kg for PCB and 3.4 for FRB.
[2] Sales based on: $120.00 per 1,000 marketable plantain fruits; and $6.00 per banana box. Each box weighed 18.14 kg.

**TABLE 4** Yearly irrigation (liters/20 ha), energy and water costs (US$) required for plantain or banana production under three irrigation regimes in a 20-hectare planting in the semiarid southern coast of Puerto Rico[1].

| Pan factor | Yearly supplied irrigation, A | Yearly energy cost, B | Yearly water cost, C | Total energy and water costs, B + C |
|---|---|---|---|---|
| | Liters/20 ha | | US$ | |
| 0.75 | 125,788,950 | 2,871.96 | 830.84 | 3,702.80 |
| 1.00 | 167,718,096 | 3,789.28 | 1,107.78 | 4,897.06 |
| 1.25 | 209,648,250 | 4,706.59 | 1.384.73 | 6,091.32 |

[1] Irrigation recommendations were based on the following assumptions: A plant population of about 39,840 plants/20-ha planted at a distance of 1.8 x 1.8 x 3.7 m; 156 irrigation applications per year, which is equivalent to three applications per week. Each plant was provided with drip lines equipped with 4 Lph emitters spaced 61 cm apart.

**TABLE 5A** Approximate costs and income per hectare for the production of a plant crop and first ratoon bananas grown under drip irrigation in the semiarid region of Puerto Rico using a plant population density of about 1,990 plants per ha.

| Item | Unit | Quantity | Cost($) |
|---|---|---|---|
| **A. Supplies and materials** | | | |
| 1.  Fertilizer | | | |
|  - Urea | Kg | 659 | 266.00 |
|  - Potassium Sulfate | Kg | 1795 | 756.00 |
|  - Phosphoric acid (6 applications) | Liters | 147 | 310.00 |
| 2.  Nematicide | kg | 113 | 560.00 |
| 3.  Herbicide | Liters | 14 | 208.00 |
| 4.  Fungicide (2 applications of Tilt)[1] | Liter | 1 | 66.00 |
| 5.  Polyethylene bunch bags | bags | 1990 | 338.00 |
| 6.  String | spools | 6.7 | 80.00 |
| 7.  Miscellaneous supplies | — | — | 124.00 |
| **Total Supplies and Materials** | — | — | **2,708.00** |
| **B. Other costs** | | | |
| Electricity | ha | 1 | 186.00 |
| Irrigation water (pan factor 1.0) | ha | 1 | 55.00 |
| Use of land and irrigation equipment | ha | 1 | 618.00 |

**TABLE 5A** *(Continued)*

| Item | Unit | Quantity | Cost($) |
|---|---|---|---|
| Use of equipment (tractor, cultivator, sprayer, drip system; includes repair maintenance and depreciation) | ha | 1 | 370.00 |
| Payroll taxes (Social Security, state, unemployment, Christmas bonus, vacations)[2] | ha | 1 | 1,120.00 |
| Financing interests through assignment[3] | ha | 1 | 529.00 |
| Crop Insurance[4] | ha | 1 | 332.00 |
| Supervision (other than owner) | ha | 1 | 495.00 |
| Fuel and oil | ha | 1 | 55.00 |
| Financing interests on equipment (9.75%) | ha | 1 | 62.00 |
| **Total other costs** | **ha** | — | **3,822.00** |

[1] Trade names in this publication are used only to provide specific information. Mention of a trade name does not constitute a warranty of equipment or materials by the Agricultural Experiment Station of the University of Puerto Rico, nor is this mention a statement of preference over other equipment or materials.
[2] About 24.5% of labor costs.
[3] Crop average costs at **10%** interest.
[4] Insurance covers 75% of the plantation (1,493 plants) at a cost of $4.77 for each $100.00 insured.

**TABLE 5B**  Approximate costs and income ($ per hectare) for the production of a plant crop and first ratoon bananas grown under drip irrigation in the semiarid region of Puerto Rico using a plant population density of about 1,990 plants per ha.

| Item | Unit | Quantity | Cost, US$ |
|---|---|---|---|
| **Labor cost, US$** | | | |
| 1.  Land preparation[1]. | man/days | 3.7 | 104.00 |
| 2.  Digging, cleaning and planting of suckers[1]. | man/days | 17.3 | 484.00 |
| 3.  Post planting cultivation (Twice). | man/days | 2.5 | 70.00 |
| 4.  Herbicide application (Three). | man/days | 3.7 | 104.00 |
| 5.  Fertigation through drip irrigation. | man/days | 4.9 | 137.00 |
| 6.  Desuckering and removal of dead leaves. | man/days | 22.2 | 622.00 |
| 7.  Supporting plants with twine. | man/days | 17.8 | 498.00 |
| 8.  Bunch bagging. | man/days | 8.8 | 246.00 |
| 9.  Spraying against Sigatoka (Two applications). | man/days | 1.2 | 34.00 |
| 10. Miscellaneous tasks. | man/days | 5.0 | 140.00 |
| 11. Harvest, transport and packing. | man/days | 76.2 | 2,134.00 |
| **Total labor cost: Plant crop** | man/days | 163.3 | 4,573.00 |
| **Total labor cost: Ratoon crop** | man/days | 142.3 | 3,985.00 |

[1] Not considered for ratoon crop.

| Item | Unit | Quantity | Cost($) |
|------|------|----------|---------|
| Income[2] | | | |
| Sale of high grade fruits[7] | | | |
| 1.  Plant crop | Boxes/ha | 2,282 | 13,692.00 |
| 2.  Ratoon crop | Boxes/ha | 3,115 | 18,690.00 |
| Net income | | | |
| 3.  Plant crop | ha | 1 | 2,589.00 |
| 4.  Ratoon crop | ha | 1 | 8,175.00 |

[2] Net income may increase by about $1,166.00 for the plant crop and by $1,022.00 for the ratoon crop if reimbursements for supplementary salaries at the rate of $7.00 for each man-day is considered. Also it is assumed that the buyer supplies the packing boxes. For income taxes, only 10% of the annual net income will be taxed at rates not exceeding 36%.

[7] $6.00/box of 18.14 kg.

On the basis of the economic analysis provided in Table 5 and the yield data for banana contained in Table 3, we estimated the net income to be $51,780 in the plant crop and $163,500 in the ratoon crop when 20 hectares are irrigated with a pan factor of 1.0. The increase in the net income for the ratoon crop was the combined result of a 36% yield increase (Table 3) and lower operational costs associated with less use of labor and operation of machinery (Table 5) .

These results provide evidence on the importance of proper irrigation management for plantain and banana production on the southern coast of Puerto Rico and show that the additional costs brought about by the increments in irrigation are compensated by improved fruit quality and higher yields. We therefore, recommend irrigation according to a pan factor of at least 0.75 for plantains and 1.0 for bananas to attain optimum growth and yields.

## 15.4  SUMMARY

Plantain and banana production in the semiarid lowlands of the southern coast of Puerto Rico has been increasing because of a greater demand for high-quality fruits, high farm-gate prices and the availability of arable land with an irrigation infrastructure. There is, however, a scarcity of information on optimum water requirements and practical irrigation recommendations for growers of these crops.

Five irrigation regimes, based on class A pan factors ranging from 0.25 to 1.25, were used to obtain fractions of the potential evapotranspiration and to evaluate their influence on yield, the crop parameters, and crop performance. Results were extrapolated to make projections on productivity, gross sales, and on irrigation costs incurred in the operation of a 20-hectare farm of drip irrigated plantains or bananas.

Increasing the amount of applied irrigation in a 20-hectare plantation from a pan factor of 0.75 to 1.25 increased the number of banana fruit boxes by 6,747 in the plant crop, and by 18,009 in the first banana ratoon. This irrigation increment resulted in gross sales increases of $40,482 for the banana plant crop, and $108,054 for the first banana ratoon, with additional water and energy cost of only $2,388. The net income for the plant crop and first banana ratoon irrigated according to a pan factor of 1.0 was estimated to be $51,780.00 and $163,500.00, respectively in a 20-hectare banana plantation.

There were no significant differences in the number of plantain fruits in irrigated plants when pan factors ranged from 0.75 to 1.25. However, irrigating plantains according to a pan factor of 1.25 significantly increased bunch yield and fruit weight.

## KEYWORDS

- **Banana**
- **bunch weight**
- **class A pan evaporation**
- **crop coefficient, Kc**
- **drip irrigation**
- **economic analysis**
- **energy cost**
- **evapotranspiration**
- **irrigation recommendations**
- **micro irrigation**
- **Mollisol soil**
- **Musa spp.**
- **pan coefficient, Kp**
- **pan factor**
- **plant crop**
- **Plantain**
- **ratoon crop**
- **semiarid land**
- **Yield**

## REFERENCES

1. Abruña, F., J. Vicente-Chandler and S. Silva, 1980. Evapotranspiration with plantains and the effect of frequency of irrigation on yields. *J. Agric. Univ. P.R.,* 64(2):204–210.
2. Agricultural Experiment Station, 1986. Technological package for the production of plantains and bananas. Publication 97. Agricultural Experiment Station, College of Agricultural Sciences, Mayaguez, Puerto Rico (In Spanish).
3. Asoegwu, S. N. and J. C. Obiefuna, 1987. Effect of irrigation on late season plantains. *Tropical Agriculture*, 64(2):139–143.

4. Doorenbos, J. and W. 0. Pruitt, 1977. *Guidelines for Predicting Crop Water Requirements*, Irrigation and Drainage Paper 24, FAO - United Nations, Rome, Italy.
5. Goenaga, R., H. Irizarry and E. Gonzalez, 1993. Water requirement of plantains *(Musa acuminata* x *Musa balbisiana* AAB) grown under semiarid conditions. *Tropical Agriculture* 70(1):3–7.
6. Goyal, M. R. and E. A. Gonzalez, 1989. Climatological data collected in the agricultural experiment stations of Puerto Rico. Publication 88–70. Agricultural Experiment Station, College of Agricultural Sciences, Mayaguez, Puerto Rico (In Spanish).
7. Goyal, M. R. and E. A. Gonzalez, 1988. Irrigation requirements for plantain in seven ecological zones of Puerto Rico. *J. Agric. Univ. P.R.,* 72(4):599–608 (In Spanish).
8. Hedge, D. M. and K. Srinivas, 1990. Growth, productivity and water use of banana under drip and basin irrigation in relation to evaporation replenishment. *Indian J. Agron.,* 35(1–2):106–112.
9. Hedge, D. M. and K. Srinivas, 1989. Irrigation and nitrogen fertility influences on plant water relation, biomass, and nutrient accumulation and distribution in banana cv. Robusta. *J. Hort. Sci.,* 64(1):91–98.
10. Lahav, E. and D. Kalmar, 1988. Response of banana to drip irrigation, water amounts and fertilization regimes. *Commun. Soil Sci. Plant Anal.,* 19(1):25–46.
11. Ortiz-Lopez, J., 1992. Plantain and banana. Pages 123–138. *In: Carmen I. Alamo (ed.) Situation and Perspectives of Agricultural Enterprises in Puerto Rico during 1990–1991.* Agricultural Experiment Station, College of Agricultural Sciences, Mayaguez, Puerto Rico (In Spanish).
12. Robinson, J. C. and A. J. Alberts, 1986. Growth and yield responses of banana (cultivar 'Williams') to drip irrigation under drought and normal rainfall conditions in the subtropics. *Scientia Horticulturae,* 30:187–202.
13. SAS Institute, Inc., 1987. *SAS/STAT Guide for Personal Computers.* Raleigh, North Carolina.
14. Tai, E. A. 1977, Banana. Pages 441–460. *In: Alvim, P. de T. and Kozlowski, T. T. (eds.) Ecophysiology of Tropical Crops.* Academic Press, New York, N.Y.
15. Young, S. and 1-Pai Wu, 1981. Proceedings of XIII Annual Meeting. Pages 51–69. Conference Research Extension Series 021, College of Tropical Agriculture and Human Resources, Univ. Hawaii.

# CHAPTER 16

# BIOMETRIC RESPONSE OF EGGPLANT UNDER SUSTAINABLE MICRO IRRIGATION WITH MUNICIPAL WASTEWATER

VINOD KUMAR TRIPATHI, T. B. S. RAJPUT, NEELAM PATEL, and PRADEEP KUMAR

## CONTENTS

## 16.1 INTRODUCTION

Worldwide, agriculture must be more proactive in managing its demand for water and improving the performance of both irrigated and rain-fed production. Availability of water for irrigation sector is declining because of increasing demand for water in domestic and industrial sector at significant rates. Production of more food to feed the burgeoning population is the big challenge. It is also vital to control the increasing prices of agricultural produce. There is a need to invest in both improved technologies and better management in order to achieve "more crops per drop." Water supply and water quality degradation are global concerns that will intensify with increasing water demand.

Worldwide, marginal-quality water is becoming increasingly important component of agricultural water supplies, particularly in water-scarce countries [18]. One of the major water resources having marginal-quality water is the municipal wastewater from urban and peri-urban areas. The wastewater has been recycled in agriculture for centuries as a means of disposal in cities such as Berlin, London, Milan and Paris [1]. However, in the recent years wastewater has gained importance in water-scarce regions. Italian legislation states that natural fresh water sources should be used as a priority for the municipal water supply, and that the recycling and reuse of water are viable alternatives for meeting industrial and agricultural needs. Wastewater reuse in agriculture requires best treatment practices, management practices and appropriate irrigation technology [8]. Wastewater treatment plants in most cities in developing countries are nonexistent or function inadequately [18]. In many cases, the quality standards for reclaimed wastewater are the same as for drinking water [7]. Therefore, wastewater in partially treated, diluted or untreated form is diverted and used by urban and peri-urban farmers to grow a range of crops [10, 15, 16].

In arid and semi arid developing countries, farmers are using municipal wastewater for irrigation by traditional surface irrigation methods (generally flood or furrow method). These methods require more water for irrigation; pose numerous problems of soil, water and environmental degradation compared to micro irrigation method. Main disadvantage of these methods are supply driven rather than crop demand driven, which cause mismatch between the need of crop and the quantity supplied [20]. Micro irrigation method may accomplish higher field level application efficiency of 80–90%, because surface runoff and deep percolation losses are minimized. Aujla et al. [4] observed a 4% higher yield of eggplant with drip irrigation compared to furrow irrigation by saving 50% water. Therefore, micro irrigation can help to give high crop yield per unit of applied water, and can allow crop cultivation in an area where available wastewater and fresh water is insufficient to irrigate with surface irrigation methods. It is prudent to reuse wastewater through subsurface micro irrigation method so that limited amount of wastewater can be applied below the ground to reduce contamination of the crop and environment. In this method, there is no risk of contamination through aerosol. Kiziloglu et al. [14], while conducting experiment with flood irrigation, concluded that primary treated wastewater can be used in sustainable agriculture in the long-term. Therefore, it is becoming increasingly important to adopt subsurface micro irrigation method [22].

Eggplant (*Solanum melongina* L.) is an extensively grown vegetable in the out-skirt of Indian cities. Its annual production is 11.9 million tons that is 27.6% of the global production [11]. Al-Nakshabandi et al. [3] observed that average eggplant yield under treated effluent was twice the average eggplant production under fresh water irrigation using conventional fertilizer application in Jordan. The application of 50% water through micro irrigation can produce 4% higher eggplant yield compared to furrow irrigation with freshwater [4]. Eggplant yield with wastewater using surface irrigation methods has been studied by many researchers [3, 5, 12, 17]. **Douh and Boujelben [9] and** Cirelli [8] used wastewater through drip irrigation system. Most of the researchers have given emphasis on heavy metal contamination in the produce and much importance has not been given to subsurface micro irrigation. Subsurface micro irrigation with wastewater can play vital role in minimizing the contamination of the produce. According to the literature review, not enough information is available to evaluate the impact of subsurface micro irrigation. This chapter summarizes the effects of subsurface micro irrigation with wastewater on the biometric parameters of eggplant compared to surface micro irrigation.

## 16.2 MATERIALS AND METHODS

### 16.2.1 EXPERIMENTAL SITE

The experiment was conducted at the research farm of Precision Farming Develop-ment Center, Water Technology Center, Indian Agricultural Research Institute, New Delhi, India (Between latitudes 28°37'22"N and 38°39'05"N and longitudes 77°8'45" and 77°10'24"E and AMSL 228.61 m) during November 2009 to May, 2010. January was the coldest month with a mean temperature of 14°C however; the minimum tem-perature dips to as low as 1°C. Frost occurs occasionally during month of December and January. The average relative humidity was 34.1 to 97.9% and average wind speed was 0.45 to 3.96 m/s.

### 16.2.2 SOIL CHARACTERISTICS

Soil samples were collected up to 60 cm soil depth with 15.0 cm intervals. Hydrometer method was used to determine the sand, silt and clay percentage of soil. The soil at the experimental site was deep, well-drained sandy loam soil comprizing mean value 62% sand, 17% silt and 21% clay. The soil bulk density was 1.56 g.cm$^{-3}$. Field capac-ity and mean value of saturated hydraulic conductivity were 0.16 and 1.13 cm.h$^{-1}$, respectively.

### 16.2.3 EGGPLANT SEEDLINGS AND CROP PRACTICES

Seedlings for the eggplant (cv: *Supriya*) were raised in the plastic trays with the mix-ture of coco peat, vermiculite and perlite in the ratio of 3:2:1. Seeds were sown in No-vember 2009 under polyhouse with partial ventilation. Light irrigation was provided frequently during warm, dry periods for adequate germination. A hand sprayer was used to spray fresh groundwater on the nursery. No wastewater was used during the growth period of nursery.

Twenty-five days old seedlings were planted in the field. Crop water requirement was met by estimating the reference evapotranspiration ($ET_0$) using the Penman-

Monteith method and the crop coefficient ($K_c$) as suggested by Allen et al. [2]. The nutritional requirement (120 kg ha$^{-1}$ N, 160 kg ha$^{-1}$ P$_2$O$_5$ and 160 kg ha$^{-1}$K$_2$O) as suggested by Chadha [6] with freshwater has been suggested to decide the amount of the fertilizer to wastewater irrigated plots. Availability of nutrients in wastewater (average value: 28 mgl$^{-1}$ N, 16 mgl$^{-1}$ P, 28 mgl$^{-1}$ $^K$) was analyzed by taking the water sample during crop period. Nutrient application was done by deducting the available of nutrients in wastewater. Therefore, 64 kg ha$^{-1}$ N, 132 kg ha$^{-1}$ P$_2$O$_5$ and 96 kg ha$^{-1}$ $^K$$_2$O were applied through fertigation system.

The following treatments with different types of filter arrangement for micro irrigation systems were considered for irrigation with wastewater (WW) in this research:

W1D1: WW through media filter and placement of drip laterals at soil surface;

W1D2: WW through media filter and placement of drip laterals at 15 cm depth below ground surface;

W1D3: WW through media filter and placement of drip laterals at 30 cm depth below ground surface;

W2D1: WW through disk filter and placement of drip laterals at soil surface;

W2D2: WW through disk filter and placement of drip laterals at 15 cm depth below ground surface;

W2D3: WW through disk filter and placement of drip laterals at 30 cm depth below ground surface;

W3D1: WW through media and disk filters and placement of drip laterals at soil surface;

W3D2: WW through media and disk filters and placement of drip laterals at 15 cm depth below ground surface; and

W3D3: WW through media and disk filters and placement of drip laterals at 30 cm depth below ground surface.

### 16.2.4   BIOMETRIC PARAMETERS OF EGGPLANT

The biometric parameters were leaf area index, root length, root density, yield and dry matter content were measured. All observations were made from center rows after border rows were discarded to avoid edge effects.

#### 16.2.4.1   LEAF AREA INDEX

Leaf area index (LAI) estimation included both assimilating area and growth. Observations for LAI were recorded during the crop season starting at 25 days after transplanting with an interval of 15 days in all treatments. LAI was determined by Canopy Analyzer (model: LAI-2000) in the experimental field. For crop production, leaf area per unit land area is more important than the leaf area of individual plants. Therefore, Leaf area index was calculated as follows:

$$LAI = \frac{LA}{LLA} \tag{1}$$

where, $LAI$ = Leaf area index, LA = Leaf area, and LAA = Land area.

### 16.2.4.2   ROOT LENGTH DENSITY

Root sampling for determination of root length density was carried out during crop season starting at 25 days after transplanting with an interval of 15 days. It was done with an auger having an internal diameter 0.15 m to collect soil cores. Samples were collected at each treatment up to a depth of 45 cm. The samples were steeped and flushed prior to measure root length and root density. It was measured using root scanner (Epson Expression model LC 1600) of make (WinRHIZO 2002c). RHIZO system measured the root length by scanning the length of the root skeleton.

### 16.2.4.3   YIELD AND DRY MATTER

Matured eggplant fruits were harvested manually in three stages at the interval 6–7 days as per availability of proper size. Weight of fruits was recorded for each treatment separately. Dry matter content is a measure of the quantity of total solids in fruit. It was determined by removing all the moisture from the fruit. Judgment of moisture removal was done by weighing the remaining solids with the help of digital electronic balance. Samples were kept in electric oven till it attained the constant weight. This was reported as a percentage weight of fresh fruit.

### 16.2.5   STATISTICAL ANALYSIS

The experimental design was split plot randomized block design, where main plot was irrigation water at three levels and sub plot was irrigation systems, that is, placement of micro irrigation laterals at three levels (three depths). Each treatment was replicated thrice. Transformed data were analyzed by the General Linear Model (GLM) procedure of the SAS statistical software [19]. The Kolmogorov-Smirnov (K-S) statistics were used to test the goodness-of-fit of the data to normal distribution. To ensure that data came from a normal distribution, standardized skewness and kurtosis values were checked. All main-effects were compared by pair wise t-tests, equivalent to Fisher's least significant difference (LSD) test using the MEANS statement under GLM procedure with mean at $\alpha=0.05$.

### 16.3   RESULTS AND DISCUSSIONS

### 16.3.1   LEAF AREA INDEX

LAI was determined by leaf area meter and mean value of 20 plant sample are presented in Fig. 1. During initial 40–50 days after transplanting, LAI was higher under surface placed lateral treatments (W1D1, W2D1 and W3D1) in comparison to subsurface placed lateral treatments. After 25 days of transplanting, LAI under treatment W1D1 was significantly ($P<0.01$) different from W1D2 and W1D3, however, W1D2 and W1D3 were not significantly different ($P<0.05$). Similar trend was observed for water treatments (W2 and W3). After 60 days of transplanting increase rate of LAI for subsurface irrigated treatments (W1D2, W1D3, W2D2, W2D3, W3D2, and W3D3) was higher in comparison to surface irrigated treatments (W1D1, W2D1 and W3D1). After 100 days of transplanting, highest LAI of 4.23 was observed in treatment W3D2 but the lowest 3.12 was under treatment W1D1. After 115 days, almost constant value of LAI was observed in all the treatments showing no crop growth. It was also observed physically. At this stage, LAI of the surface irrigated treatments (W1D1,

W2D1, W3D1) were significantly different at *P*<0.01 with subsurface irrigated treatments (W1D2, W1D3, W2D2, W2D3, W3D2 and W3D3).

Subsurface treatments (W1D2, W2D2, and W3D2) at 15 cm were significantly different at *P*<0.05 compared to subsurface treatments (W1D3, W2D3, and W3D3) at 30 cm. Overall, higher LAI was observed with subsurface (15 cm) irrigated treatments. This may be due to more volume of soil under wet condition as compared with surface placed laterals.

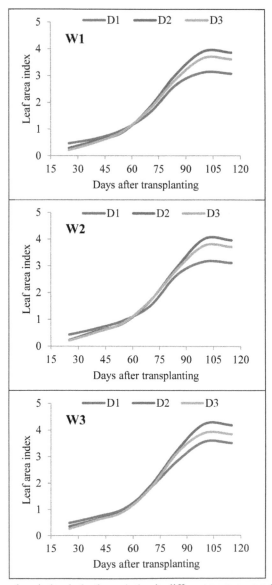

**FIGURE 1**   Temporal variation in leaf area index in different treatments of eggplant.

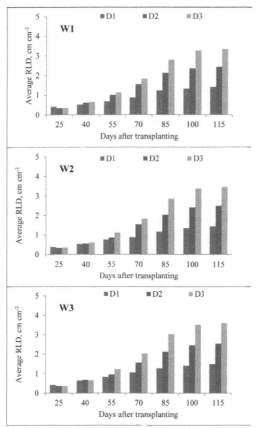

**FIGURE 2**    Increase in root length density of eggplant due to impact of lateral placement.

## 16.3.2   ROOT LENGTH DENSITY (RLD)

Measurements of RLD and LAI were done on the same days i.e., starting from 25 days after transplanting till the last harvesting at an interval of 15 days (Fig. 2). During initial 40 days after transplanting, similar values of RLD were observed in all the treatments. Afterward the growth rate of roots was faster in the sequence of W1D3>W1D2>W1D1. Similar trend was observed with all the treatments with water, W2 and W3. This trend was continued till the last harvest of the crop in all the treatments. Highest RLD value of 3.6 cm cm$^{-3}$ was observed in treatment W3D3. Under same depth of lateral placement, RLD values of 3.34 and 3.45 cm.cm$^{-3}$ were observed in treatments W1D3 and W2D3, respectively. Minimum RLD value of 1.4 cm cm$^{-3}$ was observed in treatment W1D1.

Higher root volume with subsurface placement of drip lateral at 30 cm (W1D3, W2D3, and W3D3) and lowest with surface place lateral (W1D1, W2D1, and W3D1) were observed visually. Based on RLD values, the treatments were significantly different ($P<0.01$) as per Student's $t$-Test. Good relation between RLD and LAI was observed with the value of correlation coefficient 0.72. This may be due to lower flow rate in surface place lateral and smaller soil volume wetting. Yavuz et al. [23] also observed more pronounced emission uniformity reduction in surface placed laterals due to their exposure to higher temperatures in comparison with subsurface placed laterals, which achieved the best system performance in terms of emission uniformity and flow rate reduction.

### 16.3.3 EGGPLANT YIELD

In all the treatments, eggplant yield was higher than the national average yield of 20–25 Mgha$^{-1}$. Treatment wise average fruit yield was estimated by adding the weight of harvested fruit obtained in all three stages. Yield of 44.65, 45.96 and 45.32 Mgha$^{-1}$ was recorded in treatments W1D2, W2D2, and W3D2 by subsurface placement of drip lateral at 15 cm depth, respectively. It was higher than the yield obtained in surface irrigated plots (Table 1). Minimum crop yield of 38.95 Mgha$^{-1}$ was observed in the treatment W1D1 by surface placement of drip lateral. It may be due to lower flow rate in emitters causing water stress. These results agree with those obtained by Kirnak et al. [13] who demonstrated that water stress resulted in a reduction of fresh fruit yield and fruit size of eggplants. Moderate crop yield was observed in the treatments (W1D3, W2D3 and W3D3) by subsurface placement of drip lateral at 30 cm depth. The decline in crop yield for subsurface irrigation (30 cm) may be due to the fact that higher BOD$_5$ and COD values for WW reduce oxygen availability in deeper layer soil for respiration of roots. 50% decrease in soil respiration rate was observed by Singh et al. [21] in sewage water irrigated plots without drip system.

### 16.3.4 DRY MATTER (DM)

Average dry matter content for all the treatments was estimated and presented in Table 1. Maximum DM of 8.76% was recorded in treatment W2D1 and minimum of 7.04% with W3D3. DM in treatments with surface placed laterals was higher and significantly different than the subsurface irrigated treatments. Variation in the DM values for subsurface (30 cm) was in the range of 7.04 to 7.92% but for surface it was in the range of 8.51 to 8.76%. Lower wetting volume of soil with surface placed lateral treatments may lead to water stress. This may cause higher DM content. Kirnak et al. [13] demonstrated that water stress resulted in a reduction of fresh fruit yield and fruit size of eggplants, while it caused increases in the fruit dry matter content. An induced water shortage leads high dry matter percentages for the eggplants [8].

**TABLE 1** Impact of drip lateral placement on yield and dry matter content of eggplant.

| Treatments | Average yield, Mg ha$^{-1}$ | Average dry matter content, % |
|---|---|---|
| W1D1 | 38.95 | 8.68 |
| W1D2 | 44.65 | 7.37 |
| W1D3 | 43.26 | 7.92 |
| W2D1 | 41.11 | 8.76 |
| W2D2 | 45.96 | 7.14 |
| W2D3 | 43.99 | 7.29 |

**TABLE 1** *(Continued)*

| Treatments | Average yield, Mg ha⁻¹ | Average dry matter content, % |
|---|---|---|
| W3D1 | 41.46 | 8.51 |
| W3D2 | 45.32 | 7.30 |
| W3D3 | 44.57 | 7.04 |
| LSD P=0.05) | | |
| Water (W) | 1.19 | 0.46 |
| System (D) | 0.82 | 0.27 |
| W×S | 1.64 | 0.55 |

**TABLE 2** Significance level (P-value) of the statistical model (split plot) and of each factor (main plot water and lateral depth subplot) and interaction for fruit yield and dry matter content.

| Parameter | Model | Main and interaction effect | | | t-grouping | | | | | |
|---|---|---|---|---|---|---|---|---|---|---|
| | | | | | Water (W) | | | Lateral depth (D) | | |
| | | W | D | W×D | W1 | W2 | W3 | D1 | D2 | D3 |
| Fruit yield | *** (R²=0.86) | * | *** | n.s. | B | A | A | B | A | A |
| Dry matter | *** (R²=0.94) | *** | *** | *** | B | A | A | A | B | C |

n.s. : not significant, *P>* 0.05.

* :*P*<0.05 ** :*P*<0.01 *** :*P*<0.001.

Statistical analysis with split plot design for fruit yield and DM content by ANOVA are presented in Table 2. ANOVA model for fruit yield and DM was valid with higher value of $R^2$ (0.86 for fruit yield and 0.94 for DM). Effect of water on yield was significant at $P<0.05$ but effect of lateral depth on yield was significant at $P<0.001$. Interaction effect was not significant on fruit yield. Higher least significant difference (LSD) values for the effect of water and lateral depth interaction can also be seen from Table 1. Subsurface depths D2 and D3 are in same group as per *t*-grouping. It may be explained that higher wetting volume of soil due to subsurface placement of lateral makes more availability of accumulated nutrients in the root zone. Individual and combined effect of water and lateral depth on DM content was significant ($P<0.001$) Different filtration system have significant role because emitter flow rate was varied with respect to time. As per *t*-grouping, DM was grouped separately for lateral placement depth. This shows subsurface drip can play a vital role for utilization of wastewater for irrigation.

## 16.4 CONCLUSIONS

The field study was conducted to evaluate eggplant performance through surface and subsurface micro irrigation system by using municipal wastewater. Higher LAI and

RLD were observed in subsurface irrigated plots. Good correlation was also observed among LAI and RLD. Higher dry matter content was recorded in surface irrigated plots but higher fruit yield was recorded with subsurface irrigated plots by placing drip lateral at 15 cm depth. Irrigation with wastewater partially contributed (47%, 18%, and 40% of N, $P_2O_5$, and $K_2O$ respectively) nutritional requirement of eggplant crop. With this field experiment, it can be concluded that subsurface micro irrigation might be a good alternative to use wastewater for production of eggplant. The only difficulty for subsurface micro irrigation is involvement of cost for placing drip lateral at appropriate depth. It increases the cost of cultivation. Basic information obtained about biometric response of eggplant may be used for future research.

## 16.5  SUMMARY

Utilization of municipal wastewater through subsurface micro irrigation reduces the environmental contamination. However, the impact of subsurface drip on biometric parameters of eggplant is a researchable issue. This study was conducted with wastewater. It was treated though three different types of filtration processes viz., media type, disk type and combined media and disk type filters. Eggplants were grown with wastewater. The root length density (RLD), leaf area index (LAI), fruit yield with their dry matter was recorded.

The data revealed that leaf area index was lower for subsurface drip during initial 55 days after transplanting but it was significantly higher in latter stage. Highest root length density 3.6 cm cm$^{-3}$ was recorded under subsurface placement of drip lateral at 30 cm depth. Good relation between RLD and LAI was observed with the value of correlation coefficient 0.69. Highest dry matter content (8.76%) was recorded with surface placed lateral but highest fruit yield was recorded with the subsurface placed drip lateral at 15 cm depth. Subsurface micro irrigation through laterals placed at 15 cm and 30 cm depths resulted in 12% and 8.5% higher yield, respectively, in comparison to surface placed lateral. Utilization of wastewater gave the saving of 47%, 18%, and 40% of N, $P_2O_5$, and $K_2O$ nutrients, respectively. Statistically the effect of filtered wastewater on the yield was significant. The findings of present study can be used for using wastewater through subsurface micro irrigation for protection of community living nearby wastewater-irrigated field.

## KEYWORDS

- **Crop practice**
- **Dry matter**
- **Eggplant**
- **Fruit yield**
- **Leaf area index**
- **Micro irrigation**
- **Root length density**
- **Statistical analysis**
- **Subsurface drip irrigation**
- **Wastewater**

## REFERENCES

1. AATSE (Australian Academy of Technological Sciences and Engineering), 2004.Water Recycling in Australia. AATSE, Victoria, Australia.
2. Allen, R.G., Pereira, L.S., Raes, D., and Smith, M., 1998. *Crop evapotranspiration. Guidelines for computing crop water requirements*. FAO Irrigation and Drainage Paper 56, FAO, Rome, Italy.
3. Al-Nakshabandi, G.A., Saqqar, M.M., Shatanawi, M.R., Fayyad, M. and Al-Horani, H., 1997. Some environmental problems associated with the use of treated wastewater for irrigation in Jordan. *Agricultural Water Management*, 34(1):81–94.
4. Aujla, M.S., Thind, H.S., and Buttar, G.S., 2007. Fruit yield and water use efficiency of eggplant (*Solanummelongema* L.) as influenced by different quantities of nitrogen and water applied through drip and furrow irrigation. *Scientia Horticulturae*, 112(2):142–148.
5. Brar, M.S., Arora, C.L., 1997. Nutrient status of cauliflower in soils irrigated by sewage and tubewell water. *Indian Journal of Horticulture*, 54(1):80–85.
6. Chadha, K.L., 2001. *Handbook of horticulture*. Indian Council of Agricultural Research (ICAR), New Delhi, India.
7. Cirelli, G.L., Consoli, S., and Di Grande, V., 2008. Long term storage of reclaimed water: the case studies in Sicily (Italy). *Desalination*, 218:62–73.
8. Cirelli, G.L., Consoli, S., Licciardello, F., Aiello, R., Giuffrida, F., and Leonardi, C., 2012. Treated municipal wastewater reuse in vegetable production. *Agricultural Water Management*, 104:163–170.
9. Douh, B., and Boujelben, A., 2010. Water saving and eggplant response to subsurface micro irrigation. *Agricultural Segment*, 1(2):AGS/1525.
10. Ensink, H.H., Mehmood, T., Vand der Hoeck, W., Raschid-Sally, L., Amerasinghe, F.P., 2004. A nation-wide assessment of wastewater use in Pakistan: an obscure activity or a vitally important one? Water Policy 6:197–206.
11. FAO, 2010. http:\\www.fao.org\waicent\portal\statistics.en.asp.
12. Irénikatché Akponikpè, P.B., Wima, K., Yacouba, H., and Mermoud, A., 2011. Reuse of domestic wastewater treated in macrophyte ponds to irrigate tomato and eggplant in semiarid West-Africa: Benefits and risks. *Agricultural Water Management*, 98(5):834–840
13. Kirnak, H., Tas, I., Kaya, C., Higgs, D., 2002. Effects of deficit irrigation on growth, yield, and fruit quality of eggplant under semiarid conditions. *Aust. J. Agric. Res.*, 53:1367–1373.
14. Kiziloglu, F.M., Turan, M., Sahin, U., Kuslu, Y. and Dursun, A., 2008. Effects of untreated and treated wastewater irrigation on some chemical properties of cauliflower (Brassica olerecea L. var. botrytis) and red cabbage (Brassica olerecea L. var. rubra) grown on calcareous soil in Turkey. *Agricultural water management*, 95:716–724.
15. Lai, T.V., 2000. Perspectives of peri-urban vegetable production in Hanoi. Background paper prepared for the Action Planning Workshop of the CGIAR Strategic Initiative for Urban and Peri-urban Agriculture (SIUPA), Hanoi, 6–9 June. Convened by International Potato Center (CIP), Lima.
16. Murtaza, G., Ghafoor, A., Qadir, M., Owens, G., Aziz, M.A., Zia, M.H. and Saifullah., 2010. disposal and use of sewage on agricultural lands in Pakistan: a Review. *Pedosphere, 20(1):* 23–34.
17. Papadopoulos, I., Chimonidou, D., Polycar, P., and Savvides, S. 2013. Irrigation of vegetables and flowers with treated wastewater. Agricultural Research Institute, 1516 Nicosia, Cyprus. http://om.ciheam.org/om/pdf/b56_2/00800185.pdf .
18. Qadir, M., Wichelns, D., Raschid-Sally, L., McCornick, P.G., Drechsel, P., Bahri, A., Minhas, P.S., 2010.The challenges of wastewater irrigation in developing countries. *Agricultural Water Management*, 97(4):561–568.
19. SAS, 2008.SAS/STAT® 9.2 User's Guide. Cary, NC: SAS Institute Inc.

20. Singh, D.K., and Rajput, T.B.S., 2007.Response of lateral placement depths of subsurface micro irrigation on okra (*Abelmoschuse sculentus*). *International Journal of Plant Production*, 1:73–84

21. Singh, P.K., Biswas, A.K., Singh, S.P., 2009. Effect of sewage water on heavy metal load on okra, spinach and cauliflower in vertisoil. *Indian J. Hort.* 66(1):144–146.

22. Turral, H., Svendsen, M., Faures, J.M., 2010. Investing in irrigation: Reviewing the past and looking to the future. *Agricultural Water Management*, 97(4):551–560.

23. Yavuz, M.Y., Demirel, K., Erken, O., Bahar, E., Devecirel, M., 2010. Emitter clogging and effects on micro irrigation systems performances. *African J. Agricultural Research*, 5(7):532–538.

# APPENDICES

(Modified and reprinted with permission from: Goyal, Megh R., 2012. Appendices. Pages 317–332. In: *Management of Drip/Trickle or Micro Irrigation* edited by Megh R. Goyal. New Jersey, USA: Apple Academic Press Inc.)

## APPENDIX A

## CONVERSION SI AND NON-SI UNITS

| To convert the Column 1 in the Column 2, Multiply by | Column 1 Unit SI | Column 2 Unit Non-SI | To convert the Column 2 in the Column 1 Multiply by |
|---|---|---|---|
| **LINEAR** | | | |
| 0.621 ------ | kilometer, km ($10^3$ m) | miles, mi ------------------ | 1.609 |
| 1.094 ------ | meter, m | yard, yd -------------------- | 0.914 |
| 3.28 ------- | meter, m | feet, ft ---------------------- | 0.304 |
| $3.94 \times 10^{-2}$ ---- | millimeter, mm ($10^{-3}$) | inch, in -------------------- | 25.4 |
| **SQUARES** | | | |
| 2.47 ------- | hectare, he | acre --------------------- | 0.405 |
| 2.47 ------- | square kilometer, km$^2$ | acre --------------------- | $4.05 \times 10^{-3}$ |
| 0.386 -------- | square kilometer, km$^2$ | square mile, mi$^2$ ------------ | 2.590 |
| $2.47 \times 10^{-4}$ ---- | square meter, m$^2$ | acre --------------------- | $4.05 \times 10^{-3}$ |
| 10.76 -------- | square meter, m$^2$ | square feet, ft$^2$ ------------- | $9.29 \times 10^{-2}$ |
| $1.55 \times 10^{-3}$ ---- | mm$^2$ | square inch, in$^2$ -------------- | 645 |
| **CUBICS** | | | |
| $9.73 \times 10^{-3}$ ---- | cubic meter, m$^3$ | inch-acre ------------------ | 102.8 |
| 35.3 -------- | cubic meter, m$^3$ | cubic-feet, ft$^3$ ---------------- | $2.83 \times 10^{-2}$ |
| $6.10 \times 10^4$ ---- | cubic meter, m$^3$ | cubic inch, in$^3$ ------------- | $1.64 \times 10^{-5}$ |
| $2.84 \times 10^{-2}$ ---- | liter, L ($10^{-3}$ m$^3$) | bushel, bu ------------------ | 35.24 |
| 1.057 -------- | liter, L | liquid quarts, qt ------------ | 0.946 |
| $3.53 \times 10^{-2}$ ---- | liter, L | cubic feet, ft$^3$ -------------- | 28.3 |
| 0.265 -------- | liter, L | gallon -------------------- | 3.78 |
| 33.78 -------- | liter, L | fluid ounce, oz ------------- | $2.96 \times 10^{-2}$ |
| 2.11 ------- | liter, L | fluid dot, dt --------------- | 0.473 |
| **WEIGHT** | | | |
| $2.20 \times 10^{-3}$ ---- | gram, g ($10^{-3}$ kg) | pound, -------------------- | 454 |

$3.52 \times 10^{-2}$ ---- gram, g ($10^{-3}$ kg)     ounce, oz ------------------ 28.4

2.205 ------ kilogram, kg           pound, lb ------------------ 0.454

$10^{-2}$ ------- kilogram, kg        quintal (metric), q ---------- 100

$1.10 \times 10^{-3}$ ---- kilogram, kg    ton (2000 lbs), ton ----------907

1.102 ------ mega gram, mg       ton (US), ton -------------- 0.907

1.102 ------ metric ton, t         ton (US), ton -------------- 0.907

## YIELD AND RATE

0.893 ------- kilogram per hectare     pound per acre ------------ 1.12

$7.77 \times 10^{-2}$ --- kilogram per cubic meter   pound per fanega ---------- 12.87

$1.49 \times 10^{-2}$ --- kilogram per hectare    pound per acre, 60 lb ----- 67.19

$1.59 \times 10^{-2}$ --- kilogram per hectare    pound per acre, 56 lb ----- 62.71

$1.86 \times 10^{-2}$ --- kilogram per hectare    pound per acre, 48 lb ----- 53.75

0.107 ------- liter per hectare        gallon per acre --------- 9.35

893 ---------- ton per hectare         pound per acre ---------- $1.12 \times 10^{-3}$

893 ---------- mega gram per hectare    pound per acre ---------- $1.12 \times 10^{-3}$

0.446------- ton per hectare          ton (2000 lb) per acre ----- 2.24

2.24 ---------- meter per second      mile per hour ------------ 0.447

## SPECIFIC SURFACE

10 -------- square meter per kilogram    square centimeter per gram ------------ 0.1

$10^{3}$ ------- square meter per kilogram    square millimeter per gram ------------ $10^{-3}$

## PRESSURE

9.90 ---------- megapascal, MPa       atmosphere --------------------- 0.101

10 --------- megapascal             bar ------------------------------- 0.1

1.0 ---------- megagram per cubic meter   gram per cubic centimeter ----- 1.00

$2.09 \times 10^{-2}$ ---- pascal, Pa        pound per square feet ---------- 47.9

$1.45 \times 10^{-4}$ ---- pascal, Pa        pound per square inch --------- $6.90 \times 10^{3}$

| To convert the column 1 in the Column 2, Multiply by | Column 1 Unit SI | Column 2 Unit Non-SI | To convert the column 2 in the column 1 Multiply by |
|---|---|---|---|

## TEMPERATURE

1.00 (K-273)---Kelvin, K         centigrade, °C -------- 1.00 (C+273)

(1.8 C + 32)---centigrade, °C    Fahrenheit, °F -------- (F--32)/1.8

## ENERGY

$9.52 \times 10^{-4}$ ---- Joule J         BTU ------------------ $1.05 \times 10^{3}$

0.239 -------- Joule, J            calories, cal ------------ 4.19

0.735 -------- Joule, J                    feet-pound ------------ 1.36
2.387 × 10$^5$ --- Joule per square meter    calories per square centimeter --- 4.19 × 10$^4$
10$^5$ ---------- Newton, N                  dynes ----------------- 10$^{-5}$

## WATER REQUIREMENTS

9.73 × 10$^{-3}$ --- cubic meter             inch acre --------------- 102.8
9.81 × 10$^{-3}$ --- cubic meter per hour    cubic feet per second ------ 101.9
4.40 ---------- cubic meter per hour     galloon (US) per minute ---- 0.227
8.11 ---------- hectare-meter            acre-feet --------------- 0.123
97.28 ------- hectare-meter              acre-inch ---------------- 1.03 × 10$^{-2}$
8.1 × 10$^{-2}$ ---- hectare centimeter      acre-feet --------------- 12.33

## CONCENTRATION

1 ------------ centimol per kilogram     milliequivalents per100 grams ------------- 1
0.1 --------- gram per kilogram          percents ---------------- 10
1 ----------- milligram per kilogram     parts per million --------- 1

## NUTRIENTS FOR PLANTS

2.29 -------- P       $P_2O_5$ --------------------- 0.437
1.20 -------- K       $K_2O$ -------------------- 0.830
1.39 -------- Ca      CaO -------------------- 0.715
1.66 -------- Mg      MgO ------------------ 0.602

## NUTRIENT EQUIVALENTS

| Column A | Column B | Conversion A to B | Equivalent B to A |
|---|---|---|---|
| N | $NH_3$ | 1.216 | 0.822 |
| | $NO_3$ | 4.429 | 0.226 |
| | $KNO_3$ | 7.221 | 0.1385 |
| | $Ca(NO_3)_2$ | 5.861 | 0.171 |
| | $(NH_4)_2SO_4$ | 4.721 | 0.212 |
| | $NH_4NO_3$ | 5.718 | 0.175 |
| | $(NH_4)_2HPO_4$ | 4.718 | 0.212 |
| P | $P_2O_5$ | 2.292 | 0.436 |
| | $PO_4$ | 3.066 | 0.326 |
| | $KH_2PO_4$ | 4.394 | 0.228 |
| | $(NH_4)_2HPO_4$ | 4.255 | 0.235 |
| | $H_3PO_4$ | 3.164 | 0.316 |
| K | $K_2O$ | 1.205 | 0.83 |
| | $KNO_3$ | 2.586 | 0.387 |
| | $KH_2PO_4$ | 3.481 | 0.287 |
| | Kcl | 1.907 | 0.524 |

| Column A | Column B | Conversion A to B | Equivalent B to A |
|---|---|---|---|
| | $K_2SO_4$ | 2.229 | 0.449 |
| Ca | CaO | 1.399 | 0.715 |
| | $Ca(NO_3)_2$ | 4.094 | 0.244 |
| | $CaCl_2 \times 6H_2O$ | 5.467 | 0.183 |
| | $CaSO_4 \times 2H_2O$ | 4.296 | 0.233 |
| Mg | MgO | 1.658 | 0.603 |
| | $MgSO_4 \times 7H_2O$ | 1.014 | 0.0986 |
| S | $H_2SO_4$ | 3.059 | 0.327 |
| | $(NH_4)_2SO_4$ | 4.124 | 0.2425 |
| | $K_2SO_4$ | 5.437 | 0.184 |
| | $MgSO_4 \times 7H_2O$ | 7.689 | 0.13 |
| | $CaSO_4 \times 2H_2O$ | 5.371 | 0.186 |

## APPENDIX B
## PIPE AND CONDUIT FLOW

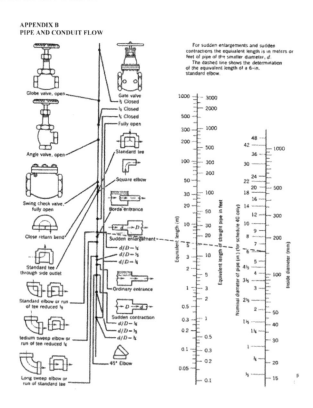

APPENDIX B
PIPE AND CONDUIT FLOW

For sudden enlargements and sudden contractions the equivalent length is in meters or feet of pipe of the smaller diameter, d. The dashed line shows the determination of the equivalent length of a 6-in. standard elbow.

## APPENDIX C

## PERCENTAGE OF DAILY SUNSHINE HOURS: FOR NORTH AND SOUTH HEMISPHERES

| Latitude | Jan | Feb | Mar | Apr | May | Jun | Jul | Aug | Sep | Oct | Nov | Dec |
|---|---|---|---|---|---|---|---|---|---|---|---|---|
| | | | | | *NORTH* | | | | | | | |
| 0 | 8.50 | 7.66 | 8.49 | 8.21 | 8.50 | 8.22 | 8.50 | 8.49 | 8.21 | 8.50 | 8.22 | 8.50 |
| 5 | 8.32 | 7.57 | 8.47 | 3.29 | 8.65 | 8.41 | 8.67 | 8.60 | 8.23 | 8.42 | 8.07 | 8.30 |
| 10 | 8.13 | 7.47 | 8.45 | 8.37 | 8.81 | 8.60 | 8.86 | 8.71 | 8.25 | 8.34 | 7.91 | 8.10 |
| 15 | 7.94 | 7.36 | 8.43 | 8.44 | 8.98 | 8.80 | 9.05 | 8.83 | 8.28 | 8.20 | 7.75 | 7.88 |
| 20 | 7.74 | 7.25 | 8.41 | 8.52 | 9.15 | 9.00 | 9.25 | 8.96 | 8.30 | 8.18 | 7.58 | 7.66 |
| 25 | 7.53 | 7.14 | 8.39 | 8.61 | 9.33 | 9.23 | 9.45 | 9.09 | 8.32 | 8.09 | 7.40 | 7.52 |
| 30 | 7.30 | 7.03 | 8.38 | 8.71 | 9.53 | 9.49 | 9.67 | 9.22 | 8.33 | 7.99 | 7.19 | 7.15 |
| 32 | 7.20 | 6.97 | 8.37 | 8.76 | 9.62 | 9.59 | 9.77 | 9.27 | 8.34 | 7.95 | 7.11 | 7.05 |
| 34 | 7.10 | 6.91 | 8.36 | 8.80 | 9.72 | 9.70 | 9.88 | 9.33 | 8.36 | 7.90 | 7.02 | 6.92 |
| 36 | 6.99 | 6.85 | 8.35 | 8.85 | 9.82 | 9.82 | 9.99 | 9.40 | 8.37 | 7.85 | 6.92 | 6.79 |
| 38 | 6.87 | 6.79 | 8.34 | 8.90 | 9.92 | 9.95 | 10.1 | 9.47 | 3.38 | 7.80 | 6.82 | 6.66 |
| 40 | 6.76 | 6.72 | 8.33 | 8.95 | 10.0 | 10.1 | 10.2 | 9.54 | 8.39 | 7.75 | 6.72 | 7.52 |
| 42 | 6.63 | 6.65 | 8.31 | 9.00 | 10.1 | 10.2 | 10.4 | 9.62 | 8.40 | 7.69 | 6.62 | 6.37 |
| 44 | 6.49 | 6.58 | 8.30 | 9.06 | 10.3 | 10.4 | 10.5 | 9.70 | 8.41 | 7.63 | 6.49 | 6.21 |
| 46 | 6.34 | 6.50 | 8.29 | 9.12 | 10.4 | 10.5 | 10.6 | 9.79 | 8.42 | 7.57 | 6.36 | 6.04 |
| 48 | 6.17 | 6.41 | 8.27 | 9.18 | 10.5 | 10.7 | 10.8 | 9.89 | 8.44 | 7.51 | 6.23 | 5.86 |
| 50 | 5.98 | 6.30 | 8.24 | 9.24 | 10.7 | 10.9 | 11.0 | 10.0 | 8.35 | 7.45 | 6.10 | 5.64 |
| 52 | 5.77 | 6.19 | 8.21 | 9.29 | 10.9 | 11.1 | 11.2 | 10.1 | 8.49 | 7.39 | 5.93 | 5.43 |
| 54 | 5.55 | 6.08 | 8.18 | 9.36 | 11.0 | 11.4 | 11.4 | 10.3 | 8.51 | 7.20 | 5.74 | 5.18 |
| 56 | 5.30 | 5.95 | 8.15 | 9.45 | 11.2 | 11.7 | 11.6 | 10.4 | 8.53 | 7.21 | 5.54 | 4.89 |
| 58 | 5.01 | 5.81 | 8.12 | 9.55 | 11.5 | 12.0 | 12.0 | 10.6 | 8.55 | 7.10 | 4.31 | 4.56 |
| 60 | 4.67 | 5.65 | 8.08 | 9.65 | 11.7 | 12.4 | 12.3 | 10.7 | 8.57 | 6.98 | 5.04 | 4.22 |
| | | | | | *SOUTH* | | | | | | | |
| 0 | 8.50 | 7.66 | 8.49 | 8.21 | 8.50 | 8.22 | 8.50 | 8.49 | 8.21 | 8.50 | 8.22 | 8.50 |
| 5 | 8.68 | 7.76 | 8.51 | 8.15 | 8.34 | 8.05 | 8.33 | 8.38 | 8.19 | 8.56 | 8.37 | 8.68 |
| 10 | 8.86 | 7.87 | 8.53 | 8.09 | 8.18 | 7.86 | 8.14 | 8.27 | 8.17 | 8.62 | 8.53 | 8.88 |
| 15 | 9.05 | 7.98 | 8.55 | 8.02 | 8.02 | 7.65 | 7.95 | 8.15 | 8.15 | 8.68 | 8.70 | 9.10 |
| 20 | 9.24 | 8.09 | 8.57 | 7.94 | 7.85 | 7.43 | 7.76 | 8.03 | 8.13 | 8.76 | 8.87 | 9.33 |
| 25 | 9.46 | 8.21 | 8.60 | 7.74 | 7.66 | 7.20 | 7.54 | 7.90 | 8.11 | 8.86 | 9.04 | 9.58 |
| 30 | 9.70 | 8.33 | 8.62 | 7.73 | 7.45 | 6.96 | 7.31 | 7.76 | 8.07 | 8.97 | 9.24 | 9.85 |
| 32 | 9.81 | 8.39 | 8.63 | 7.69 | 7.36 | 6.85 | 7.21 | 7.70 | 8.06 | 9.01 | 9.33 | 9.96 |
| 34 | 9.92 | 8.45 | 8.64 | 7.64 | 7.27 | 6.74 | 7.10 | 7.63 | 8.05 | 9.06 | 9.42 | 10.1 |
| 36 | 10.0 | 8.51 | 8.65 | 7.59 | 7.18 | 6.62 | 6.99 | 7.56 | 8.04 | 9.11 | 9.35 | 10.2 |
| 38 | 10.2 | 8.57 | 8.66 | 7.54 | 7.08 | 6.50 | 6.87 | 7.49 | 8.03 | 9.16 | 9.61 | 10.3 |

| 40 | 10.3 | 8.63 | 8.67 | 7.49 | 6.97 | 6.37 | 6.76 | 7.41 | 8.02 | 9.21 | 9.71 | 10.5 |
| 42 | 10.4 | 8.70 | 8.68 | 7.44 | 6.85 | 6.23 | 6.64 | 7.33 | 8.01 | 9.26 | 9.8 | 10.6 |
| 44 | 10.5 | 8.78 | 8.69 | 7.38 | 6.73 | 6.08 | 6.51 | 7.25 | 7.99 | 9.31 | 9.94 | 10.8 |
| 46 | 10.7 | 8.86 | 8.90 | 7.32 | 6.61 | 5.92 | 6.37 | 7.16 | 7.96 | 9.37 | 10.1 | 11.0 |

## APPENDIX D

### PSYCHOMETRIC CONSTANT ($\gamma$) FOR DIFFERENT ALTITUDES (Z)

$$\gamma = 10^{-3}\ [(C_p.P) \div (\varepsilon.\lambda)] = (0.00163) \times [P \div \lambda]$$

$\gamma$, psychrometric constant [kPa C$^{-1}$]c$_{p,}$ specific heat of moist air = 1.013

[kJ kg$^{-10}$C$^{-1}$] P, atmospheric pressure [kPa].

$\varepsilon$, ratio molecular weight of water vapor/dry air = 0.622$\lambda$, latent heat of vaporization [MJ kg$^{-1}$]

= 2.45 MJ kg$^{-1}$ at 20°C.

| Z (m) | $\gamma$ kPa/°C | z (m) | $\gamma$ kPa/°C | z (m) | $\gamma$ kPa/°C | z (m) | $\gamma$ kPa/°C |
|---|---|---|---|---|---|---|---|
| 0 | 0.067 | 1000 | 0.060 | 2000 | 0.053 | 3000 | 0.047 |
| 100 | 0.067 | 1100 | 0.059 | 2100 | 0.052 | 3100 | 0.046 |
| 200 | 0.066 | 1200 | 0.058 | 2200 | 0.052 | 3200 | 0.046 |
| 300 | 0.065 | 1300 | 0.058 | 2300 | 0.051 | 3300 | 0.045 |
| 400 | 0.064 | 1400 | 0.057 | 2400 | 0.051 | 3400 | 0.045 |
| 500 | 0.064 | 1500 | 0.056 | 2500 | 0.050 | 3500 | 0.044 |
| 600 | 0.063 | 1600 | 0.056 | 2600 | 0.049 | 3600 | 0.043 |
| 700 | 0.062 | 1700 | 0.055 | 2700 | 0.049 | 3700 | 0.043 |
| 800 | 0.061 | 1800 | 0.054 | 2800 | 0.048 | 3800 | 0.042 |
| 900 | 0.061 | 1900 | 0.054 | 2900 | 0.047 | 3900 | 0.042 |
| 1000 | 0.060 | 2000 | 0.053 | 3000 | 0.047 | 4000 | 0.041 |

## APPENDIX E

### SATURATION VAPOR PRESSURE [e$_s$] FOR DIFFERENT TEMPERATURES (T)

Vapor pressure function = $e_s = [0.6108]*\exp\{[17.27*T]/[T + 237.3]\}$

| T °C | e$_s$ kPa | T °C | e$_s$ kPa | T °C | e$_s$ kPa | T °C | e$_s$ kPa |
|---|---|---|---|---|---|---|---|
| 1.0 | 0.657 | 13.0 | 1.498 | 25.0 | 3.168 | 37.0 | 6.275 |
| 1.5 | 0.681 | 13.5 | 1.547 | 25.5 | 3.263 | 37.5 | 6.448 |
| 2.0 | 0.706 | 14.0 | 1.599 | 26.0 | 3.361 | 38.0 | 6.625 |
| 2.5 | 0.731 | 14.5 | 1.651 | 26.5 | 3.462 | 38.5 | 6.806 |
| 3.0 | 0.758 | 15.0 | 1.705 | 27.0 | 3.565 | 39.0 | 6.991 |

| | | | | | | | |
|---|---|---|---|---|---|---|---|
| **3.5** | 0.785 | **15.5** | 1.761 | **27.5** | 3.671 | **39.5** | 7.181 |
| **4.0** | 0.813 | **16.0** | 1.818 | **28.0** | 3.780 | **40.0** | 7.376 |
| **4.5** | 0.842 | **16.5** | 1.877 | **28.5** | 3.891 | **40.5** | 7.574 |
| **5.0** | 0.872 | **17.0** | 1.938 | **29.0** | 4.006 | **41.0** | 7.778 |
| **5.5** | 0.903 | **17.5** | 2.000 | **29.5** | 4.123 | **41.5** | 7.986 |
| **6.0** | 0.935 | **18.0** | 2.064 | **30.0** | 4.243 | **42.0** | 8.199 |
| **6.5** | 0.968 | **18.5** | 2.130 | **30.5** | 4.366 | **42.5** | 8.417 |
| **7.0** | 1.002 | **19.0** | 2.197 | **31.0** | 4.493 | **43.0** | 8.640 |
| **7.5** | 1.037 | **19.5** | 2.267 | **31.5** | 4.622 | **43.5** | 8.867 |
| **8.0** | 1.073 | **20.0** | 2.338 | **32.0** | 4.755 | **44.0** | 9.101 |
| **8.5** | 1.110 | **20.5** | 2.412 | **32.5** | 4.891 | **44.5** | 9.339 |
| **9.0** | 1.148 | **21.0** | 2.487 | **33.0** | 5.030 | **45.0** | 9.582 |
| **9.5** | 1.187 | **21.5** | 2.564 | **33.5** | 5.173 | **45.5** | 9.832 |
| **10.0** | 1.228 | **22.0** | 2.644 | **34.0** | 5.319 | **46.0** | 10.086 |
| **10.5** | 1.270 | **22.5** | 2.726 | **34.5** | 5.469 | **46.5** | 10.347 |
| **11.0** | 1.313 | **23.0** | 2.809 | **35.0** | 5.623 | **47.0** | 10.613 |
| **11.5** | 1.357 | **23.5** | 2.896 | **35.5** | 5.780 | **47.5** | 10.885 |
| **12.0** | 1.403 | **24.0** | 2.984 | **36.0** | 5.941 | **48.0** | 11.163 |
| **12.5** | 1.449 | **24.5** | 3.075 | **36.5** | 6.106 | **48.5** | 11.447 |

## APPENDIX F

## SLOPE OF VAPOR PRESSURE CURVE ($\Delta$) FOR DIFFERENT TEMPERATURES (T)

$$\Delta = [4098.\ e^0(T)] \div [T + 237.3]^2$$
$$= 2504\{\exp[(17.27T) \div (T + 237.2)]\} \div [T + 237.3]^2$$

| T °C | $\Delta$ kPa/°C | T °C | $\Delta$ kPa/°C | T °C | $\Delta$ kPa/°C | T °C | $\Delta$ kPa/°C |
|---|---|---|---|---|---|---|---|
| 1.0 | 0.047 | 13.0 | 0.098 | 25.0 | 0.189 | 37.0 | 0.342 |
| 1.5 | 0.049 | 13.5 | 0.101 | 25.5 | 0.194 | 37.5 | 0.350 |
| 2.0 | 0.050 | 14.0 | 0.104 | 26.0 | 0.199 | 38.0 | 0.358 |
| 2.5 | 0.052 | 14.5 | 0.107 | 26.5 | 0.204 | 38.5 | 0.367 |
| 3.0 | 0.054 | 15.0 | 0.110 | 27.0 | 0.209 | 39.0 | 0.375 |
| 3.5 | 0.055 | 15.5 | 0.113 | 27.5 | 0.215 | 39.5 | 0.384 |
| 4.0 | 0.057 | 16.0 | 0.116 | 28.0 | 0.220 | 40.0 | 0.393 |
| 4.5 | 0.059 | 16.5 | 0.119 | 28.5 | 0.226 | 40.5 | 0.402 |
| 5.0 | 0.061 | 17.0 | 0.123 | 29.0 | 0.231 | 41.0 | 0.412 |
| 5.5 | 0.063 | 17.5 | 0.126 | 29.5 | 0.237 | 41.5 | 0.421 |
| 6.0 | 0.065 | 18.0 | 0.130 | 30.0 | 0.243 | 42.0 | 0.431 |
| 6.5 | 0.067 | 18.5 | 0.133 | 30.5 | 0.249 | 42.5 | 0.441 |
| 7.0 | 0.069 | 19.0 | 0.137 | 31.0 | 0.256 | 43.0 | 0.451 |

| 7.5  | 0.071 | 19.5 | 0.141 | 31.5 | 0.262 | 43.5 | 0.461 |
| 8.0  | 0.073 | 20.0 | 0.145 | 32.0 | 0.269 | 44.0 | 0.471 |
| 8.5  | 0.075 | 20.5 | 0.149 | 32.5 | 0.275 | 44.5 | 0.482 |
| 9.0  | 0.078 | 21.0 | 0.153 | 33.0 | 0.282 | 45.0 | 0.493 |
| 9.5  | 0.080 | 21.5 | 0.157 | 33.5 | 0.289 | 45.5 | 0.504 |
| 10.0 | 0.082 | 22.0 | 0.161 | 34.0 | 0.296 | 46.0 | 0.515 |
| 10.5 | 0.085 | 22.5 | 0.165 | 34.5 | 0.303 | 46.5 | 0.526 |
| 11.0 | 0.087 | 23.0 | 0.170 | 35.0 | 0.311 | 47.0 | 0.538 |
| 11.5 | 0.090 | 23.5 | 0.174 | 35.5 | 0.318 | 47.5 | 0.550 |
| 12.0 | 0.092 | 24.0 | 0.179 | 36.0 | 0.326 | 48.0 | 0.562 |
| 12.5 | 0.095 | 24.5 | 0.184 | 36.5 | 0.334 | 48.5 | 0.574 |

## APPENDIX G

## NUMBER OF THE DAY IN THE YEAR (JULIAN DAY)

| Day | Jan | Feb | Mar | Apr | May | Jun | Jul | Aug | Sep | Oct | Nov | Dec |
|---|---|---|---|---|---|---|---|---|---|---|---|---|
| 1 | 1 | 32 | 60 | 91 | 121 | 152 | 182 | 213 | 244 | 274 | 305 | 335 |
| 2 | 2 | 33 | 61 | 92 | 122 | 153 | 183 | 214 | 245 | 275 | 306 | 336 |
| 3 | 3 | 34 | 62 | 93 | 123 | 154 | 184 | 215 | 246 | 276 | 307 | 337 |
| 4 | 4 | 35 | 63 | 94 | 124 | 155 | 185 | 216 | 247 | 277 | 308 | 338 |
| 5 | 5 | 36 | 64 | 95 | 125 | 156 | 186 | 217 | 248 | 278 | 309 | 339 |
| 6 | 6 | 37 | 65 | 96 | 126 | 157 | 187 | 218 | 249 | 279 | 310 | 340 |
| 7 | 7 | 38 | 66 | 97 | 127 | 158 | 188 | 219 | 250 | 280 | 311 | 341 |
| 8 | 8 | 39 | 67 | 98 | 128 | 159 | 189 | 220 | 251 | 281 | 312 | 342 |
| 9 | 9 | 40 | 68 | 99 | 129 | 160 | 190 | 221 | 252 | 282 | 313 | 343 |
| 10 | 10 | 41 | 69 | 100 | 130 | 161 | 191 | 222 | 253 | 283 | 314 | 344 |
| 11 | 11 | 42 | 70 | 101 | 131 | 162 | 192 | 223 | 254 | 284 | 315 | 345 |
| 12 | 12 | 43 | 71 | 102 | 132 | 163 | 193 | 224 | 255 | 285 | 316 | 346 |
| 13 | 13 | 44 | 72 | 103 | 133 | 164 | 194 | 225 | 256 | 286 | 317 | 347 |
| 14 | 14 | 45 | 73 | 104 | 134 | 165 | 195 | 226 | 257 | 287 | 318 | 348 |
| 15 | 15 | 46 | 74 | 105 | 135 | 166 | 196 | 227 | 258 | 288 | 319 | 349 |
| 16 | 16 | 47 | 75 | 106 | 136 | 167 | 197 | 228 | 259 | 289 | 320 | 350 |
| 17 | 17 | 48 | 76 | 107 | 137 | 168 | 198 | 229 | 260 | 290 | 321 | 351 |
| 18 | 18 | 49 | 77 | 108 | 138 | 169 | 199 | 230 | 261 | 291 | 322 | 352 |
| 19 | 19 | 50 | 78 | 109 | 139 | 170 | 200 | 231 | 262 | 292 | 323 | 353 |
| 20 | 20 | 51 | 79 | 110 | 140 | 171 | 201 | 232 | 263 | 293 | 324 | 354 |
| 21 | 21 | 52 | 80 | 111 | 141 | 172 | 202 | 233 | 264 | 294 | 325 | 355 |
| 22 | 22 | 53 | 81 | 112 | 142 | 173 | 203 | 234 | 265 | 295 | 326 | 356 |
| 23 | 23 | 54 | 82 | 113 | 143 | 174 | 204 | 235 | 266 | 296 | 327 | 357 |

| Day | Jan | Feb | Mar | Apr | May | Jun | Jul | Aug | Sep | Oct | Nov | Dec |
|-----|-----|-----|-----|-----|-----|-----|-----|-----|-----|-----|-----|-----|
| 24 | 24 | 55 | 83 | 114 | 144 | 175 | 205 | 236 | 267 | 297 | 328 | 358 |
| 25 | 25 | 56 | 84 | 115 | 145 | 176 | 206 | 237 | 268 | 298 | 329 | 359 |
| 26 | 26 | 57 | 85 | 116 | 146 | 177 | 207 | 238 | 269 | 299 | 330 | 360 |
| 27 | 27 | 58 | 86 | 117 | 147 | 178 | 208 | 239 | 270 | 300 | 331 | 361 |
| 28 | 28 | 59 | 87 | 118 | 148 | 179 | 209 | 240 | 271 | 301 | 332 | 362 |
| 29 | 29 | (60) | 88 | 119 | 149 | 180 | 210 | 241 | 272 | 302 | 333 | 363 |
| 30 | 30 | — | 89 | 120 | 150 | 181 | 211 | 242 | 273 | 303 | 334 | 364 |
| 31 | 31 | — | 90 | — | 151 | — | 212 | 243 | — | 304 | — | 365 |

## APPENDIX H

### STEFAN-BOLTZMANN LAW AT DIFFERENT TEMPERATURES (T):

$[\sigma^*(T_K)^4] = [4.903 \times 10^{-9}]$, MJ $K^{-4}$ $m^{-2}$ $day^{-1}$

Where: $T_K = \{T[°C] + 273.16\}$

| T | $\sigma^*(T_K)^4$ | T | $\sigma^*(T_K)^4$ | T | $\sigma^*(T_K)^4$ |
|---|---|---|---|---|---|
| | | | Units | | |
| °C | MJ $m^{-2}$ $d^{-1}$ | °C | MJ $m^{-2}$ $d^{-1}$ | °C | MJ $m^{-2}$ $d^{-1}$ |
| 1.0 | 27.70 | 17.0 | 34.75 | 33.0 | 43.08 |
| 1.5 | 27.90 | 17.5 | 34.99 | 33.5 | 43.36 |
| 2.0 | 28.11 | 18.0 | 35.24 | 34.0 | 43.64 |
| 2.5 | 28.31 | 18.5 | 35.48 | 34.5 | 43.93 |
| 3.0 | 28.52 | 19.0 | 35.72 | 35.0 | 44.21 |
| 3.5 | 28.72 | 19.5 | 35.97 | 35.5 | 44.50 |
| 4.0 | 28.93 | 20.0 | 36.21 | 36.0 | 44.79 |
| 4.5 | 29.14 | 20.5 | 36.46 | 36.5 | 45.08 |
| 5.0 | 29.35 | 21.0 | 36.71 | 37.0 | 45.37 |
| 5.5 | 29.56 | 21.5 | 36.96 | 37.5 | 45.67 |
| 6.0 | 29.78 | 22.0 | 37.21 | 38.0 | 45.96 |
| 6.5 | 29.99 | 22.5 | 37.47 | 38.5 | 46.26 |
| 7.0 | 30.21 | 23.0 | 37.72 | 39.0 | 46.56 |
| 7.5 | 30.42 | 23.5 | 37.98 | 39.5 | 46.85 |
| 8.0 | 30.64 | 24.0 | 38.23 | 40.0 | 47.15 |
| 8.5 | 30.86 | 24.5 | 38.49 | 40.5 | 47.46 |
| 9.0 | 31.08 | 25.0 | 38.75 | 41.0 | 47.76 |
| 9.5 | 31.30 | 25.5 | 39.01 | 41.5 | 48.06 |
| 10.0 | 31.52 | 26.0 | 39.27 | 42.0 | 48.37 |
| 10.5 | 31.74 | 26.5 | 39.53 | 42.5 | 48.68 |
| 11.0 | 31.97 | 27.0 | 39.80 | 43.0 | 48.99 |
| 11.5 | 32.19 | 27.5 | 40.06 | 43.5 | 49.30 |

| | | | | | |
|---|---|---|---|---|---|
| **12.0** | 32.42 | **28.0** | 40.33 | **44.0** | 49.61 |
| **12.5** | 32.65 | **28.5** | 40.60 | **44.5** | 49.92 |
| **13.0** | 32.88 | **29.0** | 40.87 | **45.0** | 50.24 |
| **13.5** | 33.11 | **29.5** | 41.14 | **45.5** | 50.56 |
| **14.0** | 33.34 | **30.0** | 41.41 | **46.0** | 50.87 |
| **14.5** | 33.57 | **30.5** | 41.69 | **46.5** | 51.19 |
| **15.0** | 33.81 | **31.0** | 41.96 | **47.0** | 51.51 |
| **15.5** | 34.04 | **31.5** | 42.24 | **47.5** | 51.84 |
| **16.0** | 34.28 | **32.0** | 42.52 | **48.0** | 52.16 |
| **16.5** | 34,52 | **32.5** | 42.80 | **48.5** | 52.49 |

## APPENDIX I

## THERMODYNAMIC PROPERTIES OF AIR AND WATER

### 1. Latent Heat of Vaporization ($\lambda$)

$\lambda = [2.501 - (2.361 \times 10^{-3})\,T]$

Where: $\lambda$ = latent heat of vaporization [MJ kg$^{-1}$]; and T = air temperature [°C].

The value of the latent heat varies only slightly over normal temperature ranges. A single value may be taken (for ambient temperature = 20°C): $\lambda = 2.45$ MJ kg$^{-1}$.

### 2. Atmospheric Pressure (P)

$P = P_o\,[\{T_{Ko} - \alpha(Z - Z_o)\} \div \{T_{Ko}\}]^{(g/(\alpha \cdot R))}$

Where: P, atmospheric pressure at elevation $z$ [kPa]

$P_o$, atmospheric pressure at sea level = 101.3 [kPa]

$z$, elevation [m]

$z_o$, elevation at reference level [m]

g, gravitational acceleration = 9.807 [m s$^{-2}$]

R, specific gas constant == 287 [J kg$^{-1}$ K$^{-1}$]

$\alpha$, constant lapse rate for moist air = 0.0065 [K m$^{-1}$]

$T_{Ko}$, reference temperature [K] at elevation $z_o$ = 273.16 + T

T, means air temperature for the time period of calculation [°C]

When assuming $P_o$ = 101.3 [kPa] at $z_o$ = 0, and $T_{Ko}$ = 293 [K] for T = 20 [°C], above equation reduces to:

$P = 101.3[(293 - 0.0065Z)\,(293)]^{5.26}$

### 3. Atmospheric Density ($\rho$)

$\rho = [1000P] \div [T_{Kv}\,R] = [3.486P] \div [T_{Kv}]$, and $T_{Kv} = T_K[1 - 0.378(e_a)/P]^{-1}$

Where: $\rho$, atmospheric density [kg m$^{-3}$]

R, specific gas constant = 287 [J kg$^{-1}$ K$^{-1}$]

$T_{Kv,}$ virtual temperature [K]

$T_{K,}$ absolute temperature [K]: $T_K$ = 273.16 + T [°C]

$e_a$, actual vapor pressure [kPa]

T, mean daily temperature for 24-hour calculation time steps.

For average conditions ($e_a$ in the range 1–5 kPa and P between 80–100 kPa), $T_{Kv}$ can be substituted by: $T_{Kv} \approx 1.01\,(T + 273)$

## 4. Saturation Vapor Pressure function ($e_s$)

$e_s = [0.6108]*\exp\{[17.27*T]/[T + 237.3]\}$
Where: $e_s$, saturation vapor pressure function [kPa] T, air temperature [°C]

## 5. Slope Vapor Pressure Curve ($\Delta$)

$\Delta = [4098. \; e°(T)] \div [T + 237.3]^2$
$= 2504\{\exp[(17.27T) \div (T + 237.2)]\} \div [T + 237.3]^2$
Where: $\Delta$, slope vapor pressure curve [kPa C$^{-1}$]
T, air temperature [°C]
$e°(T)$, saturation vapor pressure at temperature T [kPa]

In 24-hour calculations, $\Delta$ is calculated using mean daily air temperature. In hourly calculations T refers to the hourly mean, $T_{hr}$.

## 6. Psychrometric Constant ($\gamma$)

$\gamma = 10^{-3} \; [(C_p.P) \div (\varepsilon.\lambda)] = (0.00163) \times [P \div \lambda]$
Where: $\gamma$, psychrometric constant [kPa C$^{-1}$]
$c_p$, specific heat of moist air = 1.013 [kJ kg$^{-10}$C$^{-1}$]
P, atmospheric pressure [kPa]: equations 2 or 4
$\varepsilon$, ratio molecular weight of water vapor/dry air = 0.622
$\lambda$, latent heat of vaporization [MJ kg$^{-1}$]

## 7. Dew Point Temperature ($T_{dew}$)

When data is not available, $T_{dew}$ can be computed from $e_a$ by:
$T_{dew} = [\{116.91 + 237.3 \text{Log}_e(e_a)\} \div \{16.78 - \text{Log}_e(e_a)\}]$
Where: $T_{dew}$, dew point temperature [°C]
$e_a$, actual vapor pressure [kPa]

For the case of measurements with the Assmann psychrometer, $T_{dew}$ can be calculated from:
$T_{dew} = (112 + 0.9T_{wet})[e_a \div (e° \; T_{wet})]^{0.125} - [112 - 0.1T_{wet}]$

## 8. Short Wave Radiation on a Clear-Sky Day ($R_{so}$)

The calculation of $R_{so}$ is required for computing net long wave radiation and for checking calibration of pyranometers and integrity of $R_{so}$ data. A good approximation for $R_{so}$ for daily and hourly periods is:
$R_{so} = (0.75 + 2 \times 10^{-5} \; z)R_a$
Where: z, station elevation [m]
$R_a$ extraterrestrial radiation [MJ m$^{-2}$ d$^{-1}$]

Equation is valid for station elevations less than 6000 m having low air turbidity. The equation was developed by linearizing Beer's radiation extinction law as a function of station elevation and assuming that the average angle of the sun above the horizon is about 50°.

For areas of high turbidity caused by pollution or airborne dust or for regions where the sun angle is significantly less than 50° so that the path length of radiation through the atmosphere is increased, an adoption of Beer's law can be employed where P is used to represent atmospheric mass:

**$R_{so} = (R_a) \exp[(-0.0018P) \div (K_t \sin(\Phi))]$**

Where: $K_t$, turbidity coefficient, $0 < K_t \leq 1.0$ where $K_t = 1.0$ for clean air and $K_t = 1.0$ for extremely turbid, dusty or polluted air.

P, atmospheric pressure [kPa]

$\Phi$, angle of the sun above the horizon [rad]

$R_a$, extraterrestrial radiation [MJ m$^{-2}$ d$^{-1}$]

For hourly or shorter periods, $\Phi$ is calculated as:

$\sin \Phi = \sin \varphi \sin \delta + \cos \varphi \cos \delta \cos \omega$

Where: $\varphi$, latitude [rad]

$\delta$, solar declination [rad] (Eq. (24) in Chapter 3)

$\omega$, solar time angle at midpoint of hourly or shorter period [rad]

For 24-hour periods, the mean daily sun angle, weighted according to $R_a$, can be approximated as:

**$\sin(\Phi_{24}) = \sin[0.85 + 0.3 \, \varphi \sin\{(2\pi J/365)-1.39\}-0.42 \, \varphi^2]$**

Where: $\Phi_{24}$, average $\Phi$ during the daylight period, weighted according to $R_a$ [rad]

$\varphi$, latitude [rad]

J, day in the year

The $\Phi_{24}$ variable is used to represent the average sun angle during daylight hours and has been weighted to represent integrated 24-hour transmission effects on 24-hour $R_{so}$ by the atmosphere. $\Phi_{24}$ should be limited to $\geq 0$. In some situations, the estimation for $R_{so}$ can be improved by modifying to consider the effects of water vapor on short wave absorption, so that: $R_{so} = (K_B + K_D) R_a$ where:

$K_B = 0.98\exp[\{(-0.00146P) \div (K_t \sin \Phi)\}-0.091\{w/\sin \Phi\}^{0.25}]$

Where: $K_B$, the clearness index for direct beam radiation

$K_D$, the corresponding index for diffuse beam radiation

$K_D = 0.35-0.33 \, K_B$ for $K_B \geq 0.15$

$K_D = 0.18 + 0.82 \, K_B$ for $K_B < 0.15$

$R_a$, extraterrestrial radiation [MJ m$^{-2}$ d$^{-1}$]

$K_t$, turbidity coefficient, $0 < K_t \leq 1.0$ where $K_t = 1.0$ for clean air and $K_t = 1.0$ for extremely turbid, dusty or polluted air.

P, atmospheric pressure [kPa]

$\Phi$, angle of the sun above the horizon [rad]

W, perceptible water in the atmosphere [mm] = $0.14 \, e_a P + 2.1$

$e_a$, actual vapor pressure [kPa]

P, atmospheric pressure [kPa]

# APPENDIX J

# PSYCHROMETRIC CHART AT SEA LEVEL.

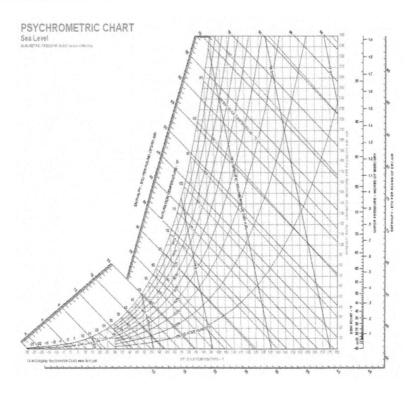

# INDEX